T0192347

Crop Responses to Environment

Crop Responses to Environment

Adapting to Global Climate Change, Second Edition

Anthony E. Hall

CRC Press is an imprint of the
Taylor & Francis Group, an informa business

CRC Press
Taylor & Francis Group
6000 Broken Sound Parkway NW, Suite 300
Boca Raton, FL 33487-2742

ISBN-13: 978-1-03-209573-8 (pbk)
ISBN-13: 978-1-138-50638-1 (hbk)

Publisher's Note
The publisher has gone to great lengths to ensure the quality of this reprint but points out that some imperfections in the original copies may be apparent.

Visit the Taylor & Francis Website at
http://www.taylorandfrancis.com

and the CRC Press Website at
http://www.crcpress.com

Contents

Preface to Second Edition

I wrote this second edition to update the material presented in the first edition of *Crop Responses to Environment*, which was published 16 years ago. I also added discussions of crop responses to environment that relate to crop adaptation to climate change. I consider climate change to be important. I began teaching that the increase in atmospheric carbon dioxide concentration would impose an increased heat load on the earth in the first class on crop ecology that I taught at the University of California, Riverside, in spring 1972. In this class I also said that it was not clear to me what impact this increased heat load would have on the earth.

Since 1972 I have been amazed by the extent of signs and impacts of global warming—increasing night temperatures in the tropics negatively correlated with decreasing yields of rice, decreasing chilling hours in winter with potential catastrophic negative effects on deciduous fruit and nut crops, a rise in sea level and increased flooding in deltas, increasing sea temperature and death of coral jeopardizing marine ecosystems, retreat of most glaciers and problems with water supplies in the warm season, and decreases in arctic ice opening up the Northwest Passage to shipping and jeopardizing the existence of polar bears.

I also have added some information concerning the need to prevent the biosphere from being substantially damaged by attempts to increase crop production.

In teaching and in writing this book, I have emphasized topics for which I either have some practical expertise or research experience or at least a reasonably sound theoretical and empirical understanding. The material is complex, however, and it is possible that I made some mistakes. I will provide a chronological summary of my experiences to alert the reader concerning the areas where I have significant expertise.

I entered grammar school in England (equivalent to high school in the United States) two years younger than the average student. I passed the Cambridge O Level examinations in English, geography, mathematics, chemistry, and physics. I then left grammar school because I was not happy there. In addition, I frequently experienced headaches. I went to work on the family farm because I enjoyed farmwork.

We had a moderate-sized farm operated by my father, me, one farmhand and occasional temporary laborers. It was a mixed farm with livestock enterprises integrated with field crops, truck crops, and pastures. The livestock included a herd of beef cattle fattened on pastures in the warm season, and brought inside a barn in the winter where they were provided with straw and feed and produced manure. We had a cow which we milked by hand to produce milk for the house. We had several sows producing piglets, which were fattened to be sold for bacon. We had flocks of hens producing eggs that we raised in movable enclosures on pastures and brought into a building during the winter.

Our most profitable field crop was Irish potatoes; the other field crops were sugar beet, wheat, and barley. We also grew red beet, carrots, and peas for the fresh vegetable market. We had fields with grass and/or clover and alfalfa for pasture or for producing hay and silage. We milled the wheat and barley grains to produce flour and mixed in maize flakes, soybean flour, sugar beet pulp supplied by a factory, and other components to produce year-round feed for the pigs and milk cow, and feed for the steers in winter. In addition, the steers were fed hay, silage, wilted sugar beet tops, and any carrots or red beet that we couldn't sell.

Crops were rotated. In the fall we spread manure on the pasture and other fields and plowed it under. We grew Irish potatoes on the fields that had been pasture with added manure so that they could benefit from the fertile soil. Crops such as Irish potato, sugar beet, and carrots, which are susceptible to plant-parasitic nematodes or other detrimental microorganisms that might build up in the soil, were only grown on the same field every four to six years after growing pastures, grain, and other crops. We had a water supply to part of the farm where we had a dam and a small reservoir from which, after unusual dry spells, we pumped water and irrigated fields using portable aluminum pipe and overhead sprinklers.

I worked on the farm for 40 months, gaining strength because of the hard physical labor and becoming healthy. I did not experience any more headaches. I gained some practical understanding of agriculture and mechanization. I gained an appreciation of mixed-farm agriculture as an ecologically sound and sustainable approach to farming. I also gained an appreciation for supplemental irrigation as a method to stabilize production.

I had no future prospects continuing to work on this farm because much of the land was leased and the lease was destined to end in a few years and could not be renewed. My parents and the government then helped me to attend college, for which I am very grateful.

I went to Harper Adams Agricultural College for two years, gaining college and national diplomas in agriculture and expertise mainly in temperate-zone agriculture. By this time I had developed an additional interest in agricultural engineering, and during the summers I took courses from farm machinery companies. The companies used the courses to train their technical staff. The courses covered the design, maintenance, and operation of farming equipment. I took three courses: one on combine harvesters, one on hay balers, and another on various types of tractors. I then went to the Essex Institute of Agriculture, Essex, United Kingdom, for one year, gaining college and national diplomas in agricultural engineering and membership in the Institute of Agricultural Engineering. I now had developed a greater interest in irrigation.

I then took a position with the Ministry of Agriculture in Tanganyika, partially because I felt a need to help poor agricultural communities in Africa and partially for the adventure. The position in Tanganyika was supposed to involve working on irrigation development but the ministry was short-staffed and I served most of my contract as a Field Officer directing a district team conducting agricultural extension. At this time Tanganyika was a Trust Territory being administered by Great Britain for the United Nations until it reached the ultimate goal of self-government—which took place while I was there.

My first district was South Mara, which is between Lake Victoria and the Serengeti National Park. Initially I worked with the current field officer who was a Tanganyikan with much practical experience. It was a pleasure working with him but when he left to take an official vacation, I was on my own. I gained some experience with agricultural extension and cotton cultivation, which was an important crop in

South Mara, and tropical-zone agriculture. I thoroughly enjoyed working with the farmers and camping on the edge of the Serengeti National Park. The savanna landscapes and wild animals were magnificent.

I then was sent to take a government course on the widely spoken local language—Kiswahili. The course was conducted at Tengeru near Arusha at the base of Mount Meru, which provided me with an opportunity to visit the Ngorongoro Crater where I saw more wild animals and beautiful scenery. After completing this course I mainly conducted my work with farmers by speaking swahili.

I was then transferred to be field officer in charge of North Mara, which has a complex ecology. In this district the elevation varied from 1,135 meters at the shore of Lake Victoria, where it often was very hot, to an elevation of about 2,000 meters in the North Mara highlands, where it often was cool. The lakeshore area was tropical and suitable for cotton and other warm-season annual crops and tropical perennials, whereas the environment in the highlands was subtropical, suitable for Arabica coffee and some cool-season annual crops. While I was working in this district, the transition to independence began and I gained some experience managing elections and establishing cooperatives for marketing maize and operating the North Mara coffee industry. The climate in the highlands of North Mara was humid in that significant rain occurred every month. But it also was very sunny in that the rain came as convective storms which only occurred in the evening.

While working in North Mara, I was given the opportunity to travel by road with two farmer representatives of the coffee cooperative to a conference in northern Tanganyika. We used a government Land Rover and traveled through Kenya via Kisii, Nakuru, and Nairobi to Arusha and then to the Coffee Research Station at Lyamungu, which is at the base of Mount Kilimanjaro in northern Tanganyika. At the conference the farmers and I were provided with new information about growing and processing coffee beans. After the conference we benefited from lessons on the best ways to prune coffee bushes.

Toward the end of my 27-month contract with the Government of Tanganyika, I worked at the Ilonga Research Station near Kilosa, where I developed an irrigated nursery for sugar cane varieties and evaluated different soybean and maize varieties to choose ones that were suitable for this area. I also was asked to visit two substations of the ministry near Mboze and Njombe to create maps of the stations to facilitate their use for conducting field plot trials. While working in these areas of Tanganyika I gained some understanding of climates influenced by monsoons, which have two rainy seasons and two dry seasons every year.

As my work in Tanganyika was ending, I did not ask to renew my contract. Tanganyika now had become independent and would soon be called Tanzania. I felt the work of Field Officer should be conducted by local people. I also felt a need to learn more about irrigation, after which I planned to return to Africa to contribute to agricultural development as a specialist involved in planning and establishing new irrigation schemes.

I then enrolled at the University of California, Davis where I obtained a BS degree in irrigation science. I took courses in plant, soil, environmental and irrigation sciences, and engineering. To help with financing my education I worked half-time as a senior engineering aid for the research program of W (Bill). O. Pruitt. In this work

I installed and operated a Soviet-style weather station with a very large evaporation pan, made weather station measurements, and worked with a weighing lysimeter to measure reference crop evapotranspiration for comparisons with the pan evaporation. This provided me with valuable experiences in micrometeorology, agricultural climatology, and predicting crop water use.

I continued on at the University of California, Davis, studying for a PhD in Plant Physiology under the direction of Professor Robert S. Loomis. Bob taught me an advanced undergraduate course in crop ecology that I thoroughly enjoyed and that influenced my future academic directions. I took many courses during my graduate program at Davis including ones in plant physiology, physical chemistry, and genetics. My dissertation concerned the effects of the sugar beet yellows virus on the photosynthesis, respiration, and transpiration of sugar beet. As a part of this study I learned about measuring different aspects of photosynthesis and transpiration, measuring the performance of healthy and diseased plants, and also learned how to rear aphids, maintain diseased plants, and transfer the virus disease to healthy plants. My dissertation research involved a complex experimental system manipulating plants, aphids, and the virus disease.

Toward the end of my dissertation studies, I decided that combining empirical studies with mathematical modeling was a powerful research methodology. I began developing a mathematical model of photosynthesis and respiration to test a hypothesis I had developed: that one enzyme, RuBP-carboxylase, was responsible for both the fixation of carbon dioxide and photorespiration. I continued on with testing this model during a short postdoctoral appointment at the Department of Plant Biology of the Carnegie Institution for Science on the Stanford University campus. My model was compared with measurements of photosynthesis made by Dr. Olle Bjőrkman and predictions made by my model were in perfect agreement with the measurements (Carnegie Institution of Washington Year Book, Volume 70, 1971).

I continued developing this model while working on my next position as a Professor of Plant Physiology at the University of California, Riverside, and during a sabbatical leave at the Research School of Biological Sciences at the Australian National University in Canberra working with Professors Ian R. Cowan and Graham D. Farquhar. I completed the model of photosynthesis and respiration during a sabbatical leave at the University of Bayreuth in West Germany working with Professor E. -D. Schulze where we tested an elegant hypothesis for optimal stomatal function developed by Professors Cowan and Farquhar. In 1979 I published the model of photosynthesis and respiration (Hall, 1979). This model was used to create the figures in Chapter 4 in this book.

In 1971 I had obtained the professorial position at the University of California, Riverside, that I occupied for 32 years until my retirement. As a professor I conducted some basic research and about every year taught an advanced undergraduate course in crop ecology and a graduate course on photosynthesis, transpiration, and the temperature and water relations of plants, essentially covering most of the topics in this book. I wrote the first edition of this book because a suitable text was not available for these classes.

One type of basic research that I conducted consisted of empirical studies of stomatal responses to environment that I mainly conducted by myself. This research made me quite skeptical in that I discovered that most of what was then known about stomatal responses to humidity, soil drought, temperature, and carbon dioxide concentration was mainly incorrect. For example, we discovered that stomatal responses to soil drought did not involve concurrent changes in bulk leaf water status but probably involved signaling between roots and leaves via hormonal messengers. During this period I made major contributions to the design of a steady-state porometer, which for many years was the most effective device for measuring stomatal conductance and was used in many countries.

While at the University of California, Riverside, I also held a joint appointment as a crop ecologist in the California Agricultural Experiment Station that enabled me to conduct applied research. In addition, I collaborated with farm advisors in the University of California system who conducted extension work.

Initially I conducted research on the responses to environment of various citrus species. The University of California, Riverside, was famous for its research on oranges, lemons, and grapefruit. I then decided to work on cowpeas because it was a minor crop that had been neglected by scientists but had some importance in California and was a major crop in Sub-Saharan Africa. First I studied how to improve the irrigation management of cowpeas because this was important in California. In this manner, I gained expertise in irrigated subtropical-zone agriculture. The major valleys in California have Mediterranean climates, which are subtropical with rain in winter and dry summers when many crops can be grown using irrigation. According to the California Department of Food and Agriculture, about 400 different crop species are grown in California.

My subsequent major basic research contributions involved studies of the physiology and genetics of cowpea responses to environment with emphasis on heat tolerance, adaptation to drought, and chilling tolerance. With respect to heat stress effects, we demonstrated the surprising fact that high temperatures in the late night and early morning can have detrimental effects on pollen development and seed and fruit set in cowpea, whereas much hotter temperatures later in the day have no impact on seed or fruit set. Other scientists have now shown that other crops including some cereals and grain legumes and annual fruit crops such as peppers and cotton also are susceptible to high late-night temperatures during pollen development.

I then decided that major contributions could be made to understanding the physiology and genetics of plant responses to environmental stresses by combining plant physiology and plant breeding. Interdisciplinary areas between classical disciplines have been neglected by scientists. Breeding-improved varieties also provided a mechanism whereby I could contribute to the well-being of farmers in California and Africa. So I became a plant breeder and developed a major program of cowpea breeding, including the creation of a major cowpea germplasm collection and breeding varieties with resistance to heat and various pests and diseases for use in California and resistance to drought for use in Africa.

By now I had decided that progress in plant breeding and agronomy in Sub-Saharan Africa could most effectively be done by local scientists. But, that we could assist them by providing educational opportunities and other support. I guided many African and U.S. students, as well as Brazilian, Japanese, and Spanish students, in masters and doctoral programs at the University of California, Riverside. I helped the African students to design research programs that would provide them with experiences which would be useful for them when they returned home.

A severe drought began in 1968 in the Sahelian zone of Africa just below the Sahara that caused major problems for herders and farmers (Hall, 2017). In 1974 the United States Agency for International Development provided the University of California, Riverside, with a major grant to develop solutions to the problems caused by droughts in Africa. Rainfall in the Sahelian zone has been low for at least 30 years. For example, the average annual rainfall in Louga, Senegal, was only 276 millimeters from 1968 through 1998, whereas it had been 442 millimeters from 1918 through 1967 (Hall, 2017).

Through research in California and Senegal I developed an approach for providing a partial agronomic solution to the Sahelian droughts: breed cowpea varieties that are erect with extra-early flowering and have vegetative-stage drought resistance, and sow them at a spacing of 20 × 50 centimeters, which is much denser than farmers were using of 100 × 100 centimeters for their prostrate cowpea varieties. From my fieldwork in California, I estimated that in the Sahel, these varieties should be able to produce substantial quantities of dry grain of about 1,000 kg/ha in 60 days, with only a small amount of rain of about 200 millimeters occurring over about 50 days. They also would produce significant quantities of fresh peas in about 50 days which was during the hungry season when little other food was available. Traditional prostrate cowpea varieties produced grain in about 90 days and required about 400 millimeters of rain.

There were other advantages from growing more cowpeas. They improve the quality of food available to people in that their grains have large quantities of a type of protein that complements the protein in cereal grains. Cowpea also produces high-quality hay for feeding livestock during the dry season. Having more livestock also improves the quality of food for people and provides more manure to improve the fertility of the soil. Cowpeas grown in rotation also improve the fertility of the soil through the nitrogen fixation of the nodules on their roots. Even moderate increases in soil fertility would substantially increase the grain yields of pearl millet, which is a staple food for the people, grown in rotation with cowpeas. Peanut is another important component in the crop rotation in the Sahel providing cash, food grains, oil for cooking and hay. Fresh peas and dry grains are the most important reasons for growing cowpea but the several other benefits when combined together could result in significant increases in pearl millet grain and peanut grain and hay production.

I designed and directed a project in which I collaborated with some of my African students and other African scientists in research and extension. We bred extra-early cowpea varieties with vegetative-stage drought resistance and tested them in the Sahel. They performed in the manner that I had predicted. The varieties that had been shown to be effective were extended to farmers. The project was successful in helping many poor farm families to gain more food and cash (Hall, 2017).

I worked on this project for about three decades. Almost every year I worked with African collaborators for a few weeks in Africa, mainly in Senegal where we developed a comprehensive cowpea research program. We bred and extended early-flowering cowpea varieties to farmers in the Sahelian zone of Senegal. These varieties produced significant quantities of food within 50–60 days from sowing in a dry environment where no other crops are able to produce significant quantities of food. I also developed a plan to increase cowpea production in the Sahelian zone of Sudan. We bred and extended early-flowering cowpea varieties to farmers in the Sahelian Zone of Sudan and they were effective. I collaborated with scientists from Ghana and this resulted in two cowpea varieties being bred for the wetter Savanna zone of West Africa. While working in Senegal, Sudan, and Ghana I gained some experience with semiarid tropical Sahelian and Savanna zones that have monomodal rainfall patterns.

For California, we bred three cowpea varieties. In all cases we developed improved management methods that complemented the new varieties.

After I retired in 2003 I worked as a professor emeritus, continuing a small agricultural consulting firm I had initiated soon after joining the University of California, Riverside. In the early 1970s I had taken a month's vacation time to work on agricultural development in Guinea-Conakry where I gained some experience with agriculture in a humid tropical zone in West Africa.

In later years I used vacation time to work for the United Nations, evaluating the International Maize and Wheat Improvement Center in El Batan, Mexico, and the International Crops Research Institute for the Semi-Arid Tropics in Hyderabad, India. These evaluations provided me with broadening educational experiences with a range of crops in different climatic zones. After retirement I continued working for agencies of the United Nations and various foundations on matters pertaining to designing research programs to improve agriculture in Africa.

I continued to think about global climate change and its implications for agricultural research. I contributed to a chapter on breeding for heat resistance I had developed earlier for a website www.plantstress.com. I also continued writing reviews concerning breeding crops for resistance to heat and breeding crops that are adapted to future climatic conditions (Hall, 2011). I collaborated in editing Crop Adaptation to Climate Change (Yadav et al., 2011). I also wrote a memoir that describes my work in Africa and philosophy concerning research and education (Hall, 2017).

In 2004 my wife, Bretta, and I moved to a cabin at 1,060 meters in a small forest in the northern Sierra Nevada in California. Since then I have learned a little about temperate-zone cow-calf ranching systems that use mountain meadows in summer. We are enjoying the environment and moved here to experience some change in seasons. In 2017 we experienced record high air temperatures for June of 41°C. It looks as if global warming is coming to northern California, we do not have and prefer not to have air-conditioning, so we may need to move to a cabin at a higher elevation.

Preface to First Edition

This book is primarily for scientists and students who are interested in developing improved crop cultivars and management methods. However, it emphasizes principles and theories concerning plant responses to environment that are relevant to plants in natural as well as agricultural systems. Many practical applications to plant breeding, agronomy, and horticulture are discussed, including some examples from my work in agricultural research and extension on irrigated systems in California and rain-fed systems in Africa, and as a farmer in England. I have included many references to key papers that describe original concepts or research observations or reviews of important topics and some addresses to websites that provide useful information.

This book is designed so that it is most easy to read in a linear sequence from the front to the back. Experienced readers will have no difficulty skipping among the chapters, in that each chapter is designed to be independent with references to critical parts of other chapters as they are needed. The reader may get the impression that some themes are repeated in different chapters. This is deliberate, in that I feel that crop responses to environment cannot be explained well in a simple linear sequence but are most effectively explained by a series of iterative cycles that bring in either additional elements or different ways of looking at the same issue.

Acknowledgments and Dedication for the First Edition

I thank Professors Carol J. Lovatt and Timothy J. Close of the University of California, Riverside, for reading an early draft of the first edition of this book and making useful suggestions. I am responsible, however, for any mistakes in the book. I dedicate this book to the many graduate students and other scientists who I have been privileged to work with and to my wife, Bretta, for her patience and support.

Anthony E. Hall

Acknowledgments and Dedication for the First Edition

[illegible]

[illegible]

1 Introduction

Plant responses to environment determine the adaptation of plants and influence the improvement of cropping systems that can be achieved through changes in management practices and plant breeding. The importance of this discipline is that, as will be shown later in this chapter, substantial increases in the efficiency of crop production will be required during the twenty-first century. My initial definition of *efficiency of crop production* is production per unit land area, that is, yield. Subsequently, I will describe some other types of efficiency, which in general terms is a ratio of output to input. An understanding of crop responses to environment will provide the fundamental basis for developing improved varieties and complementary management methods that result in increases in yield.

In many parts of the world increases in yield will be needed because there will be greater demands for agricultural products. Increased demands will occur due to increasing human populations and changes in consumption patterns, such as increased consumption of milk and other dairy products and meat from livestock reared on feed grains. There also may be increased demands for crop products to produce biofuels, such as the use of grains of maize or stalks of sugar cane to produce ethanol.

Increases in crop yield are particularly important for developing countries because this is where the greatest increases in demand for food will occur, and because improvement in agriculture can stimulate rural and urban development. This development is essential for decreasing the income gap between the rural poor and the rich people in this world. Increases in crop production efficiency are required in all countries to maintain profitability and enhance sustainability of agricultural enterprises, and to contribute to environmental health.

The alternative to increasing yield—increasing production through expansion of arable lands by reclaiming marshes and other wet lands, cutting down forests, and plowing up grass lands as was practiced in the past—should be discouraged. These wild lands should be preserved because they provide important services to the biosphere on which we all depend. *Biosphere* is a term used to describe all of Earth's living organisms interacting with the physical and chemical environment as a whole. At this time, China is planning the preservation of large areas of land for use as national parks.

For developing countries increasing crop production through expansion of arable lands could result in another problem: further increases in the human population. Evans (1998) examined relationships between human populations and agriculture from ancient times to the late 1900s. He describes two contrasting hypotheses: (1) that increases in human population have stimulated attempts to increase agricultural production, or (2) that opportunities for increasing food production have encouraged or permitted increases in human population. Distinguishing between these two possibilities is important for communities of people whose populations have a tendency to increase rapidly.

New technologies and other developments that make possible the exploitation of new lands by agriculture could tend to stimulate even greater increases in human population, especially if the new lands provide only poor living conditions for farm families. Note that, typically, the best lands already are being used for agriculture, so expanding the amount of land that is cultivated usually involves moving into areas where the soils have major problems such as acidity, alkalinity, or salinity. In general, the very rapid increases in human populations that could occur due to expanding farming into marginal soil areas would cause major problems.

New technologies that enable poor farmers to make more efficient use of their current arable land in general are more desirable. Although they lead to increased profits and improved living conditions for farm families, the new technologies can indirectly result in decreases in human birth rates. As was pointed out by Murdock (1990), "Poor parents have many children because the economic benefits of the children outweigh their economic costs. The benefits come in the form of labor, income, and security for parents in their old age. As parents' incomes rise, and inevitably also their level of education, and as the economic structure of society changes, the benefit/cost ratio of children declines. As income increases, the balance will favor smaller families." Note that improved education and emancipation of women will enhance this trend toward smaller families. Consequently, increases in crop yield can give three important benefits: provide more food for hungry people, increase profits, and encourage decreases in the growth rate of the human population.

By 2050 the human population is predicted to increase to 9.3 billion from the current level of 7.4 billion. A major analysis has been made by Fischer et al. (2014) of whether increases in crop yield will continue to feed the world up to 2050. Their book, which is available at no cost on the Internet (http://aciar.gov.au/publication/mn158), covers many topics that are relevant to crop responses to environment. The authors analyze data for the world's most important food crops: rice, wheat, maize, and soybeans. These crops either indirectly (as feed grains) or directly (as food for people) provide two thirds of the calories and protein consumed by humans. They point out that global food demand for these crops is predicted to increase by 60% from 2010 to 2050. The authors conclude that the minimum yield increases of these major staple food crops needed to feed the world in 2050, while preventing price increases, is 1.1%–1.3% per year (relative to 2010 yields). They point out that the current increases in yield in 2014 of wheat, rice and soybean were 1.0% relative to 2010 yields. Consequently, a 10%–30% increase is needed in the rate of increase in yield of these very important crops. I consider that these increases in the rate of increase in yield of wheat, rice, and soybean will be very difficult to achieve because in many cases the easiest ways for increasing yield already have been taken. Also, changes are occurring in the abiotic and biotic environment that will tend to decrease yield. In 2014 the global increase in yield of maize was 1.5% per year relative to 2010 yields. If maintained, this increase could be enough to meet future demands for maize. But much of the grains of maize are being diverted to either feed livestock or produce ethanol, and maize contributes only moderately to the food needs of poor people.

Simply maintaining yields at current levels also often requires the development of new cultivars and new management methods, since pests and diseases continue to evolve, and aspects of the chemical, physical, and social environment can change over several decades (Dobermann et al., 2000). About 50% of the current effort by rice and wheat breeders is devoted to maintenance breeding to incorporate resistances to diseases and pests and does not result in increases in yield potential (Fischer et al., 2014).

In the 1960s, many people considered pesticides to be mainly beneficial to mankind. Developing new, broadly effective, and persistent pesticides often was considered to be the best way to control pests on crop plants. Since that time, it has become apparent that broadly effective pesticides can have detrimental effects on beneficial insects, which can negate the overall effects of the pesticide in controlling pests. In addition, persistent pesticides can damage non-target organisms in the ecosystem, such as birds and people. Also, it has become difficult for companies to develop new pesticides, even those that can have major beneficial effects and few negative effects. Very high costs are involved in following all of the procedures needed to gain government approval for new pesticides. Consequently, more consideration is being given to other ways to manage pests, such as incorporating greater resistance to pests into cultivars by breeding and using other biological control methods.

Global climate change is occurring. Due to the burning of fossil fuels the carbon dioxide concentration in the atmosphere has been increasing and this impacts the photosynthesis of different crop plants in different ways, as is discussed in Chapter 4. The increases in carbon dioxide concentration and other greenhouse gases raise the heat load on the Earth, as is discussed in Chapter 7, and are causing rises in temperature. In some circumstances, the increases in temperature are reducing crop yields. For example, yields of rice crops grown under optimal management on the experimental fields of the International Rice Research Institute in the Philippines decreased during a 12-year period from 1992 to 2003 (Peng et al., 2004). From 1979 to 2003, day temperatures had increased 0.35°C while night temperatures had increased 1.13°C. One might expect high day temperatures to be most stressful, however, grain yield exhibited a 10% decrease in yield per degree C increase in night temperature with no correlation with day temperature.

Correlations do not necessarily indicate a causal relation. However, experiments where nighttime temperatures were raised in the field resulted in 4% decreases in grain yield of cowpea per degree C increase in night temperature (Nielsen and Hall, 1985b). In some hot regions there will be a tendency for grain yields of some crops to decrease with global warming if heat-tolerant varieties are not developed. In contrast, in some cold regions there will be a tendency for yields of some crops to be increased by global warming.

Chapter 5 discusses crop physiological responses to temperature, while Chapter 6 discusses crop developmental responses to temperature such as the chilling requirements of crops. Whether chilling requirements are met is being influenced by global warming—for example, winter chilling hours are decreasing in California and jeopardizing the stone fruit and nut tree industries (Baldocchi and Wong, 2008). Global warming also is raising seawater levels and could result in greater flooding of delta and low-lying coastal areas that will impact crop production, as is discussed in Chapter 11. The adaptations of different crops to global climate change in different parts of the world are discussed in a book edited by S. S. Yadav et al. (2011).

When considering methods for increasing crop yields it is useful to distinguish among the average yields obtained by farmers in an area (FY), the potential yield that can be obtained in that area by using the best variety grown with the most effective management methods (PY), and the yield gap, that is, the difference between FY and PY (Fischer et al., 2014). The yield gap can be described as a % of FY.

For maize in much of sub-Saharan Africa, the biggest opportunity for increasing FY is to decrease the yield gap by using improved management methods. For example, the yield gap for maize in East Africa is about 400% (i.e., with a FY of 2 ton ha^{-1}, the yield gap would be 8 ton ha^{-1} and PY would be 10 ton ha^{-1}). The main limitation for yield is the infertile soil. FY of maize in sub-Saharan Africa can be substantially increased by increasing the supplies of nitrogenous fertilizer together with using the best hybrid varieties. However, the ratio of the cost of a kg of nitrogen (N) in nitrogenous fertilizer in relation to the price the farmer receives for a kg of maize grain is high in East Africa illustrating why it can be uneconomic for these farmers to apply much fertilizer to their maize crop. Farmers also often can have difficulty obtaining credit to buy fertilizer or seed of improved varieties. Encouraging farmers to apply more nitrogenous (and phosphate) fertilizer to their maize crops and use hybrid varieties that are responsive to fertilizer applications will require improving infrastructure so that they can obtain credit at a reasonable price, buy fertilizer and hybrid seed at cheaper prices, and obtain higher prices when they sell their crops. Also, more extension agents are needed to advise African farmers concerning the most effective varieties and management methods for their region.

Increasing crop yields by applying more fertilizer can be complex. For example, the yield gap of other cereals grown in Africa can be high, as with the cases of pearl millet in the Sahelian zone of West Africa and sorghum in northern Sudan. The yields of these crops can be substantially increased by small applications of fertilizer (i.e., 20 kg ha^{-1} of N plus 9 kg ha^{-1} of P) even though these zones suffer from droughts, providing the crops are grown at a denser spacing (1 × 1 meter) than is the common practice of placing one plant every 2 × 2 meters. This farming practice of using very wide spacing probably was an adaptation to the very infertile soils and the need to plant a large area in a short period of time. Note that increasing the plant density without applying fertilizer can result in a decrease in yield because the plants may suffer a severe deficiency of nitrogen and turn yellow. Also, applying fertilizer but retaining the wide spacing may only result in a small increase in yield that may not be economic. Both new practices must be used together—fertilizer application and denser plant spacing.

In the U.S. state of Iowa, the yield gap for maize is about 36% (i.e., with a FY of 10 ton ha^{-1}, the yield gap would be 3.6 ton ha^{-1} and PY would be 13.6 ton ha^{-1}). FY is close to the attainable yield and there is little opportunity for closing the yield gap by extending improved management methods and varieties. In this case, the major opportunity for increasing FY is by breeding improved maize hybrids and developing complementary improved management methods that result in greater yields, thereby increasing PY. Opportunities exist for increasing PY of maize because in 2014 the average rate of increase of PY was 1.1% per year relative to PY in 2010.

For wheat and rice, attempts should be made to increase FY by both decreasing the yield gap, through improved extension, and increasing PY by breeding improved varieties and developing complementary improved management methods. The average rates of increase per year in PY in 2014 were 0.6% for wheat and 0.8% for rice relative to PY values in 2010. Consequently, increasing PY of wheat and rice will not be easy.

In addition to improved cultivars, increasing yields of cereals usually will require enhanced soil nitrogen supplies. For example, cereals with PY of 6–9 ton ha^{-1} must take up 200–300 kg ha^{-1} of nitrogen (N) if they are to achieve these yields. Deficiencies in soil N are common in the tropics and subtropics. The major available additional source of soil N for cereal crops is from the application of nitrogenous fertilizers. On a global basis, increased applications of nitrogenous and phosphate fertilizers will be needed, but injudicious use can have costs in terms of nitrate pollution of groundwater, phosphate pollution of surface waters, and pollution of the atmosphere with gaseous nitrogen oxides (NO_x).

Another source of soil N for cereal crops is the symbiotic fixation of atmospheric nitrogen by previous leguminous crops. As Graham and Vance (2000) point out, however, there has been a worldwide decline in agricultural use of leguminous crops and inoculation with rhizobia. For example, expansion in land area devoted to cereal production has been associated, in some cases, with a decrease in area devoted to grain legumes. Graham and Vance (2000) have reviewed the advantages and constraints on increasing nitrogen supplies to cropping systems by increasing nitrogen fixation. From this review, it is clear that the main opportunity for enhancing contributions to agriculture from nitrogen fixation may be with the more extensive systems such as pastures that contain legumes, and that intensive agricultural systems will continue to need large applications of nitrogenous fertilizer or manure.

In addition to increasing food quantity, there is a need to enhance the nutritional quality of the food that people eat (Welch and Graham, 1999). For example, in South Asia, where cereal production increased fourfold between 1965 and 1995, grain legume production declined about 20%. Yet, grain legumes provide certain essential amino acids, vitamins, and minerals that are not provided in sufficient quantities by cereal grains.

I directed a project that increased yields of a grain legume, cowpea, on fields of poor farmers in a very dry part of sub-Saharan Africa (Hall, 2017). Cowpea is a cheap source of protein that complements the protein in cereals grown in the Sahel, such as pearl millet, sorghum, and rice. In addition, through nitrogen fixation and uptake of phosphate, and supply of hay to animals, cowpeas enrich the soil, such that cereals grown in rotation produce greater yields.

Information from several international centers that are working to enhance yields and qualities of the major cereals and grain legumes and other food crops can be obtained from the Consultative Group on International Agricultural Research website (www.cgiar.org). The U.S. Department of Agriculture's science magazine provides information on a wide range of agricultural topics and can be found online (www.ars.usda.gov/is/AR/).

Future needs for agricultural products will be influenced by the size of the human population. There are some parts of developing countries where human populations are increasing at rates as fast as 3% per year which, if maintained, will result in a doubling of these populations within the short period of 23 years. The doubling time in years can be calculated from the following equation:

$$\text{Doubling time} = \frac{100 \times \ln 2}{\text{annual percentage increase}} \tag{1.1}$$

In cases where the percentage increase rate is constant. Note that ln 2 = 0.693. If the rate of increase in yield is only 1% per year, it would take 69 years for the production to double; and in 23 years, production would only increase about 26%. That means that the amount of food per person for the human population that doubled in 23 years would have decreased over the 23 years to only 63% of what it had been at the beginning.

Providing the additional food, housing, schools, hospitals, medicines, jobs, and so on, required by rapidly increasing human populations will be an impossible task. Agricultural development can promote rural development, which can result in reductions in birthrates. The increase in world human population is slowing down, and attempts should be made to achieve zero or negative population growth through agricultural development and other methods. The objective would be to achieve a balance between the capacity of the Earth to provide agricultural products in a sustainable manner and the demands set by the needs of the people. It is not clear whether a sustainable balance can be achieved. Large human populations usually cause substantial damage to the biosphere and have already caused substantial damage in some places, such as through overfishing. Damage to the biosphere reduces the resources that are available to all organisms, including people, and compromises the health of the biosphere.

The ability to produce agricultural products depends on the resources available for agriculture. In the United States and many other countries, urban sprawl and new highways are continuing to take away much of the best arable land. In addition, some of the land area that is not cultivated at this time is fragile, and its cultivation could result in environmental problems such as enhanced soil erosion, pollution of aquatic systems, and reductions in the area and quality of wetlands required by migrating birds. In addition to enhancing food supplies, increased efficiency of crop production can contribute to the maintenance of environmental health and biodiversity by enabling crop production to be practiced on current arable lands, permitting the other lands to continue to be used as natural habitats.

Increases in the efficiency of irrigation and use of agricultural chemicals are needed because, in addition to enhancing the profitability of agriculture, they can reduce damage to the environment. Reducing irrigation requirements can make more water available for maintaining natural aquatic systems. In many watersheds, competition with domestic, industrial, and environmental requirements will reduce supplies of water for irrigation. Development of new projects for enhancing water supplies through building dams, reservoirs, and canal systems has slowed down and is being constrained by the recognition of the complex impacts of these endeavors. Even the opposite trend is occurring in the United States, where consideration is being given to removing some dams to try to return rivers to their wild state and enhance habitats for salmon and other creatures that depend on those rivers. Increases in the efficiency whereby agricultural chemicals are used could reduce the extent to which bodies of water become polluted. The biosphere can benefit in many ways from increased efficiencies of different aspects of crop production systems.

What are the possibilities for increasing various efficiencies of crop production? Will there be technological revolutions in the twenty-first century that provide alternative methods for producing the foods, beverages, clothes, biofuels, and other important materials that we obtain from agriculture? Some simple principles provide guidelines concerning the types of approaches that will be effective in increasing various efficiencies of crop production.

First, mankind will continue to obtain most food energy requirements (carbohydrates) from crop plants growing in fields that are harvesting energy from solar radiation by photosynthesis. The reasons for this constraint are twofold: (1) there currently is no replacement for the sun as the major supplier of the massive amounts of energy required by food production systems, and (2) field crops are the most efficient mechanism for harvesting this radiant energy. In principle, nuclear fusion could provide tremendous amounts of energy, but its use on a large scale in agriculture, such as by providing artificial lighting, would subject the Earth to destructive levels of thermal pollution.

Second, during the twenty-first century, most of the food energy for mankind is still likely to come either directly or indirectly from current major crop plants, particularly the cereals: wheat, rice, and maize. By *indirectly*, I mean where these and other cereal grains, such as barley and sorghum, are fed to livestock, such as pigs, that then provide food for people. As people become more affluent, they often demand more meat or other livestock products such as eggs, milk, butter, and cheese, and this can result in a considerable diversion of cereal grains into the production of these products. About 90%–95% of the food energy available to people is lost when people eat meat from animals fed on cereals instead of eating the cereals directly, and it takes 5–6 kilograms of the proteins in cereals to produce 1 kilogram of animal protein. This means that greater crop production is required per person when people eat diets with a substantial component of animal products.

Complete vegetarianism for everyone is not a practical solution to future problems concerning food production. Many people prefer diets that contain some meat or fish or other animal products, and animal products can enhance the nutritional value of food. Future expansions in fisheries will include more fish farming, and this places greater demands on agriculture, because it involves providing the fish with supplementary food in the form of plant protein from field crops. Note that a substantial part of livestock protein involves animals that are fed on plant products, such as grass, and other plant products that humans cannot digest or do not wish to consume.

An additional reason why mankind will continue to depend on the cereal grains for major supplies of food energy (and protein) is that a large area (about 75%) of the cultivated land is being used to produce cereals at this time. Consequently, converting agriculture away from cereal production would require considerable effort. There are many other types of food crops. For example, certain Indian tribes in the United States used acorns as a staple food, after processing by leaching to remove tannins from the acorns. However, converting farms and industries to produce and process other types of crops, such as acorns, would take many years. Also, people are very conservative with respect to the foods that they prefer to eat and they do not readily adopt new staple foods. In addition, cereals are very effective as food and feed crops because they are easy to process, transport, and store. This is important since the marketing of staple foods and feed for livestock operates on a global scale.

Radical changes in approaches to field crop management have been proposed. For example, it has been suggested that *organic* methods, defined as those in which only natural products can be used as inputs, would be less damaging to the biosphere. Large-scale adoption of *organic* farming methods, however, would reduce yields and increase production costs for major crops in most cases (Stewart, 2004). In addition, food produced by organic farming methods has no scientifically proven benefits for consumers (Fischer et al., 2014). Inorganic nitrogen supplies are essential for maintaining moderate to high levels of productivity for many of the non-leguminous crop species, such as cereals, because organic supplies of nitrogenous materials often are either limited or more expensive than inorganic nitrogen fertilizers, which cannot be used on organic farms. In addition, there are constraints to the extensive use of either manure or legumes as *green manure* crops (Graham and Vance, 2000).

In many cases, weed control can be very difficult when using organic farming methods, since synthesized herbicides cannot be used. In addition, organic methods can require much hand labor, which may not be available since fewer people are willing to do hand labor as societies become more affluent.

Some methods used in organic farming, however, such as the judicious use of crop rotations and specific combinations of cropping and livestock enterprises, can make important contributions to the sustainability of rural ecosystems. Developing the most effective and sustainable systems will require a scientific synthesis of the best ideas coming from organic and other approaches to farming, but with emphasis on mainstream approaches to farming.

Why not assume that genetic engineering will make possible substantial increases in the efficiency of crop production? Refer to Chrispeels and Sadava (1994) for descriptions and a discussion of plant genetic engineering, and Miflin (2000) and Fischer et al. (2014) for some more recent information. The simple answer to this complex question is that genetic engineering is unlikely to have a large impact on potential crop yield (PY) of the major annual crops. Physiological analyses described in Chapter 4 indicate that, under optimal conditions, some current cropping systems already may be producing close to the maximum possible biomass per unit land area per day. Also refer to the analysis of limits to crop yield by Sinclair (1994). Fischer et al. (2014) concluded that genetic engineering (in this case defined as the use of transgenes) had not had a measurable effect on the yields of wheat, rice, maize, and soybeans, and virtually all other crops.

Reductionist studies at the molecular level have resulted in major advances in medicine. Could reductionist studies have the same impact on crop science? I feel this is unlikely. Improvements in medicine typically have resulted from changing the phenotype of individuals using medicines whose effects were evaluated on those individuals which was substantially assisted by reductionist information. In contrast, improvements in crop science have resulted from changing genotypes and developing improved varieties which is much more difficult. New varieties have to be evaluated as populations in a range of environments which requires much integrationist information and is only moderately assisted by reductionist information.

When combined with conventional plant breeding, genetic engineering has the potential to develop crop cultivars with greater resistance to pests and diseases. These resistant cultivars could make a major contribution to environmental health and safety if they can be grown with little or no use of pesticides. Note that varieties produced by genetic engineering are as safe as those produced by conventional plant breeding as long as the traits that are bred in by either method are safe from a human and an ecological standpoint (Stewart, 2004). Genetic engineering and conventional plant breeding also have the potential to enhance crop resistance to abiotic stresses, such as by providing cultivars with increased tolerance to freezing, chilling, or heat. But increased understanding of plant organ or whole plant responses to the environment is needed if genetic engineering is to be completely effective, as will be shown in this book.

What about using genetic engineering and conventional plant breeding to develop crop cultivars that have sufficient adaptation to drought such that they can be grown in the deserts of the world with little irrigation? Physiological analyses described in Chapters 8 and 9 indicate that it is possible to develop cultivars of some crop species that could survive in deserts with little irrigation, as do many native species, but that, similar to the native species, their production per unit land area per day would be very low (also see the analysis of Sinclair, 1994).

What about breeding crops that could be productive when irrigated with seawater? The biosphere has an ample supply of seawater. The difficulties confronting the breeding of salt-tolerant crops that could be irrigated with seawater are discussed in Chapter 11. However, progress has been made in breeding varieties of wheat and rice with resistance to salinity and there are opportunities for using halophytes as food crops, such as quinoa, that can be grown on seawater. Note that irrigation with seawater has detrimental effects on the structure of soil that contains swelling-clay particles, causing it to have very low permeability and poor aeration. Some sandy soils do not have this problem and can be irrigated with seawater without causing either low permeability or poor aeration.

Global climate change will result in increases in level of seawater and greater flooding of some delta and low-lying areas that would negatively impact production of rice and other crops. Important progress made in developing submergence-tolerant varieties of rice through marker-assisted backcross breeding to incorporate the *SUB*1 gene is discussed in Chapter 11.

An important opportunity through the combination of genetic engineering and plant breeding is the development of crop cultivars that produce harvested products with special attributes desired by mankind, such as various types of vegetable oils and starches or special proteins, including ones with pharmaceutical or industrial uses, and enhanced levels of vitamins. Substantial progress already has been made in the genetic engineering of plants to produce these special chemicals (Chrispeels and Sadava, 1994; Miflin, 2000).

Genetically engineered plants also have the potential to cause specific problems, and the potential problems and benefits must be considered on a case-by-case basis prior to their release for commercial use (Barton and Dracup, 2000). Refer to the website of the Union of Concerned Scientists (www.ucsusa.org) for a discussion of these potential problems and Miflin (2000) for a broad analysis of both the problems and opportunities that could result from crop biotechnology. Procedures have been established in the United States to try to ensure that food produced by genetically engineered crops is at least as safe and nutritious as food from conventional crops, considering toxin or allergen production, decreases in nutrient levels, and development of antibiotic resistance (Kaeppler, 2000).

In subsequent chapters, I will discuss some principles of plant responses to environment and experimental approaches, including the use of mathematical models. I will focus on plant physiological and developmental responses to light and temperature, and plant–water relations. I will point out areas where this information has relevance to the development of improved crop management practices and crop cultivars, and crop adaptation to global climate change.

I will describe how climatic zones may be defined in relation to crop adaptation and optimal land use in crop production. These definitions will be based on temperature, rainfall, and the evaporative demand of the atmosphere. I also will discuss methods for determining where specific crops can be grown.

I will describe radiation and energy balances and show how they can be related to global climate change. I will present methods for predicting crop water use. I will explain how consideration of the hydrologic budget and crop physiology and the stage of crop development can be used to optimize irrigation management.

Flooding and salinity effects on plants are discussed in Chapter 11. Flooding will be exacerbated by global climate change, so I will examine methods for breeding crops with resistance to flooding and salinity. I also will examine crop responses to limiting soil conditions that are difficult to change, such as extremes of soil texture and high soil bulk density. I will discuss plant responses to the toxic levels of boron often found in arid lands and some irrigation schemes. I will describe plant responses to toxic levels of aluminum that occur in acid soils often found in the humid tropics.

I will examine the interactions among crop responses to pests and diseases, and abiotic factors such as drought and temperature in Chapter 12. These interactions illustrate why a systems approach is needed when developing improvements to agriculture.

In the concluding chapter, I will integrate the various topics that were discussed earlier to illustrate how understanding of crop responses to environment can guide an ideotype approach to plant breeding.

For ease of reading, I mainly use the common names of crops and native plants, but the scientific names are provided in a section at the end of the book (Appendix). I have included some references of general value for additional reading at the end of each chapter. Within chapters I provide a few websites that are likely to be maintained for many years. For example, www.plantstress.com provides information on the physiology, agronomy, and breeding of crop responses to abiotic stresses. All references made in the text are provided in a section at the end of the book.

ADDITIONAL READING

Barton, J. E. and M. Dracup. 2000. Genetically modified crops and the environment. *Agron. J.* 92: 797–803.

Chrispeels, M. J. and D. E. Sadava. 1994. *Plants, Genes and Agriculture.* Jones and Bartlett Publishers, Boston, MA, p. 478.

Evans, L. T. 1993. *Crop Evolution, Adaptation and Yield.* Cambridge University Press, Cambridge, UK, p. 500.

Evans, L. T. 1998. *Feeding the Ten Billion—Plants and Population Growth.* Cambridge University Press, Cambridge, UK, p. 247.

Fischer, R. A., D. Byerlee, and G. O. Edmeades. 2014. *Crop Yields and Global Food Security: Will Yield Increase Continue to Feed the World?* ACIAR Monograph No. 158 Australian Centre for International Agricultural Research, Canberra, Australia, p. 634.

Graham, P. and C. P. Vance. 2000. Nitrogen fixation in perspective: An overview of research and extension needs. *Field Crops Res.* 65: 93–106.

Hall, A. E. 2017. *Sahelian Droughts: A Partial Agronomic Solution.* Nova Science Publishers, New York, p. 216.

Miflin, B. J. 2000. Crop biotechnology. Where now? *Plant Physiol.* 123: 17–27.

Sinclair, T. R. 1994. Limits to crop yield? In K. J. Boote, J. M. Bennett, T. R. Sinclair, and G. M. Paulsen (Eds.), *Physiology and Determination of Crop Yield.* Crop Science Society of America, Madison, WI, pp. 509–532.

Stewart, C. N. Jr. 2004. *Genetically Modified Planet: Environmental Impacts of Genetically Engineered Plants.* Oxford University Press, New York, p. 240.

ADDITIONAL READING

2 General Principles

Several general principles are relevant to plant responses to environment. The first principle is relevant to all of the biological sciences and illustrates the importance of studying environmental plant physiology at different levels of biological organization.

2.1 COMPLETE UNDERSTANDING REQUIRES INFORMATION FROM SEVERAL LEVELS OF BIOLOGICAL ORGANIZATION

Crops should be studied at several levels of biological organization, for example, community, whole-plant, cellular, and molecular levels. Studies at lower levels of organization are useful for discovering cellular or molecular mechanisms of adaptation and for developing selection criteria for use in plant breeding, since the mechanisms are closely related to gene action. Studies at higher levels of organization are needed if we are to understand the effects of changes in cultivars or management practices on productivity or crop water use or other aspects of crop community function, such as competitiveness with weeds and system sustainability. The reason for this is that these different levels are hierarchical and, in addition to molecular properties common to all of the levels of organization, higher levels of organization have their own unique emergent properties. I will provide some examples of emergent properties and higher-level effects to show why it is important to take an integrative approach as well as a reductionist approach when studying crops.

Individual alleles (genes) can have multiple effects (pleiotropy) that have different manifestations at different levels of organization. A single gene may not only affect the target process, it can also have other effects that are either beneficial or detrimental. My research group has developed a cowpea cultivar with greater yields in hot environments (Ehlers et al., 2000) by incorporating several genes that confer heat tolerance. At the cellular level, one of these genes appears to maintain membrane integrity at high temperatures (Ismail and Hall, 1999). At the organ level, this gene enhances the number of flowers that set pods under high night temperatures (Ahmed et al., 1992; Thiaw and Hall, 2004). At the whole-plant level, this gene increases grain yield under high night temperatures, but there also is dwarfing of the plant due to reductions in the lengths of the internodes (Ismail and Hall, 1998). We have evidence of associations between pod set and dwarfing, but they might be caused by close linkage of a gene conferring heat tolerance during floral bud development and another gene affecting internode length (Ismail and Hall, 1999). For communities of plants in hot environments, the heat-tolerance genes enhance productivity to a greater extent under narrow rows than under very wide rows where the dwarfing exacts a penalty on interception of solar radiation and canopy photosynthesis (Ismail and Hall, 2000).

Attenuation of effects can occur during progression up the levels of organization. The enzyme responsible for the initial fixation of CO_2 in C_4 plants (PEP carboxylase) can exhibit much greater ability to fix CO_2 at ambient levels of [CO_2] than the enzyme responsible for the initial fixation of CO_2 in C_3 plants (RuBP carboxylase or rubisco). Leaves of C_4 and C_3 plants, however, exhibit smaller differences in rate of CO_2 fixation, and canopies of these crops exhibit even smaller differences (Gifford, 1974). The explanation for this attenuation of effects is that, at the leaf level, additional factors affect the fixation of CO_2, such as stomata, and at the canopy level, more limiting factors are present, such as the canopy resistance to the transfer of CO_2 from the air above the canopy to the leaf surface.

The effects of differences in stomatal opening on transpiration are attenuated in a similar but even more complex manner (Jarvis and McNaughton, 1986). At the leaf level with small leaves, high wind speeds, and high cuticular resistance to water flow, transpiration rate is proportional to stomatal conductance and about proportional to the area of the stomatal pores (Jones, 1992). For a tall, isolated plant with small leaves subjected to a strong wind in a location with no other vegetation, the transpiration rate would be proportional to stomatal conductance, because these leaves are not influencing their environment. In contrast, where there is an extensive smooth and dense canopy of leaves and a low wind speed, the functioning of the leaves would influence their environment. In this case, differences in stomatal opening would have only small effects on transpiration rate for the following reasons. With a change in stomatal opening, the change in overall canopy conductance would be small due to the relatively large resistances to water vapor flow imposed by the boundary layer of the leaves and the canopy. In addition, with a change in transpiration rate, there would be counteracting effects due to humidification of the canopy and cooling of the leaves that would decrease the driving force for transpiration. In this case, the rate of transpiration would depend more on the supply of radiant energy necessary for providing the latent heat of vaporization than it would on factors (such as stomatal opening) that influence the potential for vapor transfer. Consequently, genes or management methods that influence stomatal apertures could have large effects on relative plant transpiration and water use in well-stirred leaf cuvettes or growth chambers and with isolated plants or the small plots used in many experiments, but have only small effects with the large areas and dense populations of plants used by farmers.

A mutant has been discovered that causes leaves of cowpea to have substantially less chlorophyll per unit leaf area, such that the leaves appear a light greenish-yellow (Kirchhoff et al., 1989b). One might expect that the photosynthetic performance of the leaves would be impaired by this mutation. The chloroplasts have few grana, but the only effect on photosynthesis that was detected was a slight reduction in net carbon dioxide exchange at low light due to leaves of the mutant absorbing less light than the wild type (Kirchhoff et al., 1989c). The negligible effects of this chlorophyll deficiency may be explained by the fact that photosynthesis can be limited by factors other than either chlorophyll content or the photosystems, especially at high light levels. With populations of plants in the field, under sunny conditions, there was no difference in performance in that the mutant produced the same shoot biomass and grain yield as the wild type (Kirchhoff et al., 1989a). A canopy of leaves that are light green and reflect and transmit more light could have a more uniform distribution of light than a canopy of dark green leaves. A canopy with a more uniform distribution of light would have greater photosynthesis per unit ground area. This canopy effect would offset any reductions in photosynthesis by individual shaded leaves that might result from the chlorophyll deficiency. A hypothesis was tested using chlorophyll mutants of soybean that selection to reduce chlorophyll investment per unit leaf area could increase grain yield by improving light penetration and distribution in the canopy, thereby increasing the photosynthetic conversion to biomass efficiency. Studies with mutants did not support this hypothesis but did indicate that there may be an overinvestment in chlorophyll per unit leaf area in current soybean cultivars (Slattery et al., 2017). It should be noted that the results of research of this type will depend on the type of chlorophyll mutant which is used, in that some types of *chlorophyll deficiency* do have detrimental effects on plant function. For example, where soil nitrogen is strongly limiting, it can cause deficiencies in chlorophyll and other components of the photosynthetic system such as rubisco. In these cases, rates of photosynthesis per unit leaf area and per unit ground area and yields are substantially reduced.

A specific example is presented of what appears to be an emergent leaf property. Where photosynthetic capacity is smaller due to genetic, environmental (e.g., due to limiting supplies of nitrogen or phosphate), or developmental (e.g., aging) causes, maximal stomatal conductance also is smaller (Schulze and Hall, 1982). The overall effect is a coordination, such that balances are maintained between the processes influencing the supply and the fixation of carbon dioxide, and between the rates of photosynthesis and transpiration.

An example of an emergent whole-plant property is that there is a degree of coordination between shoots and roots with respect to their growth rates and activities (Brouwer, 1962; Farrar and Gunn, 1998). Presumably, evolution and plant breeding favored plants whose roots grow and function at rates that enable them to provide the supplies of nutrients and water needed by the shoot as they are determined by the growth rates and activities of the shoot, and without excessive investment of carbohydrate and chemical energy in root tissue and root function. The optimal balance between root and shoot activity would depend on the soil and aerial environments. I will provide an example to illustrate this point. For many years, I wondered why farmers in the Sahelian zone of Africa grew sorghum plants at extremely wide plant spacings of about 2 × 2 meters. It had been suggested that this wide spacing represented an adaptation to the droughts occurring in this semiarid environment, but we had evidence from studies with another species (cowpea) that this might not be valid. I made an observation in southern Kordofan in the Sudan that suggested another hypothesis. I saw sorghum planted at two spacings: sparse 2 × 2 meters, and very dense 0.5 × 0.2 meters, with either no fertilizer or a moderate amount of nitrogenous fertilizer. During the middle stage of vegetative growth, the plants under dense spacing and no fertilizer had become very chlorotic, indicating a deficiency of nitrogen in their leaf tissue. In contrast, the plants in the other three treatments appeared to be healthy. The treatment with sparse spacing and no fertilizer was particularly interesting. Apparently, the sparse planting at 2 × 2 meters is an adaptation to the infertile soil in relation to the balance in root and shoot growth maintained by the sorghum cultivars that were being grown. With this wide spacing, the roots continually access sufficient nutrients from the infertile soil as they grow to meet the demand set by the amount of shoot growth per hectare of land area. With the dense spacing, and after the initial growth stage, the roots of adjacent plants are beginning to compete and do not access sufficient nutrients from the infertile soil to meet the demand set by the amount of shoot growth per hectare of land area. Shoot growth during the initial stage would have been much greater at the closer spacing than at the wide spacing due to there being 40 more plants per unit area of land.

In many cases, appropriate root/shoot balances and activities are maintained when plants are supplied with slightly suboptimal levels of soil inorganic nitrogen and phosphate. With smaller supplies of soil nutrients, plant shoots grow more slowly, but in all other respects the plants appear normal. Slightly deficient plants have much less leaf area, but the supply of protein and enzymes per unit leaf area is regulated so that the plants maintain near-normal activities per unit leaf area. Only with moderately deficient supplies of soil nitrogen or phosphate do plants exhibit symptoms of disturbed function per unit leaf area. Also, plant appearance is not effective in detecting small, or in some cases even moderate, deficiencies of soil nitrogen or phosphate unless *control* plants are available that have been provided with more abundant supplies of soil nutrients to permit comparisons of plant size. In cases where farmers doubt whether a particular treatment will enhance plant performance, they can apply the treatment to a small strip across the field. If they see a positive response in the strip of plants, they could either adopt the treatment as their normal practice or work with scientists to quantify the effect and determine whether it is profitable. This would be done by conducting field experiments with replicated strips or plots having or not having the treatment and conducting statistical and economic analyses.

The mechanisms whereby root and shoot growth rates and activities are coordinated are poorly understood at this time but are thought to involve hormones, such as abscisic acid and cytokinins, transported between the root and shoot in the xylem (and possibly also in the phloem). The coordination of root and shoot growth rates and activities also can be critical for adaptation when plants are subject to drought because the root system determines the supply of water to the plant, whereas the leaf area and extent of opening of stomata in the leaves determine the rate at which water transpires from the plant as vapor. When plants are subjected to drying soil or some other edaphic stresses, such as soil compaction, stomata partially close and leaf expansion rates slow down, and these responses may involve hormonal communications between roots and shoots. Contrasting genotypes that produce different levels of abscisic acid and split root systems provide useful approaches for studying the communication mechanisms between roots and shoots (Mulholland et al., 1999).

Amplification of effects can occur with time at the whole-plant level but with less or no effect at the canopy level. Assume that a specific gene causes more carbohydrate to be partitioned to leaves and results in faster increases in leaf area. For an isolated plant, this could have a progressively larger effect on plant biomass accumulation rate with time in that greater leaf area would result in greater interception of solar radiation, greater photosynthesis, more carbohydrate, faster leaf growth, even greater interception of solar radiation, even greater photosynthesis, and so on, compared with a control (wild type) plant. For plants in communities, this gene would only enhance biomass production rate during the early seedling stage. Once the individual plants are competing in the aerial environment, additional leaf area would not result in greater interception of solar radiation compared with a community of control plants, and there would be little effect of the gene on biomass production or amplification of the effect of the gene.

The function of plant organs may influence the structure of plant communities. In a study of two co-dominant shrubs in the Mojave Desert in California, Mahall and Callaway (1991) demonstrated that roots of the creosote bush inhibit the growth of roots of the shrub ambrosia and also other roots of creosote bush in their vicinity. In contrast, root systems of ambrosia appeared to have the ability to detect and avoid other ambrosia root systems. In a subsequent paper, Mahall and Callaway (1992) discuss how these species differences in root communication could explain the commonly observed regular distribution of creosote bush and the clumped intraspecific distributions of ambrosia in the plant community. The root-mediated allelopathy (suppression of growth of one plant by another due to the release of toxic substances) of creosote bush would inhibit the growth of young plants of both ambrosia and creosote bush in the vicinity of established individuals of creosote bush. In a comprehensive review, Schenk et al. (1999) point out that similar root-mediated allelopathy has been observed with black walnut, silk oak, apple, peach, and guayule. In contrast, the detection and avoidance manifested by the ambrosia root systems would enable individual ambrosia plants to grow close to each other without competing in the soil. Evidence for spatial segregation of root systems of plants having the same genotype has been reported for onion, soybean, liquid amber, and *Pinus taeda* (Schenk et al., 1999), but it is not known whether this segregation is due to root-mediated allelopathy or is of the type found in ambrosia that permits plants to grow close together in an efficient manner. Annual crop species that are grown at high densities would benefit from having a system of root communication that enables the plants to grow close to each other without competing in the soil (Chapter 13). Species differences in root communication could have important effects on either the intercropping of different species or crop–weed interactions.

It should be apparent that effects seen at the cellular, organ, or whole-plant level may not be seen or may be stronger or different or more complex at the population or community levels. The functions of a population of a crop species or communities of species determine whether crop productivity will by increased by changes in cultivars or management practices.

2.2 SEPARATING CAUSES AND EFFECTS CAN BE DIFFICULT

Major advances in plant physiology have resulted from the recognition that some *effects* are in fact *causes*. I will provide three examples to illustrate this point. First, when plants are subjected to lower air humidity, their stomata partially close. In earlier years, the following hypothesis was proposed to explain this phenomenon: lower air humidity results in faster transpiration that causes a decrease in bulk leaf water content, a reduction in the turgor pressure in guard cells, and thus stomatal closure. A model based on this hypothesis predicts some instability and a tendency for oscillations to occur in bulk leaf water status, transpiration, and stomatal conductance—and this can occur if plant water status is perturbed in a rapid and unnatural manner, such as by excising the roots. Then, in certain experiments, partial stomatal closure was shown to occur in drier air that was associated with either no change or an improvement in bulk leaf water status, which did not fit the hypothesis. An alternative hypothesis was proposed: the drier air causes a reduced water content of the epidermal tissue, which results in partial stomatal closure and thereby acts to prevent or reduce changes in bulk leaf water status by moderating changes in transpiration (Schulze and Hall, 1982).

The second example concerns stomatal closure with soil drought. The established hypothesis for explaining this phenomenon was that with progressive drying of soil in the root zone there was a reduction in bulk leaf water status that caused stomata to close. However, a key paper by Bates and Hall (1981) demonstrated that stomatal conductance of cowpea decreased with day-to-day depletion of soil water with no decrease in bulk leaf water status. The authors hypothesized that progressive stomatal closure with soil drying probably was caused by changes in hormonal signals between roots and leaves. According to this hypothesis, stomatal responses to soil drying are not caused by changes in bulk leaf water status but act to reduce transpiration and prevent excessive reductions in bulk leaf water status.

The third example concerns relations between vegetative and reproductive growth. On several occasions, I have discussed cowpea crops that exhibited lush vegetative growth and few pods with farm advisors and farmers. They have asked me if the interval between irrigations of these cowpea crops should be extended to subject them to drought. Their reason for suggesting this procedure was that they assumed that drought-induced reductions in vegetative growth would result in a diversion of more carbohydrates to fruiting tissue and increases in pod production. I explained to them that this procedure would not be very effective because the opposite cause–effect relationship had occurred. The crop was exhibiting too much vegetative vigor because some factor had prevented fruiting. Typical cases where this occurs are when either insect pests or high night temperatures have damaged floral buds such that few or no fruiting structures have been produced on the main stem, and additional vegetative branches have been produced instead. The solution to the problem is to take steps to ensure that the crop produces flowers and pods on the main stem, such as by either controlling insect pests or using heat-tolerant cultivars. The reproductive growth will then act to constrain vegetative growth by reducing the number of vegetative branches that are produced and by some unknown mechanism attracting carbohydrate, which is then not partitioned to vegetative organs.

2.3 LIMITING FACTORS, SYNERGISMS, AND SOURCE/SINK EFFECTS

Analysis of factors that limit productivity is useful because it can provide clues concerning approaches for increasing productivity by changing either management practices or the cultivar that is being used. In cases where crop productivity is constrained by several major limiting factors, it is important to know how they interact. Several possibilities are apparent. Productivity may be limited only by the most limiting factor, with changes in other factors having no effect until this factor is brought to a higher level. Alternatively, two or more factors may be co-limiting. In this case, increased supplies of all co-limiting factors may be needed to increase productivity, or increases in any of these factors may increase productivity, with the effects being independent and additive or interactive and synergistic. The latter case involving synergism is particularly interesting to crop scientists and farmers, because overcoming limitations of this type could result in major increases in crop productivity.

Synergism was evident in the responses of rangeland to increased supplies of water and fertilizer in the semiarid Sahelian zone of Africa (Breman and de Wit, 1983). In the wetter part of the zone, with 500 millimeters of rain falling in one season of about four months (a location similar to the one described in Figure 10.8), mean annual shoot biomass production by annual grasses was 2,000 kg dry weight/ha. With irrigation to provide optimal supplies of water, but no fertilizer, annual shoot biomass production was 5,000 kg/ha. With fertilizer to provide optimal amounts of nitrogen and phosphorus, but no irrigation, annual shoot biomass production was 10,000 kg/ha. Plant growth in the first part of the season was limited by low phosphorous, whereas growth in the last part of the season was limited by low nitrogen. With both irrigation and fertilizer application, the annual shoot biomass production was 55,000 kg/ha, which clearly is a synergistic response in that the predicted production assuming additive effects only would have been 2,000 + 3,000 + 8,000 = 13,000 kg/ha, which is much less than 55,000 kg/ha. A likely explanation for the synergism is that the additional water from the irrigation resulted in a longer potential growing season that could be fully exploited only if the nutrient supplies in the soil also were enhanced. It should be noted that for plants which cannot fix atmospheric nitrogen, such as range grasses and cereals, supplies of some macronutrients such as nitrogen often must be increased to take advantage of increases in yield potential arising from changes in other factors.

The practical significance of the research in the Sahel was the demonstration that plant productivity is strongly limited by soil infertility, even in the presence of some drought. The responses to irrigation were of little practical significance, because irrigation is not economically feasible in much of the Sahel for rangeland or most of the cereal production (although some rice is produced under irrigation in river basins). The authors proposed that the application of fertilizer to rangeland also would not be economical in the Sahel. They did propose that application of phosphate fertilizer to arable leguminous crops such as cowpea may be useful. The fertilized cowpea crop would grow more rapidly and fix more nitrogen from the atmosphere, thereby enhancing soil fertility for subsequent cereal crops that provide staple foods. In addition, the cowpea would provide both protein-rich grain as food for people and protein-rich fodder for livestock, which would produce manure that could be used to further enhance soil fertility.

Where economic yield involves fruit or seed, the determination of limiting factors can be complex. In these cases, scientists have asked what is most limiting to yield: the photosynthetic sources of carbohydrates or the reproductive sinks for carbohydrates (Evans, 1993)? Answers to this question would provide guidance concerning selection criteria that could be used in breeding improved cultivars. In optimal environments, reproductive yield of major grain crops may be co-limited by both sources and sinks such that substantial increase in grain yield requires increases in both the photosynthetic source and the reproductive sink. When environments are not optimal, such as with very hot weather, different limitations to yield occur with different species. For several warm-season crop species, including common bean, cotton, cowpea, rice, and tomato, reproductive development is damaged more by heat stress than is the photosynthetic system (Hall, 1992, 1993a; Ismail and Hall, 1998). Consequently, in these cases, heat tolerance can be enhanced by selecting to increase reproductive sink strength. In contrast for the cool-season crop, wheat, the photosynthetic system may be very sensitive to hot weather, and breeding for heat tolerance may need to address this source problem (Fischer et al., 1998). In addition, feedback linkages occur between sources and sinks that make it difficult to determine cause and effect. For example, the presence of reproductive sinks can cause the photosynthetic capacity of leaves to either increase or decrease. The decreases in photosynthetic activity are associated with the breakdown of photosynthetic enzymes in leaves and the translocation of amino acids to developing seeds, which can be pronounced in soybean cultivars that have seed with a high protein content (Sinclair and de Wit, 1975). The causes of increases in photosynthetic activity are not known, but it is possible that the presence of reproductive structures influences the hormonal balance of the plant, which then causes increases in the levels of several components of the photosynthetic system in the leaves. Another type of linkage is where higher rates of leaf photosynthesis during early floral development in wheat cause the development of a larger spike with more kernels, which subsequently generates a greater sink strength compared with plants that had slower rates of photosynthesis during early floral development.

There is an important example of synergism that involves changes in both the cultivar and the management practice. Cultivars of wheat and rice have been developed that partition more carbohydrate to developing grain and are partially dwarfed and more compact. With large supplies of nitrogen fertilizer, these semi-dwarf cultivars have much greater productivity than the older, tall cultivars. This agronomic system was responsible for the *green revolution* that produced major benefits for mankind (Evans, 1993; Evans and Fischer, 1999) but also had broad socioeconomic consequences, some of which were not beneficial (Chrispeels and Sadava, 1994). When grown with a small supply of fertilizer, the semi-dwarf cultivars provide only a moderate increase in productivity over the tall cultivars because their yields are limited by the supply of nitrogen. The tall cultivars exhibit little increase in grain yield (i.e., harvestable) when given a large supply of nitrogen fertilizer because, under these conditions, the plants usually lodge (fall over), whereas the semi-dwarf cultivars do not lodge. Lodging involves the breaking or bending of stems. Lodging results in the crop becoming horizontal, which reduces canopy photosynthesis and makes the crop more difficult to harvest, and the grains may suffer from fungal diseases due to their closer contact with moist conditions near the soil surface. It should be apparent that the development of the *green revolution* system required changes in both management and cultivars and a team approach. A plant breeder who had tested new semi-dwarf genetic lines using traditional low levels of fertilizer would not have discovered this way to substantially enhance productivity. Similarly, an agronomist who had tested the responses of traditional tall cultivars to different higher levels of fertilizer application also would not have discovered this responsive system. The *green revolution* approach is more complex than I have described. The semi-dwarf wheat cultivars also require higher plant densities and more careful weed control because they are less competitive with weeds than the tall cultivars. The semi-dwarf cultivars are most responsive when water supplies are adequate, and many are broadly adapted with resistance to many diseases (Evans and Fischer, 1999).

A review of changes in harvest index (the ratio of grain mass to total shoot biomass) has been presented by Sinclair (1998) that places the *green revolution* in a historical context. He points out that past types of tall cereal cultivars with low values of harvest index were suited to the farming systems used in earlier years, when straw had a high value for use as bedding, feed for animals, material for thatching, and fuel for cooking. Also, he argues that higher harvest index is only possible if plants can acquire greater quantities of nitrogen from the soil, because the average nitrogen content of cereal grain is about five times greater than that of mature straw on a dry weight basis (e.g., average values of N content of wheat are 22 mg/g for grain and 4 mg/g for straw). A mass-balance model shows that a cultivar with a higher harvest index would require more plant-nitrogen per unit land area but less plant-nitrogen per ton of grain than a lower-yielding cultivar with a lower harvest index but the same total biomass in the shoot and root systems.

2.4 OPTIMIZATION AND EFFICIENCY

The simple concept that "bigger things (i.e., depth of roots) or faster things (i.e., rates of photosynthesis) are better" has little relevance to the selection of traits for developing improved cultivars. (Refer to Chapter 13 for a more complete discussion of this issue.) The adaptation of plants depends on complex optimizations and holistic harmony among the various parts and processes within plants. Adaptive quantitative traits are expressed at intermediate sizes or rates. For example, the depth of a root system that would be adaptive depends on several factors:

- The benefits that would be gained in terms of acquisition of water and nutrients
- The influences of these resources on plant function
- The costs to the plant of developing and maintaining the root system

Adaptation requires that plant systems be efficient as measured by various cost/benefit ratios.

The levels of plant characteristics that would be adaptive depend on the following three factors:

1. The target environment where the crop will be grown determines the specific intermediate level that will be adaptive. For example, deeper rooting is adaptive where water supplies are limiting and where hydrologic balance analyses indicate that, in most years, significant quantities of water will be available in the deeper parts of the soil profile where they can be accessed by the deeper roots.
2. The specific intermediate level that is adaptive depends on the genetic background of the plant. For example, plants having photosynthetic systems

with higher water-use efficiency can *afford* a deeper root system, because they produce more carbohydrate per unit of water transpired.

3. The extent of useful plasticity in character expression determines the breadth of adaptation to a range of environmental conditions. An example of useful plasticity is where plants have the ability to develop deeper root systems when subjected to low rainfall and when moisture is stored deep in the soil, but shallower root systems when grown under frequent small rains that are sufficient to meet the needs of the plant.

The concepts of optimization and efficiency have relevance to stomatal function in that greater stomatal opening has both beneficial effects (such as greater CO_2 assimilation rates and evaporative cooling) and negative effects (such as greater transpirational loss of water, more extreme plant water deficits, and reduced water-use efficiency). An elegant mathematical model has been developed by Cowan and Farquhar (1977) for quantifying the extent to which stomatal function is optimal with respect to maximizing daily photosynthesis and water-use efficiency for a given daily rate of water use. (Refer to Chapter 8 for a more detailed discussion.) Stomatal and photosynthetic responses to air humidity and temperature are quantitatively consistent with the model for optimal stomatal function (Hall and Schulze, 1980). Stomatal and photosynthetic responses to solar radiation and leaf age are qualitatively consistent with the model for optimal stomatal function. Plant adaptation depends on the optimization and efficiencies of many processes, such as the distribution of proteins involved in photosynthesis within the plant canopy as they influence the photosynthetic capacity of sun and shade leaves. (Also refer to Chapter 4.)

2.5 GENETIC AND ENVIRONMENTAL INFLUENCES ON PLANTS

The genotype of a plant (for which species and cultivar names are assigned) defines the range of performance of the plant and is determined by a set of heritable traits. The phenotype (which is simply the plant) produced by a particular genotype results from the interaction of the genotypic traits with the environment in which the plant is grown. Consequently, the same genotype growing in different environments can produce different phenotypes. When a set of cultivars is grown in contrasting environments, yield is determined by genotypic effects, environmental effects, and effects attributed to genotype × environment interactions. Statistical analyses of these effects make possible the definition of the geographic boundaries of different target production environments where specific types of cultivars will be effective, as well as the choice of specific cultivars that are best adapted to individual target production environments and can be recommended for use by farmers in these locations.

I will provide two simple examples to illustrate the meaning of genotype × environment interactions. Assume that the grain yields of a new cultivar and a current cultivar have been determined in experiments conducted in two different locations (Table 2.1).

TABLE 2.1
Grain Yield of Two Cultivars in Two Locations

	Yield, kg/ha		
	Location A	**Location B**	**Mean**
New cultivar	2,400	3,800	3,100
Current cultivar	2,800	3,000	2,900
Difference			200
Mean	2,600	3,400	
Difference	3,400 – 2,600 = 800		
Interaction	3,300 – 2,700 = 600		

The genotypic effect can be calculated as being the difference between the mean yields of the new and current cultivars (= 200 kg/ha). The environmental effect can be calculated as being the difference between the mean yields at locations A and B (= 800 kg/ha). The genotype × environmental interaction is the difference between the mean of 2,800 and 3,800 and the mean of 2,400 and 3,000, which is equal to 600 kg/ha. If a suitable experimental design had been used, including use of replicated plots of the cultivars in each environment, it would be possible to use statistical procedures to determine the probability to which the differences associated with the various effects deviate from zero. Assuming that all of the various effects were significant, for example at a 5% level, we could now conclude that the new cultivar had a greater average yield than the current cultivar, but this would have no practical value due to the presence of the interaction. We also could conclude that, on average, location B was more productive than location A. Of particular importance is the genotype × environment interaction due to the new cultivar performing better than the current cultivar in location B, but worse than the current cultivar in location A. A practical result of this study might involve recommending that farmers in location A continue to use the current cultivar, whereas farmers in location B should consider using the new cultivar. Typically, these types of trials are conducted over several years and locations within the target production zone to ensure the reliability of any recommendations (predictions) made based on the results of the trials.

A more subtle type of genotype × environmental interaction may be seen in the next set of data on yield responses of two cultivars to thoroughly watered (wet) and dry environments (Table 2.2).

In this case, an interaction occurred because the yield of the new cultivar was reduced more by the drought (–600 kg/ha) than was the yield of the current cultivar (–400 kg/ha). Note that the interaction is present even though, for both cultivars, yield was reduced 20% by the drought. Assuming the major differences are significant, these data indicate that the new cultivar has greater yield than the current cultivar in both the wet environment and the dry environment and that the interaction can be ignored when making recommendations to farmers.

TABLE 2.2
Grain Yield of Two Cultivars in Thoroughly Watered and Dry Environments

	Yield, kg/ha		
	Wet Environment	**Dry Environment**	**Mean**
New cultivar	3,000	2,400	2,700
Current cultivar	2,000	1,600	1,800
Difference			900
Mean	2,500	2,000	
Difference	2,500 – 2,000 = 500		
Interaction	2,300 – 2,200 = 100		

There are four types of plant responses to environment:

1. Quantitative reversible responses of plant processes to environmental factors, such as photosynthetic responses to changes in levels of solar radiation (refer to Chapter 4).
2. Phenological responses, where developmental changes, such as the initiation of flowering, occur in response to the effects of the photoperiod (day length) or accumulated heat units (refer to Chapter 6).
3. Irreversible stress responses of plants to environmental extremes, such as hot or cold temperatures or drought, where plant responses depend on the intensity, rate of imposition, and duration of the stress. Reproductive developmental processes often are more sensitive to abiotic stresses than are growth processes (refer to Chapters 5 and 9).
4. Acclimation, where the phenotype adjusts to changing environments in a manner that is adaptive. For example, phenotypic responses to mild stresses, in some cases, can enable plants to subsequently withstand more extreme stresses (for complex induction effects, refer to Chapter 12).

ADDITIONAL READING

Brouwer, R. 1962. Distribution of dry matter in the plant. *Netherlands J. Agric. Sci.* 10: 361–376.

Evans, L. T. and R. A. Fischer. 1999. Yield potential: Its definition, measurement, and significance. *Crop Sci.* 39: 1544–1551.

Gifford, R. M. 1974. A comparison of potential photosynthesis, productivity and yield of plant species with different photosynthetic metabolism. *Austral. J. Plant Physiol.* 1: 107–117.

Jarvis, P. G. and K. G. McNaughton. 1986. Stomatal control of transpiration: Scaling up from leaf to region. *Adv. Ecol. Res.* 15: 1–49.

Schenk, H. J., R. M. Calloway, and B. E. Mahall. 1999. Spatial root segregation: Are plants territorial? *Adv. Ecol. Res.* 28: 145–180.

Sinclair, T. R. and C. T. de Wit. 1975. Photosynthate and nitrogen requirements for seed production by various crops. *Science* 189: 565–567.

3 Experimental Approaches and Quantitative Methods

When studying plants, it is important to recognize that the type of experimental approach which is used can influence the plant response that is observed in both a quantitative and, more importantly, a qualitative manner. Also, the experimental approach can influence the extent to which predictions concerning plant responses, resulting from the research, are valid for farmers' fields. Examples of these effects will be provided in this chapter. The problems resulting from these effects can be minimized by studying crop responses to environment in several types of experimental conditions using:

1. Different field environments in contrasting locations or different seasons
2. Different controlled environments in greenhouses or growth chambers
3. Different environments in the same field by imposing treatments that vary specific environmental factors

The overall research program should be designed to take advantage of the various strengths of these different types of experimental approaches (Jones, 1992). Studies often begin with observations made on crops growing in different fields with contrasting environments.

3.1 VALUE OF EXPERIMENTAL STUDIES IN DIFFERENT FIELDS OR SEASONS HAVING CONTRASTING ENVIRONMENTS

Important information on the range of plant responses to environment and their adaptation can be obtained by comparing the function and development of the same set of genotypes growing in several contrasting field environments. Different temperatures can be achieved by choosing locations for the studies with different elevations. Different day lengths can be achieved by choosing locations with different latitudes. By choosing different dates of sowing, annual crops can be subjected to differences in both temperatures and day lengths. Locations with greater continental exposure will experience more extreme temperatures during the day and between winter and summer compared with locations nearer to large bodies of water. Note that research can exploit or be hindered by unexpected variation in environmental factors, such as temperatures and incidence of pests and diseases, which can occur from year to year in the same location.

Interpretations based on plant responses that are observed in this manner often are relevant to at least a specific range of field environments and provide reliable predictions for these conditions. The physical environment is complex, however, and it varies spatially and temporally (with time of day, from day to day, and from season to season). Consequently, experiments cannot be repeated in exactly the same way to rigorously test the validity of the results. This is not a trivial problem, because a fundamental feature of empirical science is that experiments should be repeatable. A partial solution to this problem is to conduct an experiment over several locations and years. Then, if a particular response occurs on many occasions and always is associated with a particular type of environmental condition, one can have some confidence in the generality of its occurrence and predictions concerning the environmental conditions where the particular response is likely to occur.

Determining causal factors for specific plant responses to environment can be difficult in field conditions because of the co-variation that can occur among environmental factors, such as positive correlations between levels of solar radiation and tissue temperatures, and between plant processes, such as changes in rate of transpiration and leaf water status. Also, precise measurement of either the plant process or the environmental condition associated with the response is difficult in field conditions, due to variation in the age of plant tissues such as leaves, and spatial variation in environmental conditions within the canopy or root zone, temporal variation of the environment, and unplanned variations due to pests and diseases.

3.2 VALUE OF EXPERIMENTAL STUDIES IN CONTROLLED ENVIRONMENTS

In controlled environments such as greenhouses, growth chambers, or gas-exchange cuvettes, one can vary individual environmental factors while keeping other factors constant. Thus, it is possible to separate and individually manipulate factors that co-vary in natural environments, such as the levels of shortwave radiation and leaf temperature or vapor pressure deficit (saturation vapor pressure of the air – actual vapor pressure of the air) and air temperature. Through the use of experiments where single factors are varied, it is possible to get a clearer understanding of causes and effects. Also, variation due to some unplanned factors, such as pests or diseases, can be eliminated by working in controlled environments.

Where environmental conditions are kept constant, precise measurements of plant processes and conditions, such as plant water status, can be made. The amount of plant material available for making destructive measurements, however, often is limited compared with field conditions where there is space for growing large numbers of plants. Continually removing leaves or other organs from plants can have obvious and more cryptic artifactual effects. The continual removal of leaf tissue can influence the water relations of the plant through effects associated with changes in the balance of the shoot and the root. Removing or even touching tissue also can induce plant responses that are systemic in that they are transmitted to other parts of the same plant, or they can cause plants to produce ethylene, which can influence plant function if the type of controlled environment that is being used permits the ethylene to accumulate. An example of the complexity of mechanical effects on plants is that brushing (or touching, rubbing, or shaking) can cause tomato plants to be shorter and also have enhanced tolerance to chilling stress (Keller and Steffen, 1995).

A potential problem with controlled environments is that they are always unnatural, and major artifacts can occur that reduce the reliability of interpretations and predictions concerning the plant responses which may occur in natural field conditions. Plant responses to drought often have been quantitatively and even qualitatively different in controlled environments from the responses observed under most natural field conditions. For example, leaves of cowpea plants often wilt in controlled environments when irrigation is delayed, yet I have rarely seen cowpea leaves wilt under natural field conditions, even with extreme drought that was killing some of the cowpea plants. Under drought in field conditions, the leaves of cowpea and many other legumes can become more vertical and move during the day, tracking the direct beams of solar radiation in a manner which appears to be adaptive in that it minimizes the interception of solar radiation (paraheliotropism) and the heat load on the leaves (Shackel and Hall, 1979). Certain genotypes of wheat, and other grasses, exhibit marked leaf rolling in the field when subjected to drought, but Jones (1992) was unable to get these wheat genotypes to exhibit this phenomenon in growth chambers. Stomata may respond differently to drought in controlled environments, exhibiting a threshold relationship with leaf water status that is not observed in most natural field conditions under drought (Chapter 8). A partial explanation for these artifactual responses is that the root volumes and available water to potted plants are usually much smaller in controlled environments than in natural field conditions where some annual crop plants can develop root systems more than 1 meter deep in the soil (Table 10.1). Consequently, drought often develops much faster in controlled environments, and acclimation processes do not occur as they do in the field, where they tend to change the plant response and ameliorate the effects of the drought. This problem usually cannot be solved by growing plants in larger pots, because pots that are big enough to simulate field root-zone conditions usually would be either too tall to fit in the growth chamber or too heavy for people to lift.

The use of artificial light can result in artifactual plant responses. Light levels are usually constant during the *day* in controlled environments, whereas they are rarely constant in sunlit environments. Also, most of the lighting systems used in controlled environments during the twentieth century did not develop the high levels of radiation that can occur in field conditions. The spectral distribution of artificial lighting systems has been different from that of the sun in virtually all cases and can result in artifactual effects. For example, floral bud development of cowpea is arrested by high night temperatures and long days under sunlight, and also with some artificial lighting systems, but not under the artificial lighting systems used in the many growth chambers that have a high proportion of fluorescent lamps to tungsten lamps and a high red/far red ratio (Ahmed et al., 1993b). Advances in engineering may have resulted in the development of artificial lighting systems for growth chambers that produce fewer artifactual plant responses than with past systems.

The use of high wind speeds and control of humidity and leaf temperature in gas-exchange cuvettes results in stomatal opening having much greater effects on transpiration rate than occur in most farmers' fields. If the objective of the research is to study effects of genotypic differences (or other types of treatments) on stomatal conductance, then this type of experimental approach will provide precise measurements of stomatal conductance and transpiration rate. But it should not be assumed that the differences in transpiration rate that were observed between the different genotypes in the controlled environments also will occur in natural field environments, because in most cases the differences in transpiration rate will be much smaller. Mathematical models are useful for predicting how differences observed at lower levels of organization, such as the leaf level, may affect the functioning of different types of plant populations in different types of environments. Jarvis and McNaughton (1986) provide a mathematical model for effects of stomata on water use at different levels of organization.

Controlled environments can be used for conducting unique single-factor experiments that provide important information on the causes of specific plant responses to environment, and the experiments can be repeated permitting rigorous testing of the validity of the results. However, the relevance of any predictions to natural field environments should be tested under field conditions (preferably with partial environmental control, as is discussed in the next section) or evaluated using mathematical models.

3.3 VALUE OF EXPERIMENTAL STUDIES WITH DIFFERENT ENVIRONMENTS IMPOSED IN THE SAME FIELD

By imposing a degree of environmental control under field conditions, it is possible to combine the reliability (relevance and accuracy) of field experiments with the separation of factors and precision that can be obtained by controlling certain variables. Irrigation, fertilizer, and salinity experiments have been conducted in which replicated plots of plants have been provided with different levels or quality of soil water or soil supplies of plant nutrients and other chemicals, but otherwise similar field environments. Experiments of this type can provide information on plant responses to drought, plant nutrients, flooding, salinity, and toxic chemicals that is relevant to many farmers' fields, except that differences between small plots in transpiration rate may be larger than those that would occur between large fields of the same treatments.

Temporary enclosures have been used to increase the temperature of plants during the night by different degree increments under field conditions, thereby simulating a range of subtropical and tropical nighttime temperatures while keeping the same natural daytime conditions in all of the treatments (Nielsen and Hall, 1985a). Studies of this type have demonstrated that the high night temperatures of hot subtropical and tropical zones can have detrimental effects on fruit set and yields of cowpea (Nielsen and Hall, 1985b) and many other crops (Hall, 2012).

In field conditions, however, it can be difficult to control certain environmental variables while keeping other important variables at natural levels. For example, if enclosures are used during the day to control daytime temperatures, they also will influence daytime humidity and wind speeds, which can influence transpiration and photosynthesis. In contrast, variation in humidity and wind speed during the night may have little influence on physiological processes occurring at night, since stomata usually are closed at this time. Also, if shades are used in the field to vary the level of solar radiation on plants, they also will influence plant temperatures, which may or may not produce an artifactual response, depending on the objectives of the study. For example, if the objective is to discover the causes of sunburn on the fruit of the tomato, it is important to separate effects due to levels (and quality) of solar radiation and effects due to tissue temperature.

Open-top chambers consisting of transparent vertical plastic cylinders that are placed over plant communities in field conditions are useful for studying plant responses to variations in air composition, such as levels of air pollutants or carbon dioxide concentration (Drake et al., 1989). These open-top chambers subject plants to environments that approximate natural field conditions. Open-top chambers have, however, at least one artifact compared with natural field conditions: air containing pollutants is delivered by fans to the base of the canopy inside the tube, flows up through the canopy, and exits at the top of the tube. In contrast, in natural field conditions, the polluted air enters the canopy from above, and concentrations of pollutants, such as ozone, decrease substantially as they pass through the canopy to the soil. Consequently, polluted air supplied at the base of the canopy will have less effect on upper-canopy leaves, which are the most active in photosynthesis, than the equivalent concentrations of the air pollutant delivered from the air above the canopy, as it occurs in natural conditions. This artifact may not be important when using open-top chambers to study plant responses to different levels of atmospheric [CO_2].

A major advantage of open-top chambers for studies of plant responses to elevated atmospheric [CO_2], compared with controlled-environment chambers, is that the plants can be grown in a natural soil environment. An analysis by Arp (1991) indicates that the use of potted plants, especially those experiments with small pots, could have produced artifactual responses of plants to elevated atmospheric [CO_2]. When plants with the C_3 photosynthetic system are first exposed to elevated [CO_2], there often is a substantial increase in their photosynthetic rate. Then, in some cases, this is followed by a day-by-day down-regulation of photosynthetic capacity. This down-regulation may be due to an imbalance between sources and sinks for carbohydrates with greater down-regulation when sinks are small in relation to sources. Plants grown in pots that have smaller root zones could develop a smaller root sink for carbohydrates. Arp (1991) pointed out that studies of responses to CO_2 enrichment with plants in small pots tended to exhibit limited root growth and substantial down-regulation of photosynthetic capacity, whereas studies with plants in large pots or field conditions showed stimulation of root growth and no down-regulation of the photosynthetic system. An alternative hypothesis for the greater down-regulation of photosynthesis with elevated [CO_2] by plants in smaller pots is that the plants have smaller supplies of inorganic nutrients, and it is the nutrient limitation that is responsible for the greater down-regulation, not the smaller root sink for carbohydrate. Studies with plants grown hydroponically demonstrated that elevated [CO_2] resulted in down-regulation of photosynthetic capacity with low NO_3^- supplies in the root zone but not with higher supplies of NO_3^- (Harmens et al., 2000). The authors also point out that little or no decrease in photosynthetic capacity was observed with elevated [CO_2] when root growth of plants was not restricted and they had ample supplies of nutrients. Nutrient limitations and reduced rooting volume may both contribute to the down-regulation of photosynthetic capacity that can occur when C_3 plants growing in small pots are subjected to elevated atmospheric $[CO_2]$.

Cowpea plants were grown in three types of pots with major differences in volume (11, 17, and 76 liters) and hopefully adequate nutrition in all cases (Ismail et al., 1994). The objective was to study the effects of drought and the volume of the rooting medium on the inheritance of water-use efficiency and on xylem ABA concentration. I recall observing that for the well-watered treatment, plants in the larger pots grew much faster even during initial stages when the root systems of the plants in the medium-sized pots were nowhere near the edge of the pots. The mechanism for this effect is not clear to me.

Some systems have been developed for modifying and controlling specific aspects of the environment under field conditions that minimally disturb other aspects of the system. The free-air CO_2 enrichment system (Hendrey and Kimball, 1994) was developed for exposing plants to elevated $[CO_2]$ under natural conditions. It consists of vent pipes placed in a circle in the field, with the CO_2 output of individual pipes automatically controlled to maintain the desired $[CO_2]$ in the area of the field within the circle. Also, a system has been developed to elevate the temperature of plants, such as buds of deciduous trees, during winter. The system uses infrared heating lamps and a control unit to maintain a preset difference in temperature compared with untreated buds and does not appear to disturb other aspects of the environment.

When attempting to discover mechanisms whereby the environment influences plants and to relate those mechanisms to crop productivity, it is often useful to use several experimental approaches. For example, studies in different fields with contrasting environments can lead to the development of hypotheses, for example, that pod production of cowpea is damaged by high day temperature (Turk et al., 1980). Controlled environments can then be used to separate the effects of co-varying factors. This hypothesis was tested using controlled environments in which cowpea plants were subjected to different day temperatures, but there was no effect on pod production (Warrag and Hall, 1984a). During hot weather in the field, however, both day and night temperatures can be hot. An alternative hypothesis was developed that pod production in cowpea is damaged by high night temperature. Controlled environment studies demonstrated that high night temperatures can reduce pod production of cowpea by causing male sterility (Warrag and Hall, 1984b). The reliability of this prediction was confirmed by studies in which different night temperatures were imposed on cowpea plants under field conditions (Nielsen and Hall, 1985b). The overall procedure should take advantage of the different strengths of each type of experimental approach.

3.4 QUANTITATIVE METHODS

When little is known about a scientific discipline or subdiscipline, research tends to emphasize description. In earlier years, many different types of plant responses to environment were described. Patterns emerged from these studies that made possible the development of hypotheses (incompletely tested models) for potential causes and mechanisms for specific effects and for specific types of regulation and emergent properties. Mathematical models are particularly useful for the formulation of hypotheses, theories (partially validated models), and laws (more rigorously validated models), and for providing a quantitative description of plant function and predicting plant performance under the conditions of farmers' fields (Jones, 1992). Once a hypothesis is formulated as a mathematical model, predictions can be made, making it possible to test the hypothesis by attempting to falsify it. An alternative and powerful approach is to develop two different hypotheses for the same phenomenon and then test their predictions to determine which hypothesis is most consistent with reality.

I will provide an overview of mathematical modeling, because many biologists may not fully appreciate its value. Mathematical models can be useful for developing a conceptual as well as a more quantitative understanding of plant responses to environment. I will consider three types of models: equilibrium models, steady-state models, and dynamic models.

3.4.1 Equilibrium Models

This is the simplest model and is most effective where there are no net flows of matter or energy, and state variables are constant. An example is where seeds are placed in a sealed container, such as a bell jar, that also contains a saturated solution of a specific salt which is not in direct contact with the seed. After sufficient time has elapsed, the humidity of the air in the container will reach a specific value depending on the type of salt that was used (e.g., saturated solutions of NaCl, NH_4NO_3, and LiCl provide relative humidities of 75%, 62%, and 12%, respectively, at room temperature). If the seed coat is reasonably permeable, the moisture content of the seeds will become stable within a few days or weeks, with seeds having higher values under the higher humidities. For these salts, cowpea seeds develop moisture contents of 15%, 12%, and 5%, respectively, on a fresh weight basis. (Refer to Ismail et al., 1997, and Chapter 5 for a description of how these differences in seed moisture content can influence seedling emergence.) It is possible to develop an empirical linear regression model for this relationship between seed moisture content (*MC*) and relative humidity of the air (*RH*):

$$MC \text{ (on a } FW \text{ basis)} = 3.0 + 0.15 \times RH \text{ for values in \% at equilibrium} \quad (3.1)$$

Correlation analysis indicates that the extent to which the model is consistent with the data is excellent. The r^2 value of 0.99 implies that 99% of the variation in *MC* can be explained by Equation 3.1 and the solid line in Figure 3.1. Obtaining a more rigorous test of this linear model would require additional data for seed moisture contents at air humidities between 12% and 62%, since actual values (the dashed line in Figure 3.1) may deviate from the regression line in this range of values. Also, it is risky to extrapolate with empirical models of this type and attempt to predict values much outside the range of values used in developing the model. Note that linear Equation 3.1 predicts values of *MC* that probably are too low for *RH* values greater

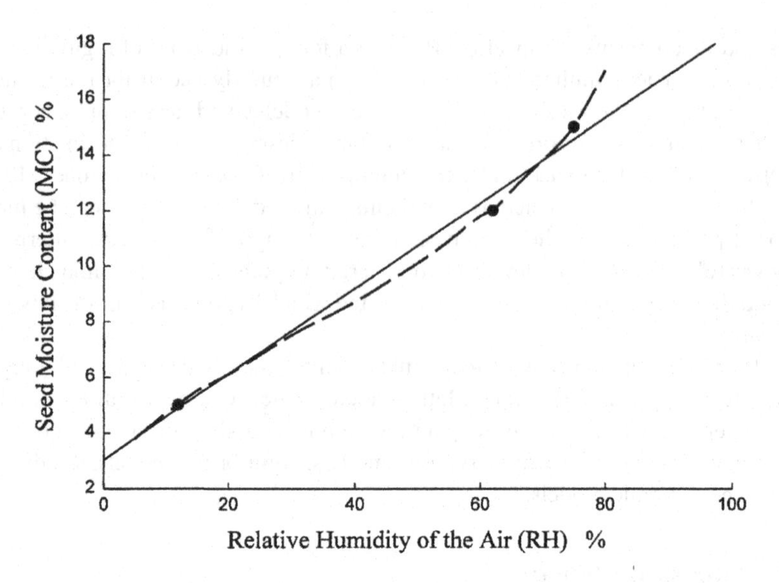

FIGURE 3.1 Moisture content of cowpea seed, on a fresh weight basis, that has been equilibrated with atmospheres having constant relative humidities at room temperature. The solid line follows Equation 3.1. The dashed line follows actual values from seed of a similar crop species. (From Vertucci, C. W. and Roos, E. E., *Plant Physiol.*, 94, 1019–1023, 1990.)

than 75% (Figure 3.1). Extrapolation is particularly dangerous when using empirical models based on polynomial equations.

Some scientists (Vertucci and Roos, 1990) have used *MC* data for seed based on dry weight (*DW*) rather than fresh weight (*FW*) as was used for the data in Equation 3.1 and Figure 3.1. The relationship between these two variables is described by the following equation:

$$MC \text{ (on } DW \text{ basis)} = \frac{100 \times MC \text{ (on } FW \text{ basis)}}{100 - MC \text{ (on } FW \text{ basis)}}$$

From this equation, it is apparent that *MC* (on a *DW* basis) would exhibit a different relationship with *RH* and curve up more rapidly at high *RH* values than would *MC* (on a *FW* basis)]. This illustrates the point that the way data are expressed can influence models and interpretations based on them.

Laws of physical chemistry can be used to determine the activity of water (a_w) in the seed based on a knowledge of the air humidity at equilibrium as shown in Equation 3.2:

$$a_w\text{(seed)} = \frac{RH\text{(air)}}{100} \tag{3.2}$$

at equilibrium with a_w having values of 0.0 to 1.0.

Knowing the activity of water in the seed enables one to calculate the water potential (Ψ, defined in Equation 8.14 in Chapter 8) of the water in the seed as shown in Equation 3.3:

$$\Psi(\text{seed}) = \frac{R \times T \times \ln a_w}{V_w} \tag{3.3}$$

where R is the international gas constant, T is the absolute temperature, and V_w is the partial molar volume of water. The relationships described by Equations 3.2 and 3.3 are used in some methods for determining the water potential of samples from plants or soils. When plant tissue or a soil sample is in equilibrium with air in a sealed container, the water potential of the water in the plant tissue or soil can be estimated based on measurements of air humidity in the container. Typically the containers and measurement systems are calibrated with samples of known water potential (Boyer, 1995).

3.4.2 Steady-State Models

This type of model is used to describe processes in which flow rates of mass or energy and state variables are approximately constant. Steady-state models are widely used in describing crop responses to environment. An example of a mechanistic model based on a law of physical chemistry is provided in Equation 3.4. The flow of gases, such as water vapor or carbon dioxide, per unit area per unit time (J_i) in air can be related to the state variable, difference in partial pressure of the gas relative to atmospheric pressure (p_i/P_{atm}) along two points in the flow path, and the porometer conductance of the gas in air (g_i) where, for leaves, the conductance describes the effects of stomata and other factors such as the depth of the boundary layer (Cowan, 1977).

$$J_i = \frac{g_i \times \Delta p_i}{P_{atm}} \tag{3.4}$$

Other types of steady-state models also are used in studying plant responses to environment, such as those based on balances involving conservation of various properties. In one case, it is assumed that energy is conserved. The steady-state energy flows of all processes involving significant energy flow between plants and their environment are evaluated in an additive manner and assumed to balance each other out (Chapter 7). In another case, it is assumed that mass (e.g., water) is conserved, and all significant steady-state flows of water are evaluated to predict soil water status in the root zone of crops (Chapter 10).

3.4.3 Dynamic Models

This type of model is used to describe processes that vary significantly with time. Dynamic models usually are much more complex than steady-state models. There is a comprehensive text that provides examples of dynamic models (Gurney and Nisbet, 1998), such as those describing how numbers of insect pests and their predators may vary with time. Dynamic models either use analytical equations to describe processes in continuous time, or they use update rules based on the premise that knowing the state of the system at a given time allows one to simulate the state of the system at some incremental time in the future. Many different types of simulation model have been developed for crop physiology (Loomis et al., 1979).

Simulation models of plant growth and development with the objectives of predicting plant productivity or time of flowering and maturity often simulate with an update increment of one day. In contrast, simulation models with the objective of predicting diurnal variation in plant properties often have shorter update increments, such as about one hour. Simple summation models and empirical models, which do not have a time element and thus are static, have been developed for describing the results of dynamic seasonal processes. This type of simple summation model is extensively used in subsequent chapters.

In attempting to understand crop responses to environment, it is useful to ask the following questions in the sequence presented:

1. Can the system be approximated by an equilibrium model? If yes, then develop such a model if only in your mind. If no, then,
2. Can the system be approximated by a steady-state model? If yes then develop such a model. If no, then,
3. Can the system be approximated better by a dynamic model? If yes, then can a static empirical model suffice to meet your objectives?

For example, you might wish to know whether hot weather is causing reductions in productivity. If you have data on productivity in different locations and years, you could then test simple or multiple correlations and regressions between productivity and different aspects of temperature (Ismail and Hall, 1998). Another environmental variable may be critical, such as the amount of solar radiation, in which case the model should combine data on both temperature and level of solar radiation (Fischer, 1985). Alternatively, one could try a simple type of simulation model, as has been developed for cereals and grain legumes. This model assumes that productivity under optimal management can be predicted by summing the product of intercepted solar radiation and factors for its conversion through photosynthesis into carbohydrate and its partitioning to grain over the daily periods when photosynthesis mainly contributes to grain formation (Equation 4.3 in Chapter 4). If static empirical or simple summation models will not provide the types of predictions that are needed, it may be necessary to develop more complex simulation models or analytical types of continuous-time dynamic models, in which case considerable effort may be required.

The overall approach that I recommend is to use the simplest model that is consistent with the objectives that you are pursuing. The reason for this approach is that models must simulate plant function in a reasonable enough manner to achieve their objectives. This means that essential features of the model must be testable. Models that are too complex are likely to have flaws of unknown character and cannot be tested adequately (Passioura, 1973).

ADDITIONAL READING

Arp, W. J. 1991. Effects of source-sink relations on photosynthetic acclimation to elevated CO_2. *Plant Cell Environ.* 14: 869–875.

Fischer, R. A. 1985. Number of kernels in wheat crops and the influence of solar radiation and temperature. *J. Agric. Sci. Camb.* 105: 447–461.

Ismail, A. M. and A. E. Hall. 1998. Positive and potential negative effects of heat-tolerance genes in cowpea. *Crop Sci.* 38: 381–390.

Ismail, A. M., A. E. Hall, and T. J. Close. 1997. Chilling tolerance during emergence of cowpea associated with a dehydrin and slow electrolyte leakage. *Crop Sci.* 37: 1270–1277.

Jones, H. G. 1992. *Plants and Microclimate*, 2nd ed. Cambridge University Press, Cambridge, UK, p. 428.

Nielsen, C. L. and A. E. Hall. 1985b. Responses of cowpea (*Vigna unguiculata* [L.] Walp.) in the field to high night temperature during flowering. II. Plant responses. *Field Crops Res.* 10: 181–196.

Passioura, J. B. 1973. Sense and nonsense in crop simulation. *J. Australian Inst. Agric. Sci.* 39: 181–183.

4 Crop Physiological Responses to Light, Photosynthesis, and Respiration

Through photosynthesis, solar radiation provides the free energy required by plants for their growth and maintenance. Photons at a wavelength between 400 and 700 nanometers have the levels of energy per photon required for eliciting the photochemical reactions of photosynthesis and are described as *photosynthetically active radiation* (PAR). The flux density of PAR photons (*PFD*) can be measured with sensors. On a clear day, at sea level, when the sun is directly overhead, it provides a *PFD* of about 2,000 μmol photon m^{-2} s^{-1} and about 50 mol photon m^{-2} day^{-1} with a day length of 14 hours. A mole of photons is 6×10^{23} photons where 6×10^{23} is Avogadro's number, the number of particles in a mole. Solar radiation data often are provided in terms of solar irradiance where R_s = energy flow per unit area per unit time for all wavelengths of sunlight (as shown in Figure 7.1 for Riverside, California), which can be converted to *PFD* using a locally determined conversion factor, which is about 2 μmol photon per joule. The conversion factor does vary with variations in light quality, especially those variations occurring with different types of artificial lighting. Some older literature gives measurements of light levels in foot candles and lux, but this practice has been discontinued because these measurements quantify the levels of light suitable for human vision and have no direct relevance to plant function. I am not providing conversion factors for the relationships between foot candles or lux and either *PFD* or irradiance because these factors are strongly dependent on the quality of the light. It can be difficult and risky to convert data from the older literature that were in either foot candles or lux into either *PFD* or irradiance.

Levels of *PFD* and the length of the growing season determine the upper limit of productivity. The photosynthetic systems of different plant genotypes can vary and have a strong influence on plant adaptation to different environments. In this chapter, I discuss studies on photosynthesis with crop species and cultivars, and native species and ecotypes. Ecotypes are genetically differentiated populations, within the same species, that are adapted to different habitats. Studies with crop species have provided much information on photosynthesis in relation to productivity. Studies with native species have provided definitive information on photosynthetic aspects of plant adaptation because much is known about the adaptation of some native species.

4.1 PHOTOSYNTHESIS AND PRODUCTIVITY

The facet of photosynthesis that is often measured by crop physiologists is the rate of carbon dioxide (CO_2) assimilation per unit of projected leaf area or per unit of ground area covered by the canopy (P_n). This measure is useful for making physiological estimates of productivity in that it is closely related to the net rate of carbohydrate accumulation by plants. P_n should not be regarded as a *pure* measure of photosynthesis, however, because it represents the balance between CO_2 uptake in photosynthesis (P), which is defined as being positive, and CO_2 release in mitochondrial respiration (R) and photorespiration (R_p), which are defined as being negative, as shown in Equation 4.1:

$$P_n = P + R + R_p \tag{4.1}$$

In crop physiological studies, it is useful to use intact plants when measuring P_n to ensure that the data have relevance to plant functioning in the natural world, because excising leaf tissue can generate artifactual responses. The type of measurement system that is ideal depends on the objectives of the study. If the objective is to compare genotypes, measuring P_n of single attached leaves can be useful because the environmental conditions to which the leaf is being subjected, such as *PFD*, leaf temperature, and boundary layer resistance, can be described and measured, and the age of the plant material can be specified. Consequently, the experiment can be repeated. Also, P_n of single leaves can be related to other parameters that influence photosynthesis, such as stomatal conductance, and levels of photosynthetic enzymes or photosystem components, making it possible to determine mechanisms for any differences in P_n. In contrast, if the objective is to determine effects of environmental factors, such as elevated [CO_2] or atmospheric pollutants, on carbon balance, it can be useful to measure diurnal curves of P_n per unit ground area for canopies in natural conditions using open-top chambers or environmentally controlled transparent enclosures with different atmospheric treatments. Measuring P_n of isolated whole plants may be useful only in special circumstances, such as when studying shoot/root interactions. For example, when working with leguminous plants, it can be useful to simultaneously measure CO_2 uptake by the shoot and ethylene production by the root/rhizobium system, with acetylene as a substrate because the nitrogenase enzyme involved in fixing atmospheric nitrogen also can reduce acetylene to form ethylene. These measurements can be used to test hypotheses concerning interactions between atmospheric nitrogen fixation by the root system and photosynthesis of the shoot. The other circumstance where it may be useful to measure P_n of a whole plant is where isolated whole plants are a component of the community, such as for young seedlings of crops in a weed-free field or an isolated barrel cactus in a desert ecosystem.

Plant photosynthetic responses to environment have been extensively modeled. Some models were developed to test hypotheses concerning the biochemistry and biophysics of photosynthesis, and the model of Farquhar et al. (1980) has been widely used for this purpose. Other models were developed to predict carbon dioxide assimilation by plants in different environments and test hypotheses concerning whole plant function. Hall's model (1979) was developed for these purposes and is used in this chapter to illustrate principles of photosynthetic responses to environment as they relate to plant productivity and adaptation. The advantage of using a mathematical model is as follows. There are many possible families of curves for photosynthetic responses to light, [CO_2], [O_2], and temperature that would be difficult to measure on the same plant. Instead, measurements could be made on key parts of a few response curves to obtain the parameters needed to calibrate the model, which can then be used to predict all of the families of response curves that are needed for the specific application. Another advantage of using models to illustrate principles is that some differences in photosynthetic responses can involve the same type of changes in components of the leaf system of photosynthesis and respiration (e.g., typical effects of genotype, leaf age, or degree of shade acclimation as is shown later, in Figure 4.3).

For individual leaves, P_n exhibits a near hyperbolic response to absorbed *PFD* that is described in Figure 4.1. The initial slope of the curve is the quantum yield (mol CO_2 taken up per mol photon of *PFD* absorbed), which strongly depends on the functioning of the photosystems. The quantum yield has a value of about 0.05 mol CO_2 per mol photon for many species under optimal natural conditions (Ehleringer and Björkman, 1977). The quantum requirement of leaves is the inverse of the quantum yield and is about 20 photon per molecule of CO_2 taken up. Quantum requirements

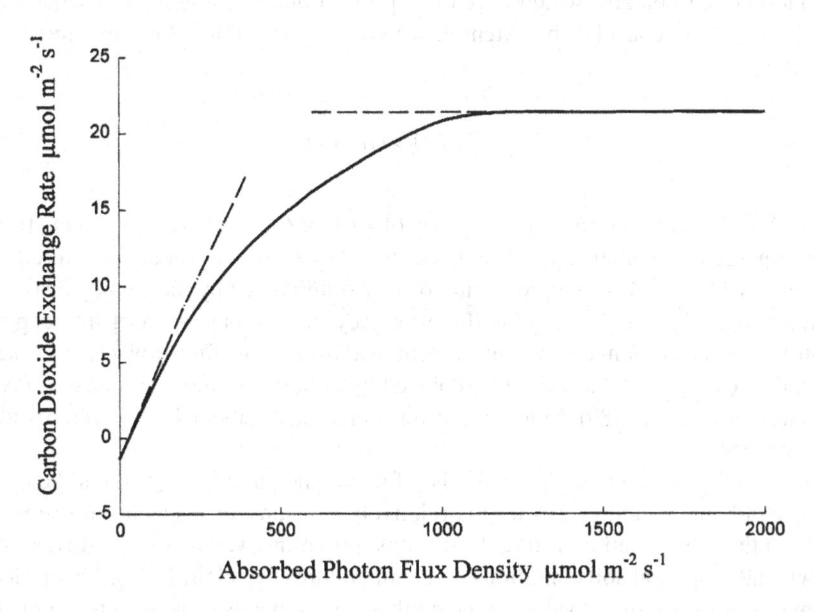

FIGURE 4.1 Leaf carbon dioxide exchange rate (P_n or R) as a function of the absorbed photon flux density based on Hall's model (1979). The carbon dioxide evolution in darkness measures the mitochondrial respiration rate (R = −1.3 µmol m^{-2} s^{-1}). The initial slope of the curve described by the lower dashed line is the quantum yield with a value of 0.05 mol CO_2/mol photon. The light-saturated rate of P_n described by the upper dashed line is 21.4 µmol m^{-2} s^{-1}.

as low as 14 have been achieved by subjecting plants to low temperatures and either low [O_2] or high [CO_2] (Ehleringer and Björkman, 1977). Light-saturated P_n strongly depends on factors influencing CO_2 transport into the leaf, carboxylation, and photorespiration and varies substantially among species and leaves of the same plant. In darkness, CO_2 evolution is observed that provides an estimate of the mitochondrial respiration which is occurring in both dark and light conditions. The response of P_n to *PFD* is linear at low *PFDs* and then curves achieving maximum levels of P_n at different *PFD* levels, depending on the crop species. The *PFD* required to saturate P_n can be less than 50% of full sunlight, which is 2,000 μmol photon m^{-2} s^{-1}, in some species to 50%–100% of full sunlight in other species.

The photosynthetic responses of canopies require more *PFD* for saturation than individual leaves (Figure 4.2). Canopies of most crop species that have achieved a leaf area index (*LAI*) of at least four projected leaf areas per unit ground area exhibit an approximately linear response of P_n per unit ground area up to full sunlight levels of *PFD*, because leaves in the lower part of the canopy are always shaded and respond in a linear manner (Sinclair, 1994). For optimal growing conditions, the production of dry plant biomass during the season (*B*), which is mainly carbohydrate, has been modeled as being proportional to the *PFD* intercepted during the growing season as shown in Equation 4.2. It should be noted that P_n of canopies is not quite proportional to the incident *PFD* (Figure 4.2), and for the model to be valid carbon losses such as mitochondrial respiration at night and root exudation of carbon also would have to be proportional to daily P_n. Consequently, it is important to consider the extent to which and conditions where this model is effective.

$$B = \sum_{i=n}^{i} PFD_i \times GC_i \times Q_i \tag{4.2}$$

where Σ is the summation of daily values of *PFD* × *GC* × *Q* over the *n* days from plant emergence to maturity, *GC* is the extent to which the ground is covered by the canopy on each day (the proportion of *PFD* intercepted by the canopy also can be used instead of *GC*), and *Q* is the efficiency of the conversion of intercepted photons into plant biomass. A similar term, *radiation-use efficiency*, often is used instead of *Q*, in which case *PFD* is replaced by either R_s or the irradiance of PAR (Sinclair and Horie, 1989). Note that, on an energy basis, about 45% of solar irradiance is PAR.

In circumstances where the model is effective, the parameter *Q* should be constant. Field studies indicate that *Q* is relatively constant for a species at different times in the season and in different locations, providing well adapted cultivars are grown under optimal soil conditions and temperatures (Sinclair, 1994). Theoretical estimates of maximum *Q* values indicate that, for C_4 plants such as maize, *Q* may approach 0.9 g/mol photon when the average apparent quantum requirement for all leaves in the canopy is 18 mol photon/mol CO_2 assimilated in photosynthesis (Loomis and Amthor, 1999). Estimates of maximum *Q* values for C_3 plants, such as wheat or rice, approached 0.6–0.8 g/mol photon for average apparent quantum requirements for all leaves in the canopy between 25 and 20 mol photon/mol CO_2

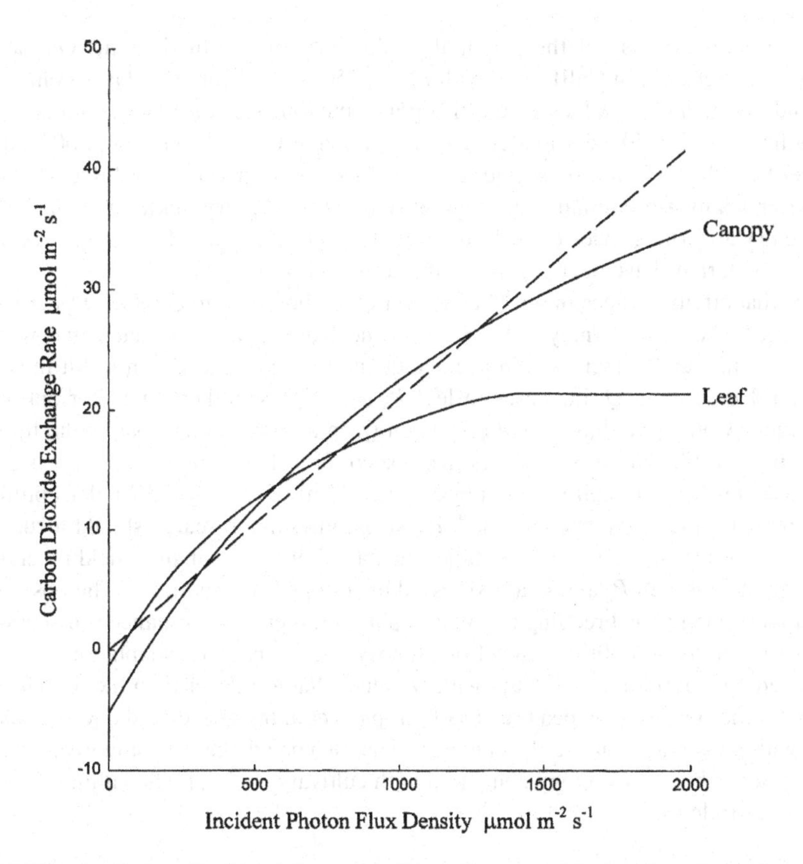

FIGURE 4.2 Carbon dioxide exchange rate P_n as a function of the incident photon flux density (*PFD*) for a canopy and a leaf based on Hall's model (1979). The dashed line illustrates a proportional response of P_n to *PFD* of 0.021 mol CO_2 taken up/mol photon.

assimilated in photosynthesis (Loomis and Amthor, 1996). Maximum values of Q obtained from measurements of shoot biomass production under field conditions were about 0.85 g/mol photon for maize, 0.7 g/mol photon for rice, and 0.6 g/mol photon for soybean (reviewed by Sinclair and Horie, 1989). Consequently, for canopies of different, well-adapted crops during active growth under optimal conditions, Q appears to be relatively constant and have maximum values of about 0.6–1.0 g/mol photon, which represents a 2.0% to 3.4% conversion of the radiant energy in sunlight into the chemical energy in the carbohydrate that is formed in the plant.

Since the model described by Equation 4.2 appears to be effective in optimal conditions, it can be used to predict the potential productivity of a growing season in a particular climatic zone. Potential productivity is the dry biomass produced per unit land area per growing season or day in environments where temperatures are optimal, water and soil nutrient supplies are not limiting, there are no problems due to pests and diseases, and plant spacing and development are such that all *PFD* is intercepted (GC = 1.0).

Here is an example of the potential productivity of a climatic zone: On clear, sunny summer days in California, with *PFD* of 50 mol photon m^{-2} day^{-1}, complete ground cover, and Q values of 0.6–1.0 g/mol photon, Equation 4.2 predicts crop growth rates of 30–50 g dry matter m^{-2} day^{-1}, which is equivalent to 300–500 kg dry matter ha^{-1} day^{-1}. Can crop species achieve these rapid growth rates? Record crop growth rates measured many years ago were up to 520 kg dry matter ha^{-1} day^{-1} for maize in California (reviewed by Loomis et al., 1971). It is possible that even greater crop growth rates have been recorded since this review was made.

In what circumstances might Q be increased by breeding to enhance P_n per unit leaf area? Theoretical analysis by Sinclair and Horie (1989) indicated that while decreases in leaf P_n (increases in average quantum requirement) can result in substantial decreases in Q, increases in leaf P_n above present-day values for optimal conditions would only have small effects on Q. In a powerful analysis of the limits to crop yield, Sinclair (1994) argues that, except for a few options that allow small increases in the yield ceiling, the physiological limit to crop yields under optimal environments may have been reached. He also argues that research should focus on breeding for resistance to stresses rather than attempting to enhance yield potential through increases in P_n and Q as expressed under optimal conditions. The possibility of increasing Q by breeding to enhance photosynthetic performance is discussed further in the section of this chapter on photosynthesis and plant adaptation.

Breeding can produce cultivars with Q values that are lower than the maximum value for the species. Cowpea breeders in tropical countries have bred cowpea cultivars with pods placed above the canopy of leaves. The advantages of cultivars with pods placed above the canopy compared with cultivars having pods within the canopy are as follows:

1. The extent of damage due to a pod boring insect pest is less because there are fewer oviposition sites for the insect to lay eggs.
2. The extent of fungal damage to pods is less because the pods dry more rapidly after rain has stopped.
3. Pods above the canopy are easier to pick than pods within the canopy.

But there is a penalty from having pods above the canopy. Pods in this position absorb many of the incident photons and, even though they are green and conduct photosynthesis, they are not as effective as leaves in this respect. Comparisons of crop growth rates of cowpea genotypes differing in pod location indicated that the pods-above trait may decrease Q by as much as 54% (Kwapata et al., 1990). This detrimental effect was confirmed by an experiment in which young pods were removed from pods-above genotypes and a substantial increase in Q to the high levels of pods-within-canopy genotypes was observed. Consequently, in regions such as California where the pod borer is not present, there is little rain when pods are present, and the crop is harvested mechanically, cultivars with pods within the canopy (e.g., variety California Blackeye 50 in Ehlers et al., 2009) should be more effective than cultivars with pods above the canopy, because they would have higher values of Q and produce more biomass and therefore more pods. For humid tropical environments in developing countries where the pod borer is present, rain occurs during pod development and maturation, and people harvest pods by hand, the ideal pod position represents a trade-off between the advantages and disadvantages of the pods-above-the-canopy and the pods-within-the-canopy traits (refer to Chapter 12 for other discussions of plant resistance to pests and diseases). As another example of where non-photosynthetic organs of the canopy can influence Q, maize hybrids released in the United States over the period from 1930 to 1992 had progressively smaller tassels (Duvick and Cassman, 1999), which could have resulted in small increases in Q due to less shading of leaves by the tassels.

The model described by Equation 4.2 can be used to make predictions about aspects of crop management, such as the extent that changes in tree spacing in orchards or row spacing for field crops influence crop biomass production under optimal conditions. Equation 4.2 predicts that B will be proportional to the amount of PFD intercepted during the growing season, which depends on the extent of ground coverage by the canopy. With denser plant spacing, GC will be greater, especially during the early stages of plant growth. One method for measuring GC is to take a photograph from above the crop and determine the extent to which the ground is covered by the canopy using the image and computer software. It is necessary to estimate GC every few days so that a curve can be generated for the growing season together with a curve for daily PFD, which would have to be measured every day. Then, it is necessary to sum $PFD \times GC$ for every day during the growing season and compare the summed values obtained for the different plant spacing treatments. An alternative approach is to measure PFD above the canopy and at ground level every day so that PFD interception can be determined from the difference in these values and then summed to determine the total intercepted PFD during the growing season.

During the previous century, substantial increases in productivity of certain row crops, such as maize, soybean, and cotton, have been achieved by progressively decreasing row width. Prior to the development of tractors, row widths of about 1 meter were used to permit a horse to walk between rows without damaging plants while pulling a cultivator to remove weeds. At the end of the twentieth century, smaller row widths were being used for growing many row crops, and they varied between 0.20 and 0.75 meters, depending on the crop species and whether surface irrigation systems were being used that require wide furrows. Dwarf trees have been bred for ease of picking and are planted at close spacing to increase productivity, especially in early years. In developing African countries, grain crops, such as sorghum, pearl millet, and cowpea, often are grown at very wide spacing, for example, as wide as 2 × 2 meters for sorghum and 1 × 1 meter for cowpea. For sorghum, this may represent an optimal system where soil nutrients, such as nitrogen (N), are limiting (refer to Chapter 2), whereas the wide spacing used with cowpea may be optimal where supplies of seed and labor are limited at sowing. Once these constraints are removed, closer plant spacing would be optimal for these crops and result in greater biomass production and grain yield per unit land area. The row spacing that is optimal also depends on the type of crop cultivar which is being grown. High grain yields per unit land area of cowpea have been achieved by developing semi-dwarf compact cultivars and growing them at high plant populations using rows 51 centimeters apart and a spacing of 10 centimeters within rows (Ismail and Hall, 2000). The optimal row spacing was wider, 76–102 centimeters, for cowpea cultivars that were taller and more spreading. Maize hybrids have been bred that are adapted to overcome stresses associated with high plant densities. The newer hybrids have more vertical leaves which enhances light penetration into the canopy and reduces barrenness and the tendency to lodge with high plant densities (Duvick and Cassman, 1999).

Equation 4.2 has been modified to produce Equation 4.3 for predicting yields (Y) of crops that produce grain, such as cereals and grain legumes (and crops that produce useful fruit, such as tomato and cotton):

$$Y = \sum_{i=d}^{i=1} PFD_i \times GC_i \times Q_i \times CP_i \tag{4.3}$$

In this case, the equation is summed over the period of days (d) when photosynthesis significantly contributes carbohydrate to the developing grain (or fruit), which often is from an early stage of flowering to the date when all grains (or fruit) are physiologically mature. The term CP describes the proportion of the carbohydrate that is partitioned to grain on each day and is conceptually related to harvest index (HI), which is the ratio of total grain yield (Y) to total shoot biomass at harvest.

Equation 4.3 can be used to estimate the increases in yield under optimal growing conditions that are possible through plant breeding and to provide explanations for differences in yield that occur in different environments. There is an intermediate value of CP (and HI) that is optimal, because some carbohydrate must be partitioned to produce and maintain the photosynthetic source tissue, stems, and roots. Increases in productivity through plant breeding in wheat, rice, oat, cotton, peanut, soybean, and sunflower during the twentieth century were mainly attributed to increased HI with little increase in total shoot biomass (Gifford, 1986; Evans, 1993; López Pereira et al., 2000). Modern cultivars of crops that have benefitted from substantial breeding, such as wheat, rice, soybean, and sunflower, may now have optimal values of CP and HI; thus, it may be difficult to achieve further increases in yield potential by breeding to increase HI. For optimal growing conditions, either no change or only small increases in photosynthetic efficiency (Q) have occurred in this century for many crops. It appears difficult to increase the value of Q through breeding. In some cases, breeding may have prevented Q from decreasing by incorporating resistance to stresses, as was observed with maize (Duvick and Cassman, 1999) and rice (Peng et al., 1999).

Equation 4.3 can be useful for explaining the differences in grain yield observed in different climatic zones. For well-adapted cultivars grown in different locations and with different sowing dates, but under optimal conditions, much of the variation in yield is due to variation in either or both daily PFD during reproductive development and the duration of reproductive development. For example, on farmers' fields with good management, rice has produced very high grain yields (about 9 tons/ha) when grown under irrigation in subtropical climates such as Davis, California in the Sacramento Valley (Figure 10.7), intermediate grain yields (about 6 tons/ha) under irrigation during the dry season in equatorial tropical climates (e.g., in the Senegal River Valley, which is adjacent to the location in Figure 10.5), and low yields (about 3 tons/ha) under rain-fed conditions in the tropics (e.g., Casamance, Senegal, in Figure 5.3). In the California rice environment, daily PFD is very high due to long days of 15 hours and clear skies, and d is large due to cool nights that slow down reproductive development and cause the period of reproductive development to be long. In the equatorial tropics during the dry season, daily PFD is moderate because, even though the skies are clear, the day length is short, about 12 hours, and d is moderate because of warm nights. During tropical rainy seasons, daily PFD is low because of the cloudiness and short days, and d is moderate because of the warm nights. These variations in daily PFD and d can explain most of the variations in grain yield I described for rice. Other warm season cereals and grain legumes have exhibited greater grain yields in subtropical zones (i.e., the Central Valleys of California) as compared with tropical zones due to the longer values of d where nights are cooler and the higher daily PFD in the subtropics. As an example of variations in d, individual pods of cowpea take three weeks to develop in cool, night conditions in California, but only two weeks to develop in the hotter night temperatures of the tropics (Nielsen and Hall, 1985b).

Models such as Equations 4.2 and 4.3 can be used to obtain an estimate of whether certain pests present in a field will reduce yield. For example, caterpillars' consumption of leaves causes two types of problems for a forage crop or a grain crop during the vegetative stage:

1. It removes some of the leaf area.
2. By reducing ground cover, it reduces the interception of PFD, and thus it will reduce the future growth rate of the crop.

The damage caused by the hairy caterpillar to cowpea in Senegal is strongly dependent on the stage of growth when the plant becomes infested with these caterpillars. If the plant is at the seedling stage, feeding by a single caterpillar can totally defoliate and kill the plant. At later stages of development, when the total leaf area is rapidly increasing, feeding by several caterpillars can have little influence on the extent of ground covered by the plant, and its accumulation of new biomass is barely affected by the caterpillars, such that they may not influence either plant survival or subsequent grain yield (also refer to the discussion of this topic in Chapter 12). For a grain crop during the reproductive period, partial defoliation by caterpillars may significantly reduce grain yield only if it causes a significant reduction in percentage of ground cover. During this period, the grain crop may have an LAI of four or more, so a considerable amount of leaves will have to be consumed by the caterpillars for it to have effects on the extent of ground cover or Q. However, nitrogen balance also must be considered. Removal of leaves by caterpillars would reduce the amount of protein that is available for translocation, as amino acids, to developing grains.

Some of the detrimental impacts of weeds can be evaluated using the models described by Equations 4.2 and 4.3. Under optimal soil conditions, yield reductions due to weeds often result from competition in the aerial environment. This competition will be serious only where weeds significantly reduce the PFD intercepted by the crop, which usually only occurs if the leaves of the weeds grow above and shade the leaves of the crop.

There are circumstances where models of the type described by Equations 4.2 and 4.3 are not effective for predicting biomass production or grain yield, such as where soil conditions are not optimal. If there is considerable variation in soil nitrogen supply, and it is limiting, there will be considerable variation in leaf nitrogen level, P_n and Q (Sinclair and Horie, 1989). In this case, it is difficult to use Equations 4.2 and 4.3 without making many measurements of Q or developing submodels to predict the variation in Q with variation in nitrogen supply, which requires substantial effort. The models described by Equations 4.2 and 4.3 mainly are effective where productivity is limited by the interception and levels of *PFD*. In dry environments where irrigation is not possible, there can be clear skies and abundant *PFD* but limiting supplies of water. For rain-fed cropping in an environment with a short rainy season, a cultivar with a short cycle length that fits the season can produce more grain than a cultivar with a longer cycle length. The cultivar with the longer cycle from sowing to maturity will intercept more *PFD* but may only produce similar amounts of biomass because of the water supply limitation. In addition, the longer-cycle cultivar could produce much less grain yield than the shorter-cycle cultivar if it experiences more extreme drought during stages of reproductive development when the crop is sensitive to drought. Where water supply is limited, other types of mathematical models are more effective for predicting productivity, as will be discussed in Chapter 9.

4.2 PHOTOSYNTHESIS AND ADAPTATION

Crop plants are needed that are adapted to different environments. Differences among species in photosynthetic responses to environment are closely related to differences in adaptation and biomass production in natural environments (Lambers et al., 1998). Photosynthesis can be related to adaptation by considering relative rates of P_n per unit leaf area by different species, ecotypes, or cultivars in specific environments, efficiency of photosynthesis (where efficiency is a specific ratio that has relevance to a hypothesis concerning adaptation), effects of various stresses on the photosynthetic system, and the extent of acclimation as this influences breadth of adaptation.

Different species of vascular plants exhibit one of four basic photosynthetic systems (Loomis and Connor, 1992). Most crop species have the C_3 system of photosynthesis in which stomata open only during the day, and the enzyme rubisco is responsible for the initial fixation of CO_2, which occurs during the day (Table 4.1). Some important tropical grasses have the C_4 system in which stomata also open only during the day, but in this case the initial fixation of CO_2 is by the enzyme PEP carboxylase, which acts to concentrate CO_2 inside the leaf during the day where it is then re-fixed by rubisco during the day. One or two crop species have photosynthetic systems that are intermediate between C_3 and C_4. A few crop species have the CAM photosynthetic system in which stomata open at night, CO_2 is fixed into organic acids during the night, which are subsequently decarboxylated during the day when stomata are closed, and the CO_2 that is released is re-fixed by rubisco during the day. The photosynthetic system of some CAM plants (e.g., pineapple) is only in the CAM mode when the plants are droughted and acts similar to the C_3 system when the plants are well watered, with stomata opening during the day and the initial CO_2 fixation being done by rubisco.

TABLE 4.1
Photosynthetic Systems of Different Crop Species

Photosynthetic System	Crop Species
C_3 system	All annuals adapted to cool seasons (e.g., wheat, barley, oats, rye, Irish potato, and sugar beet), all legumes (e.g., cool-season adapted species such as garbanzo bean, lentil, and fava bean, and warm-season adapted species such as soybean, common bean, cowpea, peanut, and pigeon pea), all woody perennials (tree and vine crops), and some non-leguminous warm season annuals (e.g., cotton, tomato, cucurbits, rice, sweet potato, sesame, and sunflower)
C_4 system	A few tropical grasses (e.g., maize, sorghum, pearl millet, and sugar cane) and a few warm-season-adapted herbaceous dicotyledons (e.g., grain amaranth)
Intermediate C_3/C_4 system	Cassava
CAM system	Pineapple, sisal, and prickly pear cactus

Most crop plants are adapted to sunny conditions, with only a few being adapted to shade. Among those adapted to sunny conditions, those with the C_4 photosynthetic system have the highest light-saturated rates of P_n, and some can require up to full sunlight for a single leaf to achieve this rate. Among the C_3 species, herbaceous annuals have higher light-saturated P_n than perennial woody species, deciduous trees can have higher light-saturated P_n than evergreen trees, annual species adapted to warm seasons have higher light-saturated P_n than those adapted to cool seasons, and non-leguminous species often have higher light-saturated P_n than leguminous species that fix atmospheric nitrogen. Species differences in light-saturated P_n tend to be positively correlated with species differences in the level of *PFD* required to saturate the P_n of single leaves (Figure 4.3), canopy efficiency for converting intercepted *PFD* to carbohydrate (*Q*) (Sinclair and Horie, 1989), and biomass production.

Since there are species differences in light-saturated P_n of single leaves that are positively correlated with biomass production in sunny environments, it is of interest to ask whether there are similar differences among cultivars within species that could be exploited by plant breeding. As was pointed out by Evans (1993), variation in light-saturated P_n has been observed in several species, is highly heritable, and can be selected for, but it seems not to have contributed much to increase in yield potential so far. In contrast to genetic manipulation, environmental treatments that enhance P_n, such as CO_2 enrichment of C_3 species, have enhanced yield in many cases. This apparent paradox can be partially explained by the fact that genetic manipulation to achieve increases in light-saturated P_n can involve a major cost to the plant in terms of greater investment per unit leaf area in the components of the photosynthetic system, whereas environmental manipulations that increase P_n do not necessarily have major costs that constrain their exploitation by the plant. Of particular importance for C_3 species is the cost of increased investment in the enzyme rubisco, which

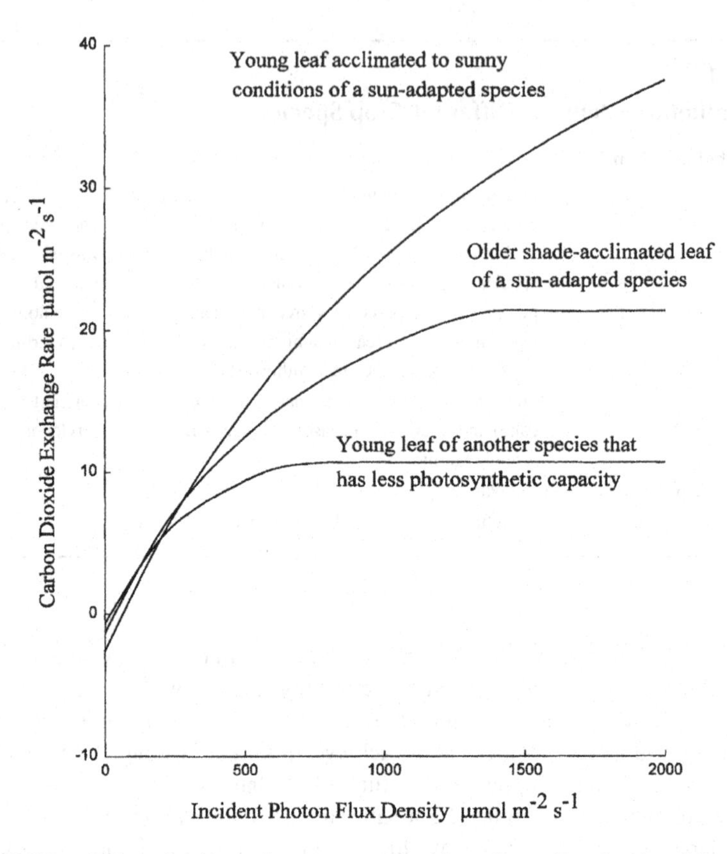

FIGURE 4.3 Carbon dioxide exchange rates to incident photon flux density of three leaves having contrasting photosynthetic capacity based on Hall's model (1979). Differences in capacity can be present in different species or occur with age or acclimation to sunny or shady conditions.

represents a large proportion of total leaf enzymes and protein. Another factor is the negative correlation that frequently, but not always, has been observed among cultivars between light-saturated P_n and the area of individual leaves. Some cultivars with higher P_n have smaller leaves and less total leaf area. During early stages of vegetative growth, cultivars with faster development of leaf area intercept more solar radiation and have a greater rate of biomass production. In contrast, once full ground cover has been achieved, cultivars with greater P_n per unit leaf area, but less leaf area, may have greater canopy P_n and a greater rate of biomass production. The impact on seasonal biomass production of cultivar differences in P_n per unit leaf area and total leaf area would depend on the relative biomass production during early and later stages of plant development. An additional complication is that, in some cases, photosynthetic capacity may vary depending on the activities of various plant *sinks* for carbohydrate. (Refer to the discussion in Chapter 3 of the down-regulation of photosynthetic capacity that can occur when C_3 plants that are growing in small pots are subjected to elevated atmospheric [CO_2].)

Sinclair et al. (2004) presented a theoretical analysis of the potential impact on grain yields of soybean of breeding to develop a genotype that synthesized 38% more rubisco per unit leaf area. The authors concluded that this could result in an 18% increase in crop biomass but depending on various assumptions concerning the plant nitrogen accumulation associated with the increase in photosynthesis, grain yield might be either increased by 6% or decreased by 6%.

The question of optimal leaf orientation also is complex. Compare a canopy with mainly horizontal leaves and a canopy with mainly vertical leaves. Assume clear sunny days with the sun vertically overhead. During early vegetative growth, the horizontal-leaved canopy would be more effective in that it would reach full ground cover sooner and with less leaf area. Also, it would be more effective than the vertical-leaved canopy in shading out and competing with weeds. In contrast, once there is substantial ground cover, the mainly vertical-leaved canopy would be more effective in that it would have greater canopy P_n, because the intercepted *PFD* would be more uniformly distributed within the canopy. Note that the second and lower layers of leaves in canopies with horizontal leaves often are in deep shade environments because leaves transmit very little *PFD* (leaf optical properties are described in Chapter 7). Many modern cultivars of wheat and rice have leaves that are more vertical and have greater potential productivity, but they are less competitive with weeds than the traditional cultivars, which have lax leaves that are more horizontal. A more ideal canopy would involve a plant producing horizontal leaves during the seedling stage and then leaves that are more vertical as the plant achieves full ground cover. Modern maize hybrids grown in the United States have leaves that are more vertical (Duvick and Cassman, 1999). The adaptive significance of this trait for maize is that it increases light penetration into the canopy, which could increase Q, decrease barrenness (promoting the tendency to produce two cobs on each stem), and decrease the tendency to lodge.

Among the shade-adapted species, there are those with an obligate requirement, such as some house plants and cocoa trees, that must be grown in shade or they will be damaged by intense sunlight. Others, such as coffee trees and Engleman spruce, are facultative shade species. They require shade as young trees but with age become adapted to sunny conditions. The ecological significance of this facultative response is that the plants evolved so that they could become established in shade conditions and then grow tall, emerge, and occupy the sunny part of the canopy when space became available due to the death of old trees. Facultative species, such as Engleman spruce, do not respond well to clear-cutting practices of forest management. Possibly this is because they experience difficulty in establishing seedlings in open sunny conditions due to damage to their photosynthetic system by intense sunlight. An additional explanation for replanting problems is that after clear-cutting coniferous forests, the predominant form of inorganic nitrogen in the soil changes from ammonium to nitrate due to changes in soil microbial activities, which then favors the establishment of aspen trees over spruce.

Coffee trees are propagated as young plants in shaded nurseries and are hardened by progressively removing the shade prior to transplanting them into plantations where they are grown to produce coffee beans. Original plantations were shaded by planting overhead trees to simulate the natural understory habitats in Africa where coffee bushes evolved (see the review by DaMatta, 2004). In many situations, however, coffee trees grow well without shade and even outyield shade-grown coffee trees. Shading was therefore abandoned in several regions, especially in Brazil, the world's largest coffee producer. However, unshaded coffee plantations require more soil nutrients and they also have a tendency for more biennial bearing and more branch dieback than shaded coffee plantations.

Coffee can grow only in a narrow temperature range and global warming has been predicted to substantially reduce the area of land in the world that is suitable for coffee production, maybe by 50% by the year 2050. Arabica coffee is responsible for producing the highest-quality coffee but requires cooler growing conditions and typically is grown at higher elevations and may suffer a greater loss of crop land than Robusta coffee. In addition, global warming is predicted to increase damage caused by coffee rust in that the fungus that causes this disease is favored by warmer conditions. Robusta coffee could be affected less by this disease because it has greater resistance to leaf rust than Arabica coffee. Growing Arabica coffee in shaded conditions, which are cooler, could counteract the effects of global warming.

Plantations that do not have shade trees provide a less desirable habitat for migrating birds. Conservation societies are currently paying higher prices for Arabica coffee beans grown in certified shaded coffee plantations that attract migrating birds, which helps the coffee farmers who adopt this program. For example, American Birding Association Song Bird Coffee comes from coffee plantations that are certified as being bird-friendly by the Smithsonian Migratory Bird Center.

Responses of leaf carbon dioxide exchange rate to incident *PFD* of a sun-adapted and a shade-adapted species (Figure 4.4) were predicted by Hall's model (1979). The main differences between the photosynthetic characteristics of shade versus sun species are as follows:

1. Sun species are better adapted to sunny conditions than shade species because they have higher P_n per unit leaf area under sunny conditions than shade species. The higher P_n results from the photosynthetic systems of sun species having higher capacity for electron flow and CO_2 assimilation than those of shade species, because they have more chloroplasts, electron transfer components, and photosynthetic enzymes per unit leaf area. Consequently, leaves of sun species reflect a greater investment of protein and nitrogen per unit area and are thicker than shade leaves.
2. Shade species are better adapted to shady conditions than sun species because, under deep shade conditions, they can have higher P_n per unit leaf area than sun species, which may evolve CO_2 in these conditions. The higher P_n of shade species under very shady conditions is mainly due to their lower rates of mitochondrial respiration (*R*) than sun species. Recall from Equation 4.1 that P_n is the balance between CO_2 fixation in photosynthesis and CO_2 release due to respiration. Rates of mitochondrial respiration are determined by rates of biosynthesis and the extent of maintenance activities within leaves, and both of these processes occur at lower levels per unit leaf area in shade-adapted leaves than in sun-adapted leaves. Also, under shady conditions, leaves of

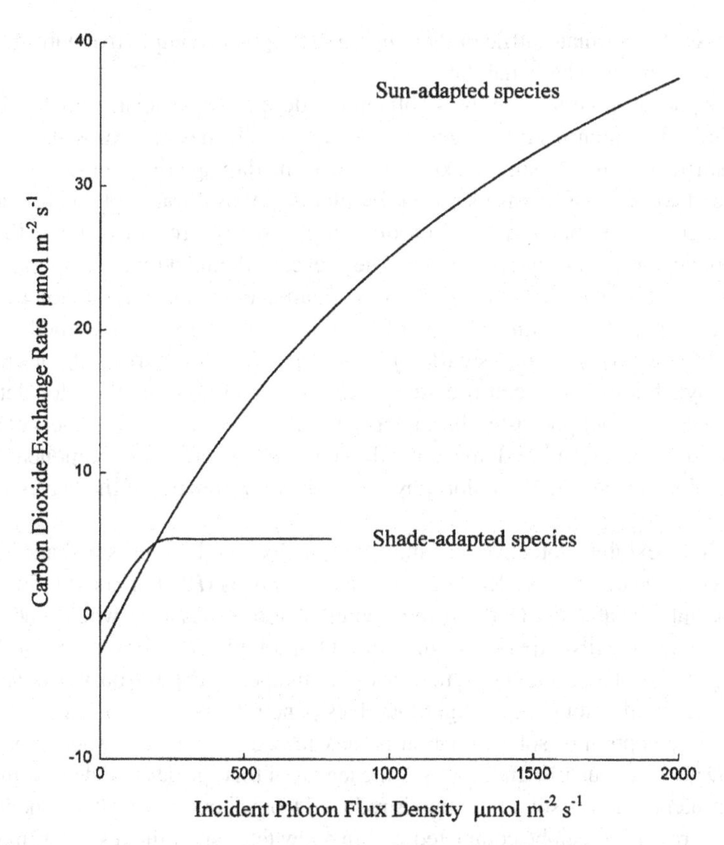

FIGURE 4.4 Leaf carbon dioxide exchange rate responses to incident photon flux density of a sun-adapted and a shade-adapted species based on Hall's model (1979).

shade species are much more efficient than leaves of sun species with respect to the ratio of net carbon gain (P_n) per unit of investment of chemical energy. Basically, shade plants are well designed for intercepting and using low levels of *PFD* through efficient investment of the relatively small amount of biomass and chemical energy that is available to them by producing thin low-cost leaves. Note that the quantum yields, the initial slopes of the curves in Figure 4.4, are very similar for sun and shade plants.

3. The photosynthetic systems of sun species can, to a certain extent, acclimate to function more efficiently under shady conditions. This acclimation involves the development of leaves with less photosynthetic capacity per unit leaf area and therefore also less *R* and under shady conditions, higher P_n, and greater efficiency (P_n per unit of investment) than sun-adapted leaves. The intermediate curve in Figure 4.3 provides an example of how leaves of a sun-adapted species might perform if they had been subjected to shade for several days and had acclimated to the shade condition. This acclimation can occur as leaves become older and occupy lower and more shaded positions in the canopy. The photosynthetic systems of sun species do not,

however, acclimate sufficiently to make these plants competitive with shade species in very shady habitats.

4. The photosynthetic systems of obligate shade species do not acclimate when exposed to intense sunlight and can be damaged if they are exposed to it for a sufficient time. A simple explanation for the damage that can result from the absorption of intense sunlight by plants is that through photochemical reactions their photosystems produce high energy products, and if these products are not used fast enough, they react with and damage components of the chloroplast. The first indication of damage involves a reduction in the quantum yield (the initial slope of the curve relating P_n to absorbed *PFD* in Figure 4.1), and this is called *photo-inhibition*. Measurements of chlorophyll fluorescence can provide an effective and rapid method for determining whether photo-inhibition has occurred (Jones, 1992; Lambers et al., 1998). More extensive damage results in photo-oxidation of components in the chloroplast, such as chlorophyll, and thus a yellowing of the leaves.

Some species exhibit protective leaf movements. Redwood-sorrel is a shade species that grows in the densely shaded floor of redwood forests (*PFD* of about 0.5% of full sunlight), but its habitat extends to the border of forest clearings where understory plants can occasionally experience full sunlight and high *PFD* for one or two hours. When exposed to shady conditions, the leaves of this species exhibit diurnal movements tracking the small amount of sunlight that does penetrate the redwood canopies, such that their interception of solar radiation is maximized. In contrast, when exposed to high *PFD* above about 400 $\mu mol\ m^{-2}\ s^{-1}$, the leaves of this species fold downward such that their interception of solar radiation is reduced from about 90% to 10%. The folding response is rapid and can be completed within a few (e.g., six) minutes (Björkman and Powles, 1981) and has been shown to enable the leaves to avoid photo-inhibition and other damage to the photosynthetic system (Powles and Björkman, 1981).

What is responsible for the ability of the many leaves that are sun adapted to withstand intense sunlight? First of all, sun leaves have the capacity to use most of the high-energy products coming from the photosystems in the process of CO_2 fixation. Also, for C_3 species, photorespiration enhances the capacity for using these high-energy products, since it acts to recycle carbon compounds. In addition, all higher plant species have the xanthophyll cycle, which acts to use any excess high-energy products from the photosystems (Lambers et al., 1998). Finally, both sun and shade leaves have the ability, during the night, to repair some of the damage to the chloroplasts that occurred during the day. The extreme sensitivity of leaves of shade plants to intense sunlight has not been well explained. They do have low photosynthetic capacity for using the high-energy products from their photosystems, but they also have photorespiration, which acts to use some of the high-energy products of the photosystems of these C_3 species shade plants.

Of particular significance for adaptation is the fact that some stresses cause plants to be more sensitive to high-*PFD*-induced photo-inhibition. For example, stresses such as soil drought and salinity cause stomata to close, thereby reducing the supply of CO_2 to the leaf mesophyll cells and reducing the capacity of these cells to use the high-energy products of the photosystems. A more complex case involves the effects of chilling (0°C–18°C) temperatures on species adapted to tropical climatic zones or warm seasons in other zones. The combination of chilling temperatures and intense sunlight can cause serious photo-inhibition in these species. The mechanism may involve chilling-induced disturbances to membranes in chloroplasts, which then in some way makes the chloroplasts more susceptible to the photo-inhibitory effects of intense sunlight. Photo-inhibition damages photosystem II. The extreme sensitivity of photosystem II may be viewed as being either damaging and a weak link in the adaptation of these species or analogous to an electrical fuse that protects the chloroplast from producing even more of the high-energy products, which would then cause more extensive leaf damage through photo-oxidation. The photosynthetic system of some C_4 species is particularly sensitive to the combination of chilling and high light. Progress has been made in breeding maize hybrids with resistance to the damage caused by chilling plus high light (Greaves, 1996).

Photosynthesis responds to changes in carbon dioxide concentration [CO_2], and this has some significance to present and future plant adaptation. During the 220,000 years prior to the year 1800, atmospheric [CO_2] (C_a) fluctuated between 180 and 290 parts per million by volume (equivalent to µmol mol^{-1} or µbar bar^{-1}). Since 1800, C_a has exhibited accelerating increases from 280 to 300 parts per million by 1900 and 360 parts per million by 2000 and is expected to continue to increase during the twenty-first century up to about 700 parts per million. Crop species differ in their photosynthetic responses to [CO_2], with C_3 species exhibiting increases in P_n with increases in C_a up to about 1,000 parts per million and C_4 species exhibiting maximum P_n at about 360 parts per million (Figure 4.5). Under optimal temperatures, the P_n of C_3 species responds to increases in C_a below 1,000 parts per million at all natural levels of *PFD* (Figure 4.6).

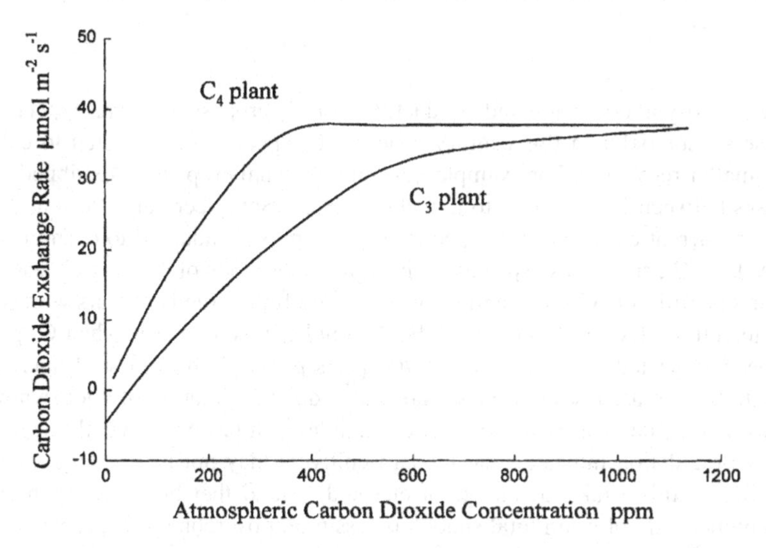

FIGURE 4.5 Leaf carbon dioxide exchange rates as a function of atmospheric carbon dioxide concentration with full sunlight for a C_3 plant and a C_4 plant based on Hall's model (1979).

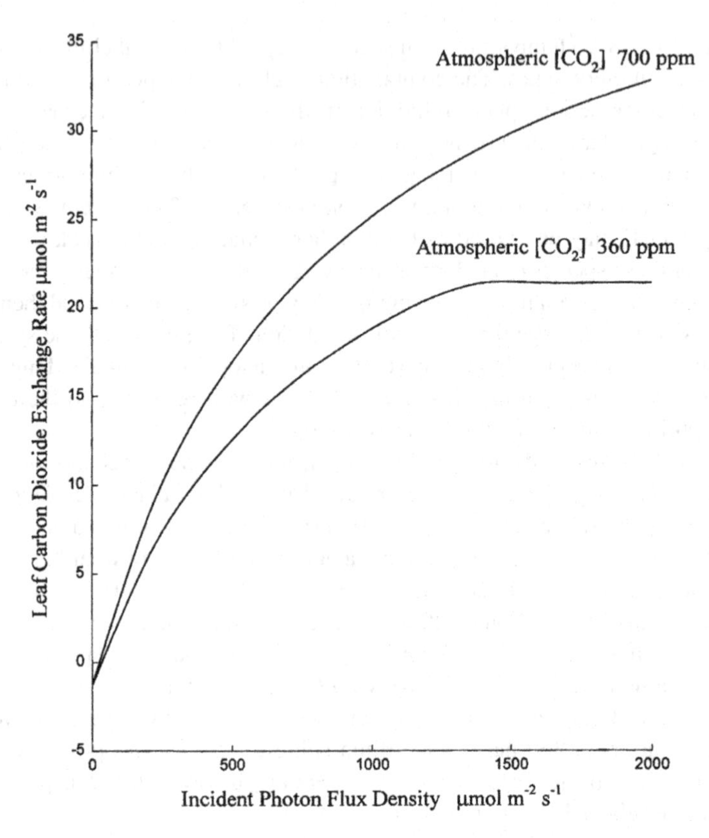

FIGURE 4.6 Leaf carbon dioxide exchange rate responses to incident photon flux density at 360 and 700 parts per million atmospheric carbon dioxide concentration of a C_3 plant based on Hall's model (1979).

The photosynthetic rates and productivity of C_3 crop species are expected to increase substantially in this century, whereas C_4 species are expected to exhibit much smaller responses. For example, several C_3 annual crops have exhibited yield increases between 5% and 40% to a doubling of C_a (Seneweera and Norton, 2011). However, current cultivars of C_3 species may not be well adapted to even present-day levels of C_a, since these species evolved over thousands of years at C_a less than 290 parts per million. Consequently, it would be useful to breed cultivars with adaptation to future elevated levels of C_a. For example, in some cases when C_3 plants have been subjected to elevated C_a of 700 parts per million, at first P_n increased substantially, but after several days there was a down-regulation of photosynthetic capacity, and P_n decreased to about the original level. It has been hypothesized that the down-regulation indicates that modern cultivars may not have a large enough sink/source ratio to take advantage of elevated C_a and that breeding grain crops with a higher ratio of grain/total shoot biomass might overcome this problem (Hall and Ziska, 2000). Alternatively, the down-regulation may have been an artifact that resulted from growing plants in pots, which restricted their root growth, or it may be a response to limited supplies of nutrients such as inorganic nitrogen. (Refer to the discussion of this topic in Chapter 3.)

At current levels of C_a, the C_4 system confers an advantage over the C_3 system in high *PFD*, high-temperature environments with adequate water and soil nutrients (maize and sugar cane are very productive in these environments), and in high *PFD*, high-temperature environments where water and nitrogen are limiting (sorghum and pearl millet can be productive in these environments). Note that stomata partially close under drought and the $[CO_2]$ inside leaves is low, and C_4 plants are more effective than C_3 plants in these conditions. Also, the C_4 system requires less rubisco per unit leaf area and less nitrogen per unit leaf area than the C_3 system (Lambers et al., 1998). Consequently, C_4 plants can be more efficient than C_3 plants with respect to P_n/T_r, where T_r is the transpiration rate per unit projected leaf area and P_n/unit N invested in leaves.

Daytime closure of stomata by droughted CAM plants, along with their nocturnal fixation of $[CO_2]$, enables them to achieve very high P_n/T_r. CAM plants are well adapted to very dry environments with a little rain every month but nights that are not too hot. Refer to Chapters 8 and 9 for a discussion of species and cultivar differences in transpiration efficiency and water-use efficiency.

Photosynthesis responds to the $[CO_2]$ inside leaves (C_i), which is influenced by stomata and the leaf conductance to transfer of carbon dioxide (g_c) as described in Equation 4.5, which is obtained by rearranging Equation 4.4. Note that C_i is the dependent variable in these equations. Essentially, C_i is determined by the demand for CO_2, which is determined by the photosynthetic capacity and the supply of CO_2, which is determined by C_a and g_c.

$$P_n = g_c \times (C_a - C_i) \tag{4.4}$$

$$C_i = C_a - \frac{P_n}{g_c} \tag{4.5}$$

Equation 4.4 is based on the law described by Equation 3.4. (Note that the partial pressure of a gas divided by atmospheric pressure is equivalent to the concentration of the gas on a volumetric basis.) Estimating C_i requires that in addition to a knowledge of the rate of net photosynthesis, we also know the value of g_c. A powerful model is available for estimating tissue conductance to one gas when the conductance of another gas in the same flow pathway is known. Equation 4.6 provides an example for the diffusion of water vapor and carbon dioxide in air:

$$g_c = g_w \times \frac{D_c}{D_w} \tag{4.6}$$

where g_w is the conductance to water vapor, and D_c and D_w are diffusion coefficients for carbon dioxide in air and water vapor in air, respectively. The values of the diffusion coefficients can be obtained from Massman (1998) and then converted to the appropriate atmospheric pressure (P) using the relation $D_{P2} = D_{P1} \times P_1/P_2$ and to the appropriate temperature (with T in kelvins) using the relation $D_{T2} = D_{T1} \times (T_2/T_1)^{1.81}$. For example, at 1 bar atmospheric pressure and 25°C, D_c is 16.4 mm^2 s^{-1}, and D_w is 25.9 mm^2 s^{-1}.

When comparing isotopes, such as $C^{12}O_2$ and $C^{13}O_2$, the ratio of their diffusion coefficients (e.g., ratio = $D(C^{12}O_2)/D(C^{13}O_2)$) can be estimated from Equation 4.7, but also refer to a discussion of this issue by Massman (1998):

$$\text{Ratio} = \left[\frac{\dfrac{MW(C^{13}O_2)}{MW(C^{13}O_2)+MW_a}}{\dfrac{MW(C^{12}O_2)}{MW(C^{12}O_2)+MW_a}} \right]^{0.5} \tag{4.7}$$

where MW are molecular weights, and MW_a is the molecular weight of air (average of 29). Equation 4.7 is derived from a law stating that the values of diffusion coefficients are inversely proportional to the square root of their reduced masses. Molecules with greater mass diffuse more slowly. Equation 4.7 should not be used to compare the diffusion coefficients of radically different molecules such as CO_2 and H_2O.

The models described by Equations 3.4, and 4.6 can be very useful, since g_w can be estimated from leaves (and other tissues) based on measurements of transpiration (T_r) using Equation 4.9, which is obtained by rearranging Equation 4.8. Note that Equation 4.9 provides a definition of g_w, whereas the dependent variable is T_r.

$$T_r = g_w \times (H_i - H_a) \tag{4.8}$$

$$g_w = \frac{T_r}{(H_i - H_a)} \tag{4.9}$$

Equation 4.8 also is from the law described by Equation 3.4, with H_a being the volumetric concentration of water vapor in the air (equivalent to air vapor pressure/atmospheric pressure) and H_i being the volumetric concentration of water vapor in the air spaces in the leaf. The latter can be obtained if leaf temperature is known by assuming either that the humidity inside the leaf is saturated, with values of the saturated vapor pressure being available in handbooks, or by determining the relative humidity inside the leaf using Equations 3.2 and 3.3 if the water potential of the leaf is known (this latter correction is necessary only if the plants have been subjected to drought and the leaf water potential is very negative).

I stated that the models described by Equations 3.4 and 4.6 can be very useful, and I will provide some examples. The obvious case is the estimation of C_i using Equations 4.5 and 4.6. Another case is the estimation of the oxygen concentration within leaves (O_i) based on a knowledge of P_n and relations between the carbon dioxide and oxygen exchanges in photosynthesis and respiration and of g_w. If sugars are being formed in photosynthesis and used in respiration, then the net oxygen exchange (P_{no}) is equivalent to $-P_n$, and O_i can be estimated using Equations 4.10 and 4.11.

$$O_i = \frac{O_a - P_{no}}{g_o} \cong \frac{O_a + P_n}{g_o} \tag{4.10}$$

where O_a is the atmospheric oxygen concentration, which is 21% (210,000 parts per million), and g_o is the leaf conductance to transfer of O_2, which can be estimated from Equation 4.11:

$$g_o = g_w \times \frac{D_o}{D_w} \tag{4.11}$$

where D_o is the diffusion coefficient for oxygen in air, which has a value of 21.6 mm^2 s^{-1} at 1 bar atmospheric pressure and 25°C (Massman, 1998). Using typical values of P_n and g_w in Equations 4.10 and 4.11, it can be shown that, during the day, leaves will have an O_i which is only a few ppm greater than 21%. The O_i can be estimated for nighttime conditions, providing the respiration rate and g_w are known, and it will be a few parts per million less than 21%. Consequently, O_i of leaves in air remains at 21% ± 0.01%. For plant tissues that are submerged in liquid water, such as roots in waterlogged soil, O_i can deviate substantially from 21% because the diffusion of gases in liquid water is 10^{-4} slower than in air ($D_o = 0.0020$ mm^2 s^{-1} for O_2 in water).

Another use of the models concerns ethylene. If the ethylene production rate by plant tissue (F_{eth}), external ethylene concentration (Eth_a), and g_w are known, then the ethylene concentration within the tissue (Eth_i) can be estimated using Equations 4.12 and 4.13.

$$Eth_i = Eth_a + \frac{F_{eth}}{g_e} \tag{4.12}$$

where g_e is the leaf conductance to the transfer of ethylene, which can be estimated from Equation 4.13.

$$g_e = g_w \times \frac{D_e}{D_w} \tag{4.13}$$

where D_e is the diffusion coefficient for ethylene in air, with a value of 23.2 mm^2 s^{-1} at 1 bar atmospheric pressure and 25°C (Massman, 1998). Knowledge of the ethylene concentration within tissues (Eth_i) can be useful because this concentration most likely will determine effects of ethylene on plant function.

Another use of the model concerns ozone. If the ozone concentration in the atmosphere (Oz_a) and g_w are known, we can estimate the rate at which ozone is being taken up by leaves (F_{oz}) using Equations 4.14 and 4.15, because it is likely that the ozone concentration within leaves is very small and can be assumed to be zero.

$$F_{oz} \cong g_{oz} \times Oz_a \tag{4.14}$$

where g_{oz} is the leaf conductance to the transfer of ozone, which can be estimated from Equation 4.15.

$$g_{oz} = g_w \times \frac{D_{oz}}{D_w} \tag{4.15}$$

where D_{oz} is the diffusion coefficient for ozone in air. Experimental values for D_{oz} may not be available, because ozone is reactive on surfaces, but an estimated value has been provided by Massman (1998) of 17.1 $mm^2\ s^{-1}$ at 1 bar atmospheric pressure and 25°C. Estimating values of the ozone flux into leaves is useful. Damage to plants by ozone probably is related to the rate of uptake of ozone by leaves. Differences in ozone uptake rates by different types of vegetation has been related to the stomatal conductances of leaves, because ozone is more reactive with the wet surfaces within leaves than it is with the dry external surfaces of plants (Grantz et al., 1994). Irrigated cotton and corn canopies took up ozone faster than irrigated grapes and fruit trees, which usually have lower stomatal conductances than cotton or corn, and dry rangeland had a very slow uptake of ozone. The importance of these observations is that they demonstrate the value to society of irrigated crop land in dry environments as an important factor in removing the pollutant ozone from the air that we have to breathe.

Photosynthetic responses to temperature provide an indication of the environments to which plants are adapted. Photosynthetic responses to temperature exhibit a non-asymmetric optimal type curve (Figure 4.7). Temperature extremes must not be used when measuring these curves to ensure the data are reversible. For example, the response in Figure 4.7 indicates that a measurement at 28°C would give a P_n of 21 μmol $m^{-2}\ s^{-1}$, and a measurement at 32°C would give a P_n of 17 μmol $m^{-2}\ s^{-1}$, and if one then returned to 28°C, a P_n of 21 μmol $m^{-2}\ s^{-1}$ would be obtained. If the leaf had been subjected to a temperature that is much higher than 32°C, the leaf would have been stressed and, on returning to 28°C, a value of P_n that is less than 21 μmol $m^{-2}\ s^{-1}$ would have been obtained. The shape of these response curves can vary depending on the humidity conditions that are

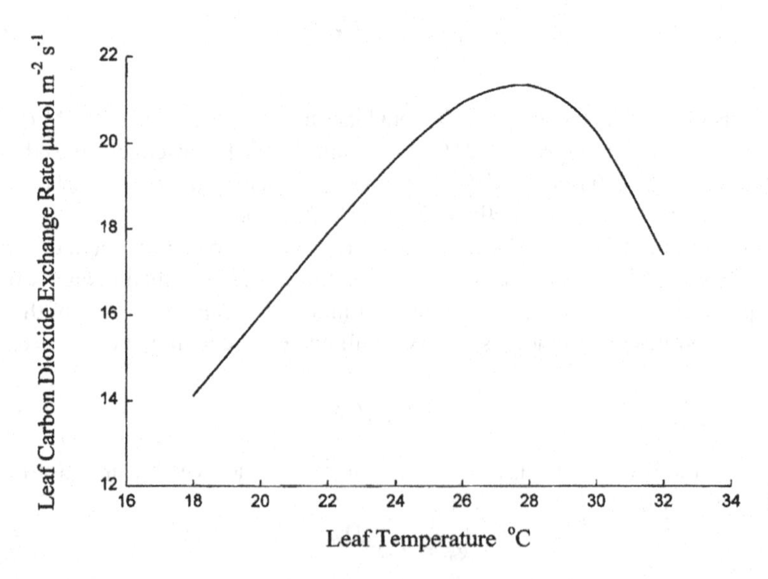

FIGURE 4.7 Reversible leaf carbon dioxide exchange rate responses to leaf temperature at saturating photon flux density based on Hall's model (1979).

used while making the measurements because humidity can influence stomatal conductance, as is discussed in Chapter 8.

The range of temperatures where plants exhibit active reversible P_n also are the day temperatures where they exhibit active growth. Therefore, P_n responses to temperature are related to the daytime temperatures to which plants are adapted. This is very useful information because it is difficult and time consuming to determine the temperatures to which plants are adapted based on studies of plant growth at different temperatures. In principle, such studies could be conducted by measuring plant growth in controlled-environment chambers with different temperatures. In practice, such studies can have many artifacts, which reduce their ability to predict plant performance in natural conditions. Using constant temperatures in the chambers would simplify data analysis but would generate artifacts because natural field environments have extremely variable temperatures with different temperatures in the root and shoot zones. Different constant day and night, root and shoot temperatures could be used, but this still could generate artifacts since temperatures are not constant during either the day or the night in field conditions. Thermal regimes with diurnal sinewave variation in temperature could be used in the controlled environments that approximate some natural environments. Analysis of such experiments to determine plant responses to temperature, however, would be difficult, and it may be more difficult to repeat them than when using chambers where temperature is kept constant.

In general, annual plants are adapted to a modest range of temperatures, with some species being adapted to hotter temperatures (warm-season species) and other species being adapted to cooler temperatures (cool-season species), as discussed in more detail in Chapter 5. Generally, but not always, species with higher optimal temperatures for photosynthesis also have higher maximum rates of P_n. Photosynthetic acclimation to temperature can occur. It is most pronounced in perennial evergreens, such as oleander, and possibly various *Citrus* species that are adapted to environments with large seasonal changes in temperature but mild winter temperatures, such as Mediterranean climates (e.g., the locations at Figures 5.2 and 10.4). Acclimation results in plants that have been subjected to hotter temperatures having higher P_n at hotter temperatures than plants that were subjected to cooler temperatures, and vice versa.

Usually, it is assumed that plant growth rate is a function of P_n per plant, and this often is the case. There are some circumstances, however, where internal regulation of photosynthesis occurs such that P_n/leaf area is itself a function of plant growth. A hypothesis for short-term (minutes) regulation has been proposed. (This hypothesis is discussed in Lambers et al., 1998.) If export of the products of photosynthesis is slow, phosphorylated intermediates of the pathway leading to sucrose accumulation build up, causing a shortage of inorganic phosphate, which results in less RuBP being regenerated and slower P_n. If this short-term regulation occurs to a significant extent, removing it would provide a mechanism for enhancing P_n through genetic engineering to reduce bottlenecks to leaf export of sucrose. However, to date, clear evidence has not been presented for the substantial occurrence of short-term feedback effects of this type. In contrast, there is abundant evidence that long-term (days) regulation of photosynthetic capacity can occur in which, for C_3 plants, decreases or increases in rubisco and other photosynthetic components occur. Down-regulation can occur when leaves are shaded or as they become older or, in some cases, when they are subjected to elevated C_a, as was discussed earlier. Also, up-regulation can occur such as with the upper leaves in a canopy if the lower leaves are subjected to deep shade for several days. The presence of nearby fruits can result in either up- or down-regulation of leaf photosynthetic capacity.

4.3 MITOCHONDRIAL RESPIRATION

A model for mitochondrial respiration of a population of plants (*R* is in μmol release of CO_2 ground-area^{-1} time^{-1}) is presented in Equation 4.16:

$$R = a \times B + b \times C + D \tag{4.16}$$

where *B* is total plant biomass per unit ground area, $a \times B$ is the extent of respiration involved in maintenance activities, *C* is the growth rate per unit ground area, $b \times C$ is the extent of respiration involved in biosynthesis and transport processes, and *D* is the extent of respiration that is uncoupled from the production of reductants such as NADH and $FADH_2$ and ATP.

Cultivar differences in rate of mitochondrial respiration have been reported but, in most cases, their relation to crop yield is not clear (Evans, 1993). For young plants with a small amount of biomass, respiration involved in biosynthesis is the dominant component of the model, and a positive relation between crop growth rate and respiration rate may be expected. At later stages, when maintenance respiration is the dominant component, the relation between crop growth rate and total respiration rate may be negative, as it also might be if cultivars differed substantially in extent of uncoupled respiration. In two specific cases, regrowth of perennial rye grass and sugar storage in the root of sugar beet, plant performance was negatively correlated with respiration rate (Evans, 1993). Further understanding is needed, however, before selection for either a lower or higher respiration rate can be recommended as being useful in plant breeding for enhancing crop performance.

4.4 PHOTORESPIRATION

Photorespiration results from the oxygenase activity of rubisco and occurs to a substantial extent in photosynthetic tissues of C_3 but not C_4 plant species. Photorespiration appears to have evolved as a negative consequence of the evolution of rubisco, which was driven by its effectiveness in fixing CO_2. But this fixation occurs at the same enzymatic site where the oxygenase activity occurs. This was not a problem during the early stages of plant evolution in that the ratio of the atmospheric $[CO_2]/[O_2]$ was much higher than current levels, and therefore the oxygenase and photorespiration rates were much slower than the carboxylase activity in the far past. It has been hypothesized that photorespiration may contribute to adaptation by reducing the extent of photo-inhibition through internal recycling of CO_2. The importance of this contribution to the adaptation of C_3 plants is not known, and it clearly is not essential for the adaptation of C_4 plants, because they exhibit very low rates of photorespiration. The extent of photorespiration can be determined by subjecting plants to 1% [O_2], which inhibits photorespiration (but has little effect on mitochondrial respiration). This treatment results in a very large increase in net photosynthesis compared with P_n under ambient [O_2] of 21%. At 1% [O_2], the P_n of C_3 plants is similar to that of C_3 plants at high [CO_2] (Figure 4.6) or that of C_4 plants at 21% [O_2] (Figure 4.5), because, in all of these cases, there is very little photorespiration. The extent of the increase in P_n at 1% [O_2] is related to the extent of photorespiration that was occurring under ambient [O_2] following Equation 4.1. For many years, scientists have pursued the hypothesis that the yield potential of C_3 plants may be enhanced by developing plants with leaf rubisco that has greater specificity for CO_2 than for O_2 and thus much slower photorespiration (discussed by Austin, 1999). This approach assumes that photorespiration has no major beneficial effects, which may or may not be true, and that it is possible to change the specificities of rubisco in the desired manner. It is not yet clear, however, whether rubisco can be modified such that P_n is enhanced and there are no negative side effects. In future environments with elevated atmospheric [CO_2], photorespiration of C_3 plants will be slower, and there will be less need to attempt to decrease it by plant breeding.

4.5 GROWTH ANALYSIS

A system for analyzing growth has been developed that is useful when studying seedlings in isolation but is not of much use for examining the performance of closed canopies. With this method, the relative growth rate (*RGR*) is defined using Equation 4.17:

$$RGR = (1/B) \times \frac{dB}{dt} \tag{4.17}$$

where *B* is biomass and *dB/dt* is the change in biomass with time. Note that crop growth rate equals *dB/dt* on a unit ground area basis. *RGR* may be partitioned into two components in Equation 4.18:

$$RGR = LAR \times NAR \tag{4.18}$$

where the net assimilation rate (*NAR*) equals *dB/dt* on a projected leaf area basis and is similar to average P_n integrated over time, and the leaf area ratio (*LAR*) equals projected leaf area/*B*; this can be further partitioned by Equation 4.19:

$$LAR = LMR \times SLA \tag{4.19}$$

where the leaf mass ratio (*LMR*) equals total leaf mass/*B*, and the specific leaf area (*SLA*) equals projected leaf area/leaf mass.

Growth analysis has been used to investigate mechanisms for differences in seedling growth rates of different species. In general, species differences in *RGR* mainly were due to differences in *LAR* that were due to differences in *SLA* (Poorter and van der Werf, 1998), but with positive correlations with photosynthetic capacity per unit leaf mass (Evans, 1998). Basically, faster-growing seedlings partitioned larger amounts of biomass into new leaf area that was efficient with respect to it having a larger area and ratio of photosynthetic cell tissue to the combination of epidermal, vascular, and sclerenchymatous tissues. Thus, the faster growth rate resulted from the amplification effect of greater partitioning into photosynthetic tissue that is capable of intercepting more light.

ADDITIONAL READING

Björkman, O. and S. B. Powles. 1981. Leaf movement in the shade species *Oxalis oregana*. I. Response to light level and light quality. *Carnegie Institution Washington Year Book* 80: 59–62.

Ehleringer, J. and O. Björkman. 1977. Quantum yields for CO_2 uptake in C_3 and C_4 plants. *Plant Physiol.* 59: 86–90.

Evans, J. R. 1998. Photosynthetic characteristics of fast- and slow-growing species. In H. Lambers, H. Poorter, and M. M. I. Van Vuuren (Eds.), *Inherent Variation in Plant Growth: Physiological Mechanisms and Ecological Consequences.* Backhuys Publishers, Leiden, The Netherlands, pp. 101–119.

Hall, A. E. 1979. A model of leaf photosynthesis and respiration for predicting carbon dioxide assimilation in different environments. *Oecologia* 143: 299–316.

Kwapata, M. B., A. E. Hall, and M. A. Madore. 1990. Response of contrasting vegetable-cowpea cultivars to plant density and harvesting of young pods. II. Dry matter production and photosynthesis. *Field Crops Res.* 24: 11–21.

Lambers, H., F. S. Chapin III, and T. L. Pons. 1998. *Plant Physiological Ecology.* Springer-Verlag, New York, p. 540.

Loomis, R. S. and D. J. Connor. 1992. *Crop Ecology.* Cambridge University Press, Cambridge, UK, p. 538.

Powles, S. B. and O. Björkman. 1981. Leaf movement in the shade species *Oxalis oregana*. II. Role in protection against injury by intense light. *Carnegie Institution of Washington Year Book* 80: 63–66.

Sinclair, T. R. 1994. Limits to crop yield? In K. J. Boote, J. M. Bennett, T. R. Sinclair, and G. M. Paulsen (Eds.), *Physiology and Determination of Crop Yield.* Crop Science Society of America, Madison, WI, pp. 509–532.

Sinclair, T. R. and T. Horie. 1989. Leaf nitrogen, photosynthesis, and crop radiation use efficiency: A review. *Crop Sci.* 29: 90–98.

5 Crop Physiological Responses to Temperature and Elevated Atmospheric $[CO_2]$—Climatic Zone Definitions, and Methods for Determining Where Crops Can Be Grown

Since I must provide many specific temperatures and changes in temperature when describing crop responses to temperature, I decided to use one temperature scale: the Celsius scale with the freezing point of water at 0°C and the boiling point of water at 100°C under atmospheric pressure at sea level. In addition, I use degrees Kelvin, which are degrees C + 273, for describing absolute temperature. Virtually all current scientific papers and scientists, as well as the general populations of most countries, use degrees Celsius for describing temperature; the United States of America is the major exception that still uses the old British system, which the British no longer use.

Due to increases in greenhouse gases, average global temperature has increased 1°C since the preindustrial period and is predicted to increase a further 2°C–5°C by the end of the twenty-first century depending on the extent of steps taken to mitigate global warming (World Bank, 2010). These increases in temperature may appear to be small, but they could have major detrimental effects on crop yield (Singh et al., 2011). Crop physiological responses to temperature largely determine plant adaptation to different climatic zones and seasons and can influence crop yield. Crop responses to high temperature are discussed in the Heat Stress section of the website www.plantstress.com, which I wrote, and in Hall (2012).

Most annual crop plants can be described as being adapted to either cool seasons or warm seasons (Table 5.1). For example, in a Mediterranean climatic zone, which is subtropical and has winter rainfall (Figures 5.2, 10.4, 10.6, and 10.7), spring wheat and pea grow well in the fall–winter–spring season, whereas maize and cowpea

TABLE 5.1
Examples of Annual Crop Species Adapted to Different Seasons

Cool-Season Annuals	Warm-Season Annuals
Barley, canola, fava bean, flax, garbanzo bean, Irish potato, lentil, lettuce, lupine, mustard, oat, pea, radish, rye, spinach, triticale, turnip, vetch, and wheat	Common bean, cotton, cowpea, cucurbits, finger millet, lima bean, maize, melon, mung bean, okra, pearl millet, pepper, pigeon pea, rice, sesame, sorghum, soybean, sunflower, sweet potato, tepary bean, tobacco, and tomato

grow well in the spring-summer season, provided they are irrigated. Note that safflower is unusual in that in the vegetative stage it grows well during cool conditions, and in the reproductive stage it grows well during hot conditions.

With respect to tree crops, subtropical zones have temperatures that are well suited to the commercial production of certain evergreen trees. Orange trees grown in subtropical parts of California produce orange-colored fruit, whereas in the tropics the same varieties of orange trees produce greenish-yellow-colored fruit that consumers do not favor. Some other evergreen trees, such as conifers, and varieties of deciduous fruit trees with a high chilling requirement grow best in temperate zones (Figure 5.6). Avocados grow well in either warm subtropical zones or tropical zones. Chilling-sensitive evergreen trees, such as mango, grow best in tropical zones (Figure 5.3).

In this chapter, I discuss the quantitative reversible and irreversible stress responses of plants to temperature that are responsible for these differences in adaptation. I then define the different climatic zones. Temperature effects are examined for different stages of plant growth: seed germination, resumption of active growth by perennials, vegetative stage, and reproductive stage and present methods for determining where crops can be grown. The thermal environment is complex with spatial, diurnal, and seasonal variations in temperature. Plant function must be examined in the context of these variations. Influences of increases in atmospheric carbon dioxide (CO_2) concentration on plant responses to temperature also are considered.

5.1 SEED GERMINATION, STORAGE, AND DORMANCY

Temperature affects seed germination through at least three separate processes (Roberts, 1988). First, seeds continuously deteriorate and ultimately die. Second, many seeds are initially dormant. Third, once seeds have lost dormancy, their rate and extent of germination are influenced by temperature.

Seeds have been put into two categories with respect to their deterioration and storage characteristics: *orthodox* and *recalcitrant*. After harvest, orthodox seeds that are dried to lower moisture contents usually store better, whereas recalcitrant seeds can be damaged by drying. I will begin by discussing optimal long-term storage conditions for orthodox seeds.

Empirical studies to determine optimal storage conditions for seeds are seriously constrained by the fact that some orthodox seeds may remain viable for several hundred—or even thousands—of years when stored under low temperatures and low moisture conditions. Scientists have attempted to overcome this research problem by conducting studies of seed storage at high temperatures and high humidities where deterioration occurs more quickly. Unfortunately, the models based on these studies may not be reliable for predicting what happens to seeds stored at low temperatures and especially at low humidities.

Based on theoretical analyses, Vertucci and Roos (1990, 1993) have proposed that there is an optimal water potential for long-term storage of orthodox seeds of about –180 megapascal (water potential is defined in Chapter 8, Equation 8.14). Equations 3.2 and 3.3 predict that seeds stored at an RH of 27% and temperature of 25°C would achieve this water potential at equilibrium. At a temperature of 5°C, an RH of 25% would result in a water potential of –180 megapascal at equilibrium. The optimal moisture content for the storage of cowpea seeds probably is between 7% and 12% since at moisture contents of 5%–6% cowpea seeds exhibit poor germination under chilling conditions, whereas they retain their vigor when stored at 10% (Ismail et al., 1997). The relative humidities of 25%–27% predicted by the model of Vertucci and Roos (1990, 1993) would result in an equilibrium seed moisture content for cowpea of about 7% (Equation 3.1 and Figure 3.1).

Vertucci et al. (1994) have obtained empirical data that support their theoretical model of an optimal water potential of −180 megapascal for the long-term storage of orthodox seed. The U.S. Department of Agriculture's (USDA) National Seed Storage Laboratory, located in Fort Collins, Colorado, is responsible for the long-term storage of plant germplasm for the United States and follows the recommendations of Vertucci and Roos (1990, 1993). In 1998, the laboratory used the following general guidelines for the storage of orthodox seeds: Seeds are brought to equilibrium, with respect to moisture content, in a room at 25% RH and 5°C and are then sealed in moisture-resistant containers and placed in cold vaults having a temperature of −18°C (equivalent to a freezer). Note that mature seeds which contain little water can withstand the low temperature of cryotanks (which use a liquid nitrogen vapor phase with a temperature of –150°C to –160°C), where they may retain their viability for many more years.

The Svalbard Global Seed vault was started in 2008 in an old coal mine inside a mountain on a remote Norwegian island in the arctic. It is intended as insurance to provide a seed store for the international community of virtually all varieties of agricultural crop species in the event of a large-scale catastrophe that destroys traditional seed banks. Samples of about 500 seeds each are stored at −18°C in sealed moisture-proof foil packages. Seed of some species are estimated to retain their viability for thousands of years in these conditions. The store is relatively secure in that, due to its location, it would remain cold even if the current small refrigeration unit failed. As of 2017 there were about one million packages of seed in the store.

For working collections it is more convenient to store seeds at moderate temperatures, such as in a refrigerator. Seeds of many species retain good viability for several years at 5°C. Before storing orthodox seeds in a refrigerator, they first should be dried to an optimal level and then placed in moisture-resistant containers. Seeds that are not placed in moisture-resistant containers can be damaged if the refrigerator malfunctions and water condenses on the seed packets.

For short-term storage, it is useful to know the highest moisture content that should be used because drying seeds can be expensive. The critical upper moisture content at which respiration of seeds began to increase rapidly was at a water activity of 0.91 (Vertucci and Roos, 1990), which, according to Equations and 3.1 and 3.2, would be 16% on a fresh weight basis for cowpea seeds and, according to Equation 3.3, would result in a water potential of –14.5 megapascal. An effective moisture content for commercial storage of cowpea seeds is lower than this at about 12%, which can be attained by equilibrium with air at an RH of 62%, resulting in a water activity of 0.62 and a water potential in the seed of −66 megapascal. Damage to the seeds due to fungi should be minimal at a water activity of 0.62 since respiration typically is inactivated when the activity of water is less than 0.85.

Dry seeds with a water potential of –66 megapascal are much "drier" than other living tissues of vascular plants. The lowest water potential developed by living leaves of cowpea is about –1.8 megapascal and the lowest water potential observed in living leaves of other vascular plant species is about –14.5 megapascal (Hall, 1982b).

The life span of orthodox seeds varies among species from several hundred or even thousands of years for some wild plants, as determined from herbarium specimens of known age or by carbon dating, to a few years for some crop plants under commercial storage conditions, to as little as a few months for a few crop species, especially those whose seeds have a high oil content (Mayer and Poljakoff-Mayber, 1989).

Recalcitrant seeds have a short life span. They often occur in woody species, are large and fleshy, and are damaged by even slight desiccation. Because they cannot be dried they cannot be cooled to freezing temperatures without being killed, due to ice formation within the tissues. Furthermore, recalcitrant seeds of tropical species, such as the rubber tree and cocoa, experience chilling injury at temperatures between 0°C and 10°C.

Some aquatic species have recalcitrant seeds that are not damaged by chilling. Under natural conditions, seeds of annual wild rice (*Zizania aquatica*) fall to the bottom of shallow lakes in the fall, remain in water at temperatures just above 0°C during the winter, and then germinate when the water warms up. Optimal seed storage requirements for cultivated varieties of wild rice (*Zizania palustris*) were similar to this set of natural conditions. Seeds were stored by submerging them in water at about 2°C for several months during the late fall and winter season. This procedure maintained their viability, but it severely constrained the operations of commercial seed companies and breeders. Note that wild rice discussed here is a crop that is native to North America and produces an exotic food eaten in expensive restaurants. It has no relation to the wild rice plants that belong to the genus *Oryza* and are related to the cultivated rice *Oryza sativa*.

Improved methods for storing seeds of North American cultivated wild rice were developed following the studies of Kovach and Bradford (1992). They established that wild rice seeds can retain viability at moisture contents ≥30% (on a fresh weight basis) for at least a year at temperatures below 20°C. The seeds have a dormancy, however, that is not broken when they are stored at temperatures above 10°C. Storage under moist chilling conditions for six months or longer (this has been called stratification) is required to break the dormancy of the seed of wild rice.

Seeds of many species, especially wild species, have various dormancy mechanisms. For a period after harvest dormant seeds will not germinate. The ecological and evolutionary basis for seed dormancy is that many seeds mature at a time in the season when immediate germination would not permit the plants to complete their life cycles. For example, some species produce seed at the end of the summer or in the fall just prior to a season in which cold temperatures would not permit adult plants to grow successfully. In these cases, seeds are dormant at harvest, and the dormancy only can be broken by subjecting the seeds to moist, cool conditions for several weeks or months (stratification), essentially simulating the conditions the seeds would experience in the ground in their native habitat during winter. During this time metabolism does occur and germination-inhibiting compounds are progressively leached from the seed such that when the soil warms up and moistens in the spring the seeds germinate.

Hard seed coats are present on some seeds of some species, such as clovers, that block the penetration of water and prevent germination. The seed coats remain hard until there has been sufficient weathering of the seed coat to permit water uptake. Seeds with hard seed coats have an interesting mechanism whereby they become dry inside: a hygroscopic valve in the seed coat opens when external environmental conditions are dry, permitting moisture to leave from the inside of the seed such that they become drier. When external environmental conditions become humid or wet, the valve closes to prevent the entrance of moisture into the seed. The significant weathering of the seed coat that permits seeds to take up moisture may be accomplished during the transportation of seeds in streams through the abrasive action of rocks. In this manner, hard seed coats act to enhance the success of genotypes by increasing dissemination of seed and providing more opportunities for successful germination and establishment in seasons conducive for growth and seed production by the plant.

In most cases plant breeders have selected against seed dormancy or hard seed coats to facilitate sowing operations. An exception to this is peanut because its fruits mature in the soil, and dormancy is a useful trait to prevent recently matured seed from sprouting when rains occur just prior to harvest. After harvest, seed dormancy of peanut cultivars gradually decreases and disappears after several months of storage so that they emerge well after sowing in the next season.

The quality of seeds can be determined by various tests of germination and vigor. From a practical standpoint, it is useful to know whether a seed lot will achieve the minimum acceptable plant emergence within a specific number of days. The minimum plant emergence that is acceptable depends on the plasticity of the crop. For example, cowpea plants are very plastic. In California, about one cowpea seed is sown every seven centimeters in the row, but a field with only one plant every 20 centimeters can adjust to fill the available space and produce maximum yields. Consequently, cowpea seed with potential emergence of at least 85% is adequate for commercial use. In contrast, a crop such as head lettuce is not plastic and, when direct seeded, every seed that does not germinate can represent a loss in yield. In seed viability tests, the maximum number of days for emergence should be specified, because a seed lot with low vigor may have plants that continue to emerge over an extended period, but the farmer cannot rely on plants that emerge late. Plants that take too many days to emerge have a high probability of being damaged by soil pests or diseases. For cowpea, it is desirable that plants emerge within at least seven days after sowing into moisture.

Plant germination and emergence exhibit an optimal-type response to soil temperature (Figure 5.1). Minimum and maximum threshold values for soil temperature in the seed zone are those values that will just provide adequate emergence in the maximum allowed number of days as determined by the type of crop species (Figure 5.1).

The minimum threshold temperature in the seed zone for germination and emergence of cool-season annuals is from 3°C to 8°C, depending on the species. In contrast, germination and emergence of warm-season annuals is damaged by chilling temperatures. For cowpea, the minimum threshold soil temperature for emergence is 18°C. At cooler temperatures, germination will be slowed and maximal emergence of seedlings will be reduced (Ismail et al., 1997). For upland cotton, the minimum threshold temperature is about 16°C; for maize, the minimum threshold is about 14°C with some variation among cultivars. These minimum thresholds influence the earliest dates that these crops can be sown in the spring in subtropical zones. Farmers in the San Joaquin Valley of California often try to sow these crops early because early-sown crops usually have greater yields, possibly because they begin flowering before the hottest weather, which normally occurs in late July. Typically,

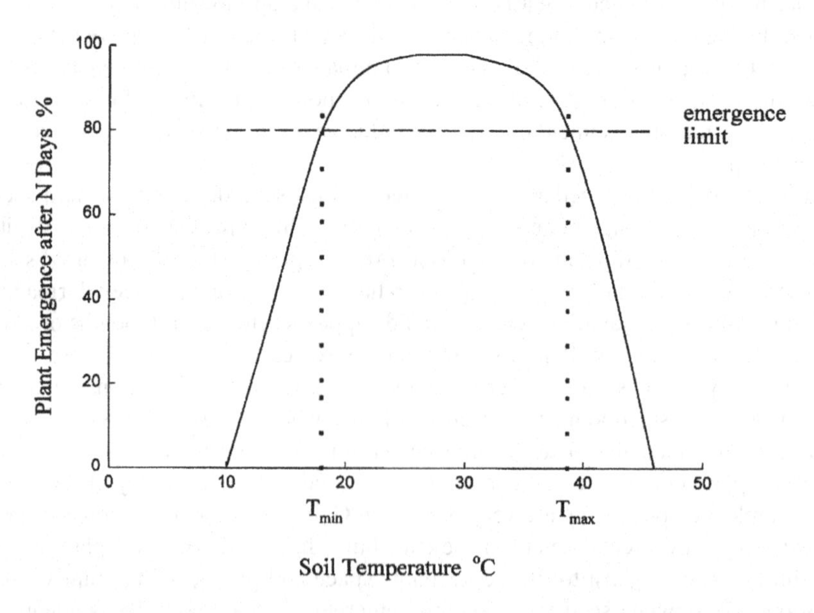

FIGURE 5.1 Plant emergence after N days where N is a practical time limit for emergence that minimizes damage due to soil-borne diseases and pests, which varies among crop species. The dashed line indicates the minimal level of emergence that will not reduce yield. The specific minimal level varies among crop species. The dotted lines indicate the lower (T_{min}) and upper (T_{max}) threshold levels of soil temperature, which also vary among crop species.

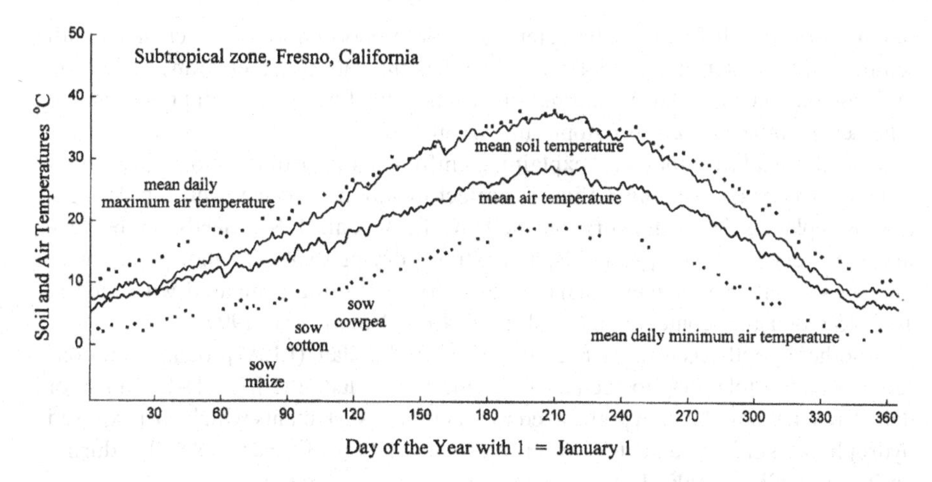

FIGURE 5.2 Daily mean temperature 10 centimeters deep in a bare, sandy loam soil (solid line), daily mean air temperature (solid line), and maximum and minimum daily air temperatures (dots) for 1961 through 1990 in the subtropical zone at Fresno, San Joaquin Valley, California, U.S.A. (36°41′N, 119°43′W, elevation 100 meters). Earliest sowing dates are indicated for maize, cotton, and cowpea.

the earliest sowing dates in Fresno in the San Joaquin Valley are March for maize, early April for cotton, and late April for cowpea (Figure 5.2).

When sowing early, it is important to choose a date when the soil temperature is at or above the minimum threshold, and weather forecasts predict that the next few days will be sunny and warm to hot so that the soil does not cool down. Note that darker soils, those with lower moisture content, and soils that slope toward the sun will warm up faster, and that mean air temperature may not provide effective predictions of soil temperature (Figure 5.2). In more temperate climatic zones (e.g., as shown in Figures 5.5 and 5.6), early sowing often is advantageous because it permits annual crops to produce fruit or seed prior to the onset of damaging cold or wet weather in the fall. Global warming will make it possible to sow earlier in the warm seasons of subtropical and temperate climatic zones and in some cases could lengthen the seasons that are suitable for crop growth and increase crop yields.

Progress has been made in elucidating mechanisms whereby warm-season annuals tolerate chilling during germination and emergence (Bedi and Basra, 1993). As seeds of sensitive genotypes imbibe water under chilling temperatures, there can be a rapid loss of electrolytes, indicating malfunction of plasma membranes. It has been proposed that genotypic differences may be explained by a positive relation between the level of unsaturated fatty acids in membranes and chilling tolerance (Lyons, 1973; Bartkowski et al., 1977). Membranes with more unsaturated fatty acids are thought to have a lower threshold temperature for their transition from a gel-like to a solid state where they become non-functional. Evidence obtained using transgenic plants supports this proposal for specific lipids in chloroplasts (Nishida and Murata, 1996). Effects of membranes on responses to chilling may not, however, be due to changes in bulk membrane fluidity in that the plasma membrane appears to contain domains having differing diffusional characteristics, the proportion of which changes with temperature (Koster et al., 1994). Some studies indicate that membrane effects on chilling tolerance may have maternal inheritance, but these studies did not establish whether the inheritance was cytoplasmic or nuclear.

An additional hypothesis for explaining chilling and dehydration tolerance is that sugars, such as sucrose and raffinose, protect membranes in seeds during desiccation by replacing water in hydration shells (Caffrey et al., 1988). Seeds can be very dry, with very low water potentials, and extreme drying to 5%–8% moisture content on a fresh weight basis can enhance sensitivity to chilling compared with seed at 10%–15% moisture content (Bedi and Basri, 1993; Ismail et al., 1997).

Another hypothesis is that late embryogenesis abundant (LEA) proteins may confer desiccation tolerance to seeds. These proteins include the LEA D-11 family of dehydrins that have been hypothesized to function as surfactants which coat exposed hydrophobic surfaces and thereby prevent coagulation of macromolecules during desiccation (Close, 1996, 1997).

A hypothesis has been proposed to explain genotypic differences in chilling tolerance during germination in cowpea. The hypothesis states there are two independent additive effects: a positive effect associated with the presence of a specific dehydrin protein inherited by a dominant nuclear allele, and a maternal effect associated with slow electrolyte leakage under chilling conditions (Ismail et al., 1997). Studies with almost isogenic lines of cowpea (Ismail et al., 1999) confirmed this hypothesis and showed that the maternal leakage effect does not persist through subsequent generations and exhibits nuclear rather than cytoplasmic inheritance.

Two broad categories are identified within *Oryza sativa*: (1) Japonica-type rice varieties that are mostly grown in subtropical areas, such as the Sacramento Valley, Arkansas, Japan, and high-elevation equatorial areas, and (2) Indica-type rice varieties that are grown in tropical areas. Some 202 accessions from the USDA rice mini-core collection have been screened for five criteria that confer adaptation under chilling conditions during germination, and growth and survival during the seedling stage (Schläppl et al., 2017). These criteria were deemed important for the success of rice cultivation in California and Arkansas. Significant differences were observed among the accessions for all five criteria with a tendency for Japonica accessions to be more chilling tolerant than Indica accessions.

When studying seed germination responses to high temperatures, it is important to use temperatures that can occur in field conditions. If extremely hot conditions are imposed, all processes can be damaged and the results usually are not relevant to field conditions. High soil temperatures that occur in farmers' fields can inhibit seed germination and emergence. For example, cool-season crops such as lettuce are grown in the lower-elevation deserts of California (Figure 10.2) to accommodate fall markets. In this case, they must be sown in the late summer, when soil temperatures can be high enough to inhibit germination. Laboratory studies demonstrated that temperatures greater than 25°C–33°C during the first 7–12 hours after seed has begun imbibing water inhibited germination of lettuce, whereas high temperatures after this period of sensitivity did not inhibit germination (Borthwick and Robbins, 1928). Field studies took advantage of this laboratory information and demonstrated that satisfactory germination and emergence could be achieved in environments with hot days by sowing lettuce into beds having dry soil and then irrigating the beds with sprinklers during the evening to permit seeds to imbibe water during the cool nighttime period. The sprinkling also cools the soil by evaporation.

Another method for avoiding detrimental effects of extreme temperatures on germination is "seed priming" (Bedi and Basra, 1993). Priming involves placing seeds in an osmotic solution such as KNO_3 at –0.8 to –1.5 megapascal solute potential (solute potential is defined in Equation 8.21) at moderate temperatures for several days. During this period, seed metabolism goes through the initial temperature-sensitive stages of germination but with the osmoticum reducing the availability of water and thereby preventing radical emergence. The seeds are then dried and subsequently can be sown with normal methods and will emerge well from hot soil. Priming mainly is useful for crops such as carrot, various onions, and lettuce, and conditions such as too hot or too cool soil temperatures, where it is difficult to achieve uniform emergence with normal seed. This approach has disadvantages in that primed seed often has a shorter shelf life and is more expensive than normal seed.

The maximum threshold temperature for germination and emergence (T_{max} in Figure 5.1) is hotter for warm-season annuals than for cool-season annuals. For example, the threshold maximum seed-zone temperature for cowpea is about 37°C, compared with about 33°C for lettuce (Argyris et al., 2005). In tropical zones (Figure 10.8), inadequate plant emergence and establishment can limit the productivity of several warm-season annual crops. In these environments, upper seed-zone temperatures can exceed 45°C. For crops such as sorghum and pearl millet, which have small seed and therefore are sown shallow, these high soil temperatures can occur in the seed zone and substantially reduce emergence (Wilson et al., 1982; Soman and Peacock, 1985). Note that germination studies in controlled environments with constant temperatures may not provide accurate predictions of germination in field conditions where soil temperatures vary with time. For example, Soman and Peacock (1985) demonstrated that sorghum lines failed to germinate at a constant temperature of 40°C but germinated and emerged when subjected to diurnally varying temperatures having an average of 40°C.

Germination, emergence, and seedling survival are relatively simple biological systems. Methods have been developed that could be useful in plant breeding to screen large numbers of plants to detect genotypic differences in heat tolerance at emergence (Wilson et al., 1982; Soman and Peacock, 1985). This type of selection, however, only will confer a level of heat resistance to varieties that is useful for farmers in specific circumstances, for example, field conditions where heat stress has detrimental effects on plant stands that reduce crop yield and where heat-resistant varieties exhibit smaller reductions in plant stands and greater crop yield. Many crops have some capacity to exhibit compensatory growth when the population of seedlings is low.

Depth of sowing may be critical in hot environments. Hot soils retard hypocotyl elongation of cowpea, and the detrimental effect on emergence is aggravated by deep sowing of seeds (Onwueme and Adegoroye, 1975; Warrag and Hall, 1984a). Seed must be sown into moisture at a depth that is neither too deep nor too shallow. Data on optimal sowing depths are provided for some crop species by Isom and Worker (1979). They point out that smaller-seeded crops and dwarf cultivars should be sown at shallower depths. The shoot structure (coleoptile) of cereals that determines ability to emerge from the soil also is dwarfed by the genes that cause the plant stem to be shorter. They also recommend shallower sowing on clayey compared with more sandy soil.

5.2 RESUMPTION OF ACTIVE GROWTH BY PERENNIALS

In temperate zones (Figures 5.5 and 5.6), the leaves of deciduous trees fall and the buds remain dormant during the cold fall and winter season. Then during the late winter or early spring, the trees resume active growth. The buds of evergreen trees such as citrus also are dormant during the cool fall and winter season in subtropical climates (Figures 5.2, 5.4, 10.6, and 10.12) but resume active growth in the late winter or early spring. For both types of trees, the first sign of the resumption of active growth is bud break, and then either leaf expansion occurs or flowers are produced.

What change in environment during the late-winter period is responsible for the resumption of bud break? The plants probably are responding to the warming that is occurring. But is the plant responding to increases in temperature of the roots or the shoots? Studies in which dormant orange trees were subjected to different soil and air temperatures during the winter provide some clues (Hall et al., 1977). Warmer soils resulted in greater total bud break than cooler soils, whereas warmer air had no influence on the total number of shoots that were initiated (but resulted in fewer flowers being produced). The authors hypothesized that control of bud break by warming soil may have adaptive significance in climatic zones with a winter season where plant function is limited by low temperatures. They proposed that temperature of the soil in the root zone at a depth below 20 centimeters provides a dampened and more reliable indicator of seasonal trends in incident energy, in these zones, than air temperature, which is more variable during the day and from day to day. A hypothetical tree that responded to increasing air and shoot temperature could be induced to break buds by occasional warm days during the middle of winter, which would not be adaptive.

If the resumption of bud break is caused by warming soil, then there must be a mechanism that links root activity with shoot activity. A likely mechanism is the production of hormones by roots that are transported to the shoots in the xylem transpiration stream and then cause bud break. This is another example of the coordination between the activities of roots and shoots described in Chapter 2 and provides an additional justification for the role of whole-plant studies in obtaining a more complete understanding of plant function and adaptation.

5.3 VEGETATIVE GROWTH

The reversible and acclimation effects of daytime shoot temperatures on vegetative growth largely reflect their effects on photosynthesis, which were described at the end of Chapter 4. Temperature extremes, such as freezing (less than –1°C) and chilling (0°C–18°C), can have irreversible stress effects on photosynthesis and other processes that affect vegetative growth and kill leaves and plants.

5.3.1 Freezing Stress during Vegetative Growth

Low temperatures that result in the freezing of solutions within plants can cause damage to all plant species. Usually, it is the extracellular solution that freezes for most crop species under natural conditions.

There are two reasons why intracellular solutions usually do not freeze. First, the presence of solutes depresses the "freezing point" by 1.86°C per osmolal of solutes, which is equivalent to a decrease of 0.82°C per MPa of solute potential. (Refer to Chapter 8 for a discussion of osmolal solutions and solute potential.) Most crop species have "freezing point" depressions of 1°C–2°C, whereas a few halophytes, which accumulate many salts, have "freezing point" depressions as large as 14°C. Second, cellular solutions often do not freeze as temperatures drop to the "freezing point" due to an absence of suitable ice-nucleation sites within cells. This is called *supercooling*. Consequently, the theoretical "freezing point" really is the temperature at which melting occurs, but is not necessarily the temperature at which plant tissue freezes, because almost all plants exhibit several degrees of supercooling.

Initially, ice forms in extracellular solutions and may not be lethal to the plant. But when plants are exposed to freezing temperatures for an extended period, significant water moves from the symplast within living cells to crystals of ice forming outside cell walls, causing the crystals to increase in size, which can mechanically disrupt and destroy the tissue. In addition, the protoplasm in the symplast becomes dehydrated. In many cases, the lethal effects of freezing may result from the dehydration of the protoplasm and destruction of the plasma membrane. The mechanical damage to tissue caused by expansion of ice crystals makes the plant more susceptible to infection by pathogens.

Damage due to sub-zero temperatures tends to be cataclysmic, with extreme damage occurring when specific thresholds are reached. Plants differ with respect to their ability to survive temperatures below –1°C. A one-night minimum temperature of about –2°C to –3°C in late winter in Florida can kill leaves and plants of warm-season annuals, such as tomato, pepper, and sweet corn that are being grown to achieve early markets for these crops. The morning after this "killing frost," the injured foliage of tomato and pepper appears flaccid, dark, and water-soaked because the plasma membranes are leaking. Leaves and plants of some frost-sensitive cool-season annuals, such as Irish potato, also can be killed by a one-night freeze of about –3°C. Commercial Irish potatoes typically are grown from small potatoes. In temperate zones the planting of these small potatoes in spring must be delayed until the probability of a frost occurring after plant emergence is very low because this could kill all of the crop. Leaves of tropical perennials, such as mango, also can be killed by a one-night freeze of about –2°C to –3°C. One-night minimum temperatures of about –3°C to –4°C can occur in warm subtropical Mediterranean zones and kill leaves of perennials such as avocado and citrus commonly grown in these zones. Young actively growing tissue on young trees are most sensitive to sub-zero temperatures. Lemons are more frost sensitive than oranges.

Cool-season cereals, such as wheat, have both spring-type and winter-type varieties. In temperate zones there are reasons why growers might prefer to grow winter wheat varieties. They are sown and become established in the fall and their roots and leaves protect the soil against wind and water erosion. They make most of their growth during cool weather and thus make more efficient use of fall and early spring moisture than spring wheat varieties that are sown in the spring and grow in the summer. Winter wheat varieties can provide greater grain yields than spring wheat varieties if the winter is not too harsh. In some temperate zones that have very cold winters, it can be risky to sow winter wheat varieties because they may not survive the winter frosts.

For spring-type cereal varieties death of leaves occurs with spring frosts of –5°C to –10°C. The frost sensitivity of winter-type cereals is more complex. They are sown in the fall and during their early growth have frost tolerance similar to that of spring-type cereals, but when subjected to chilling temperatures they progressively acclimate (harden) and during early winter can withstand temperatures as low as –14°C to –30°C. The mechanisms responsible for the increase in freezing tolerance could include increased desaturation levels of fatty acids in membrane phospholipids, accumulation of sugars such as sucrose, thought to contribute to stabilization of membranes, and induction of genes that encode sets of novel proteins that may confer tolerance to freezing (Thomashow, 1998). Genetic mapping and correlation studies suggest a role for specific dehydrin proteins in freezing tolerance of cool-season cereals (Campbell and Close, 1997). Exposure to warm spring temperatures results in dehardening of winter-type cereal varieties.

Some cool-season vegetables, such as cabbage, can survive winter temperatures as low as about –25°C. Herbaceous plants can escape freezing by either producing very dry seed that withstands very low temperatures or by having storage organs and potential growing points in the soil where temperatures do not become as cold in winter as above-ground air temperatures.

Some deciduous fruit and nut trees and ornamental woody plants exhibit considerable variation in their tolerance to cold temperatures. Deciduous trees harden in response to short days and chilling temperatures in the fall. During this hardening they can acclimate such that they survive temperatures as low as –20°C to –45°C partially because of supercooling. However, tissues usually can supercool only to about –40°C, at which temperature ice formation occurs spontaneously (even if ice nucleation sites are not present). In dormant buds of some woody plants deep supercooling occurs defined as where water is maintained in tissues in the liquid phase at temperatures as low as –50°C despite the presence of extracellular ice. This requires a barrier that prevents growth of ice crystals into dormant buds and also prevents the rapid loss of water from these buds to the extracellular ice. When deciduous trees resume active growth in spring, growing buds can be very susceptible, freezing at –2°C to –3°C.

Supercooling and deep supercooling may account for the low-temperature limits of survival of many alpine and subarctic species. Frost hardiness limitations may explain the northern limit of the deciduous forest in the northeastern United States and altitude of timberline (tree line) in the alpine zone, which occur where minimum temperatures are about –40°C. An alternative explanation for timberline in alpine zones is that it occurs where the average air temperature of the warmest month is less than about 10°C. The minimum threshold temperature for growth by cool-season-adapted vascular plants is considered to be an average temperature of about 10°C.

The correlation of timberline with summer temperature suggests that the height of vegetation may be a key factor in determining plant success through its effect on the potential for tissue temperature to be warmer than air temperature (Jones, 1992). The height of vegetation influences the boundary layer resistance which, as discussed in Chapter 7, can have a strong influence on the difference in temperature between canopy and air. In alpine zones, shorter trees typically are warmer than taller trees. The fact that the timberline can be higher on soils that slope toward the sun, compared with soils that slope away from the sun, also supports the hypothesis that the timberline is determined by minimum temperatures during the growing season.

Some woody species native to northern Canada, Alaska, and high mountains survive prolonged exposure to temperatures below –40°C and minimum temperatures below −60°C. When fully acclimated, some of these plants have survived immersion in liquid nitrogen at –196°C. Their survival does not involve deep supercooling. Instead, ice formation begins between cells after only a few degrees of supercooling (e.g., at –5°C). The ice crystals continue to grow as it becomes cooler, progressively removing water from the symplast, which remains unfrozen but becomes very dry. For these species, resistance to freezing temperatures depends on the capacity of the extracellular spaces to accommodate large crystals, such that mechanical damage does not occur, and on the ability of the symplast and the plasma membrane to withstand severe dehydration.

5.3.2 Chilling Stress during Vegetative Growth

Photosynthesis of tropical perennials and warm-season annuals, especially those with the C_4 photosynthetic system, can be damaged by the combination of chilling temperatures (0°C–18°C) plus intense sunlight. Membrane function in chloroplasts is disturbed by chilling, resulting in photo-inhibition of the photosynthetic electron transfer system of photosystem II and photo-oxidation of components such as chlorophyll, leading to loss of green coloration. The inhibition of photosystem II may be viewed as being both a weak link and damaging to the plant by reducing photosynthesis, and as being analogous to an electrical fuse that protects the chloroplast by reducing excess production of high-energy products that, when not used for productive purposes, could cause extensive photo-oxidation and be very destructive.

Maize is a C_4 species that evolved and was domesticated in the tropical zone of Central America. Cultivars have been selected for use by farmers in more extreme latitudes and altitudes with cooler environments. During the late twentieth century, cool temperatures still were the primary factor limiting the productivity of maize hybrid cultivars being grown for either grain or silage production in the northern United States, Canada, and northern Europe. Substantial progress has been made, however, in breeding maize cultivars with tolerance to chilling compared with other warm season annuals, such as cowpea, for which there has been a much smaller investment in breeding. Maize hybrids have been bred that emerge from cool soil, and the challenge facing breeders is to incorporate ability to tolerate chilling during vegetative growth and not exhibit photo-inhibition and photo-oxidation (Greaves, 1996) and to tolerate chilling during reproductive growth (Farooq et al., 2009).

5.3.3 Heat Stress during Vegetative Growth

High daytime temperatures can have both direct inhibitory effects on vegetative growth and indirect effects, due to the high evaporative demand causing more intense water stress. Among the cool-season annuals, pea is very sensitive to high daytime temperatures, with death of the plant occurring when air temperatures exceed about 35°C, whereas barley is very heat tolerant, especially during grain filling. Warm-season annuals usually can withstand higher temperatures than cool-season annuals. During the vegetative stage, cotton and cowpea can withstand about the highest temperatures experienced in crop production zones (maximum daytime temperatures in a weather station shelter of 50°C). During these very high temperatures, these crops can produce substantial amounts of vegetative biomass providing they have an adequate supply of water. For monocotyledons, including both cool-season and warm-season annuals, one symptom of damage due to high temperature is leaf firing, which involves necrosis of leaf tips, but this symptom also can be caused by drought.

Maximum threshold temperatures at which high temperatures kill warm-season seedlings, such as soybean, can depend on plant preconditioning. Seedlings subjected to high but sublethal temperatures for a few hours subsequently can survive higher temperatures than seedlings maintained at moderate temperatures. This acclimation to heat can be induced by the gradual diurnal increases in temperature that occur in hot natural environments. During the period of acclimation, there is a repression of the synthesis of many normal proteins and the initiation of the synthesis of a small set of novel proteins that have been called heat-shock proteins (Vierling, 1991). Studies with loss-of-function mutants using the model cool-season plant species *Arabidopsis* demonstrated that the enhanced thermo-tolerance can be associated with at least three independent effects: the synthesis of heat-shock proteins (e.g., Hsp101), protection of membrane integrity, and molecular chaperones which aid recovery of protein activity/synthesis after the heat shock experience (Lee and Vierling, 2000; Queitsch et al., 2000; Hong et al., 2003). The potential for enhancing heat tolerance of crop plants by genetic engineering for constitutive or overexpression of specific heat-shock proteins has been discussed by Gurley (2000), who suggests that Hsp101 and similar types of heat-shock proteins are attractive targets for this type of research.

Heat-shock proteins also may be involved in providing cross-tolerance to some other stresses (Sabehat et al., 1998). For example, when tomato fruits are subjected to high temperature (55°C) for a few minutes or low temperature (2°C) for a few days, they develop injuries. However, if tomato fruit are first subjected to moderately high temperature (38°C), several heat-shock proteins are synthesized, and the fruit develop some tolerance to both the high temperature and the chilling temperature (Sabehat et al., 1998). The adaptive significance of this type of cross-tolerance is not clear.

The photosynthetic system can be damaged by heat. Sensitivity of photosynthesis to heat may be due to damage to components of photosystem II (PSII) located in the thylakoid membranes of the chloroplast (Al-Khatib and Paulsen, 1999). The extreme sensitivity of PSII may be due to effects of temperature on the membranes in which it is embedded. In a study by Murakami et al. (2000), transgenic tobacco plants were developed in which the gene encoding chloroplast omega-3 fatty acid desaturase was silenced. The transgenic plants had less trienoic fatty acids and more dienoic fatty acids in their chloroplasts than the wild type. The transgenic plants had greater photosynthesis and grew better than wild-type plants in hot but highly artificial environments.

In comparisons of contrasting species, PSII of the cool-season species wheat was more sensitive to heat than PSII of either rice or pearl millet both of which are warm-season species adapted to much hotter climates (Al-Khatib and Paulsen, 1999). Portable instruments are available to rapidly measure chlorophyll fluorescence parameters of intact leaves that could be used to screen many plants to quantify the extent that PSII has been damaged and hopefully detect genotypic differences in heat tolerance. Selection of this type with grain crops only will enhance heat resistance where grain yield is being limited by heat-stress effects on the supply of carbohydrate through effects on PSII.

Cultivar differences in grain yield of spring wheat in hot, irrigated environments have been positively associated with photosynthetic carbon dioxide fixation rate (Reynolds et al., 1994b). Even stronger positive associations were observed between grain yield and stomatal conductance and canopy temperature depression. Note that canopy temperature depression provides an indirect measure of stomatal conductance (Chapter 7). Grain yield also was negatively correlated with electrolyte leakage from leaves subjected to high temperature, which provides a measure of membrane thermostability. This indicates that the more open stomata of the heat-resistant cultivars may be enhancing photosynthesis both by facilitating the diffusion of CO_2 into leaves and by enhancing transpirational cooling bringing leaf temperatures below damaging thresholds. However, cultivar differences in grain yield of spring wheat growing in a hot, irrigated environment also have been positively correlated with the number of kernels per spike (Shpiler and Blum, 1991). This could be explained by either cultivar differences in heat tolerance during reproductive development or the possibility that processes determining kernel number may be linked to photosynthesis. Cultivar variation in kernel number has been correlated with spike dry weight at anthesis and the ratio of solar irradiance to air temperature for the 30-day period prior to anthesis (Fischer, 1985). Consequently, there are circumstances where damaging effects of heat on photosynthesis can reduce both the source of carbohydrates and the reproductive sink for carbohydrates, making it difficult to determine which of them is most limiting for grain yield.

The source versus sink issue concerning limiting effects of heat stress on yield is further complicated by the possibility that photosynthetic capacity and stomatal behavior may be influenced by the size of the reproductive sink through long-term feedback effects. Pima cotton cultivars were bred that have greater boll yields in hot environments by selecting plants with the ability to set more bolls at low nodes under very hot irrigated field conditions. Subsequent studies showed that these heat-resistant cultivars also had greater carbon dioxide assimilation rates and higher stomatal conductances than cultivars with lower boll yields under hot conditions (Cornish et al., 1991). Plants that have higher photosynthetic capacity often have greater maximum stomatal conductance, and the mechanism of this long-term regulation is not known

(Schulze and Hall, 1982). Possible causes for the higher photosynthetic rates of the heat-tolerant cotton (and wheat) cultivars include a feedback effect from a stronger sink strength due to heat-tolerance during reproductive development resulting in more fruiting, more open stomata enhancing the conductance for carbon dioxide diffusion into the leaf and also greater transpiration resulting in cooler leaves that are operating closer to the optimum for photosynthesis, PSII with greater heat tolerance, and slower senescence of the leaves.

When grains are developing, leaves begin to senesce and this can be accelerated by late-season heat stress. Delayed-leaf-senescence traits have been sought as a means to enhance grain filling and lengthen the reproductive period under late heat stress or late drought. This trait can be easily screened for visually in field nurseries providing one only selects plants that have both delayed leaf senescence and abundant fruit and/or grain because plants that have low fruit and/or grain set typically also exhibit a type of delayed leaf senescence that has limited agronomic value. Cultivars and genetic lines with delayed leaf senescence have been bred for sorghum (Reddy et al., 2011), cowpea (Gwathmey and Hall, 1992), and wheat (Farooq et al., 2011).

Clones of Irish potato have been bred with differences in resistance to heat, in terms of tuber yield. Under hot conditions, several processes are inhibited that influence tuber production: the rate of photosynthesis, induction to tuberize, and tuber development and growth. Controlled-environment studies demonstrated that these processes are influenced differently by root and shoot temperatures (Reynolds and Ewing, 1989). High soil temperature inhibited tuber development and growth under either hot or more optimal shoot temperatures. In contrast, high shoot temperatures caused leaf rolling and accelerated leaf senescence and reduced the induction to tuberize under either hot or more optimal root-zone temperatures. The inhibition of tuber development and growth of Irish potato by hot soil temperatures has implications for crop management methods. Frequent overhead sprinkler irrigation may have an advantage over frequent drip irrigation when growing Irish potato in hot environments because it would cool the soil beds by evaporation. Near Bakersfield, California (Figure 10.3), some fields of Irish potato are planted as late as March, which means that the crops grow during very hot weather in the spring and are then harvested in the early summer. These potato crops might benefit from overhead sprinkler irrigation that would cools the soil beds more than would drip or furrow irrigation.

The Irish potato example illustrates the importance of considering both root-zone and shoot-zone temperatures when developing techniques for screening for heat tolerance and when developing management methods for hot environments. When plants growing in pots are subjected to high air temperatures, both the shoot and the roots are subjected to hot conditions. In contrast, when plants growing in the field are subjected to high air temperatures, the shoot is subjected to more extreme temperatures than the root system, because the temperature of the soil, below about 10 centimeters, is buffered and does not heat up as much or cool down as much as the air. Consequently, using plants in pots when studying heat stress effects can subject roots to unnaturally high temperatures and generate artifacts.

Finally, field studies with Irish potato conducted in a location in South Korea, where the current temperatures are near the upper limit, showed reductions in marketable tuber yield of 11% per 1°C increase in day and night temperature (Kim et al., 2017). The reduced yields were caused by reduced tuber bulking rates. These experimental results indicate a possible minimum effect of climate change on Irish potato production in South Korea.

5.4 REPRODUCTIVE DEVELOPMENT

5.4.1 Freezing Stress during Reproductive Development

Plants are particularly susceptible to temperature extremes during reproductive development. When cool-season cereals begin flowering, temperatures just below 0°C can cause them to produce no grain. When barley and wheat are grown under rain-fed conditions in Mediterranean climatic zones (Figures 5.2, 10.4, and 10.6), very early flowering varieties can be useful because they can escape end-of-season drought, but they also are more likely to be damaged by early frosts if they occur during the initiation of flowering. Deciduous fruit and nut trees and evergreen subtropical trees such as citrus and coffee are very sensitive to and are damaged by temperatures just below 0°C during flowering.

5.4.2 Chilling Stress during Reproductive Development

Warm-season annuals are particularly sensitive to chilling temperatures at flowering. Pollen development of Japonica-type rice varieties is damaged when minimum temperatures decrease below a threshold of 13°C, and this is an important problem in the Sacramento Valley of California (Figure 10.7). Growers can reduce the extent of damage to rice pollen by several methods (Board and Peterson, 1980):

1. By sowing early or by using varieties that flower early to escape the chilling air that comes from the San Francisco Bay in late summer
2. By using semi-dwarf rice varieties whose panicles are close to water as they develop in the stem, since water is warmer than air at night
3. By maintaining deeper water in the fields 7–21 days before panicle emergence and using warming basins to heat the water prior to it entering the rice field
4. By avoiding too high nitrogen fertilizer rates that can cause excessive floret sterility due to unknown mechanisms

Note that Indica-type rice varieties that are grown in hot tropical areas are even more sensitive to chilling than Japonica-type rice varieties. Damage to pollen development of Indica-type rice varieties occurs when temperatures are less than a 15°C threshold compared with less than 13°C for the Japonica-type rice varieties.

Tomato, which is a warm-season annual, exhibits low seed and fruit set when subjected to chilling temperatures at night, and the low fruit set can be overcome by sprays with certain plant hormones. For the cool-season annual garbanzo bean, chilling night temperatures between 5°C and 0°C can cause malfunctions in pollen development and reduce seed set (Srinivasan et al., 1999).

5.4.3 Heat Stress during Reproductive Development

In very hot environments many crop plant species produce significant amounts of biomass but few flowers, fruits, or seeds. In these cases, reproductive development clearly is more sensitive to heat than photosynthesis and biomass production.

Surprisingly, the reproductive development of several warm-season annuals, such as cowpea, can be damaged by moderately high night temperatures greater than about 20°C, even though their reproductive development can withstand day temperatures as high as 40°C (Warrag and Hall, 1984a, 1984b). For tomato, common bean, cowpea, cotton, rice, and sorghum moderately high night temperatures cause damage to pollen development such that few seeds or fruit are produced. Possible mechanisms for this sensitivity to heat have been discovered. Artificial pollination studies demonstrated that the female part of the flower, the pistil, was not damaged by high night temperatures but that the pollen was infertile (Warrag and Hall, 1983). In studies where cowpeas were transferred between growth chambers having high or optimal night temperatures, Ahmed et al. (1992) demonstrated that the stage of floral development most sensitive to high night temperature occurs nine to seven days prior to anthesis and coincides with release of the pollen microspores from the tetrads. This sensitive stage is after meiosis which occurs 11 days before anthesis. Damage due to high night temperature was associated with premature degeneration of the tapetal layer that provides nutrients to developing pollen. Field studies by Mutters et al. (1989a) showed that the transfer of proline from the tapetal layer to the pollen is inhibited by hot conditions in heat-sensitive but not in heat-tolerant cowpea genotypes. Large amounts of proline are required for pollen development, pollen germination, and pollen tube growth. Tapetal malfunction probably is responsible for some of the genetic male sterility occurring in plant species (Dundas et al., 1981). Pollination in the cool-season crop wheat is particularly sensitive to high temperatures (Saini et al., 1984) and also may involve tapetal malfunction (Dolferus et al., 2011). It is likely that pollen development of many crop species is sensitive to high temperatures.

Growth chamber studies by Mutters and Hall (1992) demonstrated that there is a distinct period during the 24-hour cycle when pollen development in cowpea is sensitive to high night temperatures. Plants subjected to high temperatures during the last six hours of the night exhibited substantially decreased pollen viability and pod set, whereas plants subjected to high temperatures during the first six hours of a 12-hour night exhibited no damage. For peanut, high temperatures during the morning reduced fruit set, whereas higher temperatures during the afternoon had no effect on fruit set (Prasad et al., 2000).

Why is seed set of many warm-season crops sensitive to high temperatures occurring during the late night or early morning, but not sensitive to much higher temperatures occurring during midday and afternoon? Mutters and Hall (1992) hypothesized that there is a heat-sensitive process in pollen development that is under circadian control. Note that reproductive events, such as anthesis (pollen shedding), meiosis, and flower opening have a degree of circadian control occurring at a particular time in the 24-hour cycle. They hypothesized that natural selection would have favored plants in which the heat-sensitive process in pollen development takes place in the coolest part of the 24-hour cycle, which is the late night and early morning.

For different genotypes within the genus *Oryza* the time of day for flower opening varies widely from early morning to evening with many rice cultivars exhibiting anthesis between 10 a.m. and noon. An early-morning flowering trait was introgressed from a wild rice that begins flowering at 6 a.m. into a rice cultivar with later flowering to breed an early-morning flowering line of rice. In a hot location the line was found to flower a few hours earlier and exhibit less spikelet sterility than the parental cultivar, which flowered later when temperatures were hotter (Ishimaru et al., 2010).

Based on earlier research it had been proposed that selection for slow leaf-electrolyte-leakage could provide a method to breed for tolerance to several stresses (Blum, 1988) but a linkage between leaf-electrolyte-leakage and yield or processes directly influencing yield under heat stress had not been established. Subsequently, progress was made in breeding wheat cultivars that were heat resistant, that is, they produced more grain yield than other cultivars in hot environments, by selecting for slow leaf-electrolyte-leakage under hot conditions (Blum et al., 2001).

Definitive genetic selection studies with cowpea demonstrated that slow leaf-electrolyte-leakage under hot conditions is associated with heat tolerance during pod set. Lines of cowpea that were heat tolerant during pod set were shown to have less leaf-electrolyte-leakage under heat stress than a set of heat-sensitive lines (Ismail and Hall, 1999). Thiaw and Hall (2004) selected two of these lines, a heat-sensitive cultivar and a heat-tolerant genetic line with similar genetic background, and used them as parents in a selection program. They crossed the parents and then divergently selected one population for high and low pod set under heat stress and another population for slow and fast leaf-electrolyte-leakage under heat stress. Several cycles of selection were conducted. When stable lines were evaluated: lines selected for slow leaf-electrolyte-leakage under heat stress also had high pod set under heat stress; and lines selected for high pod set under heat stress also had slow leaf-electrolyte-leakage under heat stress. These results indicate that there is a strong association between ability to set pods under heat stress and maintenance of plasma membrane function which may be pleiotropic, that is, influenced by the same gene. Both methods of selection produced lines that had greater grain yield under hot conditions.

Global warming has been predicted to increase daytime and nighttime temperatures by 2°C–5°C by the end of this century (Singh et al., 2011). Several studies on the extent to which increases in temperature can effect yield of grain crops are relevant to global warming. Variations in rice yield over 12 years under best management practices on the experimental farm of the International Rice Research Institute were compared with variations in temperature (Peng et al., 2004). The authors used partial correlation analysis and established that grain yield decreased 10% per 1°C increase in night temperature with no significant association with day temperature. A meta-analysis and review of 95 papers showed that higher night and higher day temperatures both strongly reduced rice grain yield mainly through decreased seed set with higher night temperature having more complex effects than higher day temperature (Xiong et al., 2017).

A more direct method for determining effects of increases in temperature on crop yield is to subject plots of plants growing in the field to increases in temperature without otherwise disturbing the environment, which is easy to achieve when increasing night temperature. Night temperature of sorghum plots was increased 5°C for one week during floret differentiation and there was a 28% decrease in grain yield associated with a 30% decrease in number of grains (Eastin et al., 1983). Plots of cowpea plants were subjected to only elevated night temperatures during early flowering using enclosures (Nielsen and Hall, 1985a). The cowpea plants exhibited a 4.4% decrease in grain yield per 1°C increase in night temperature above a threshold daily minimum temperature of 15°C (Nielsen and Hall, 1985b). The reductions in grain yield were due to reductions in the proportions of flowers producing pods. When rice plots were subjected to a 4°C increase in day and night temperature using open-top chambers, there was a decrease in grain yield of 18% (Moya et al., 1998) due to reductions in spikelet and pollen fertility (Matsui et al., 1997).

Since current high temperatures have been shown to reduce grain or fruit yield, breeding programs have focused on incorporating heat tolerance at flowering. In a program I directed for the cowpea, we initially screened hundreds of diverse cowpea accessions for heat tolerance during early flowering in field environments with extremely hot night and day temperatures (El Centro, Imperial Valley [Figure 10.2], and the Coachella Valley, California, which have minimum/maximum daily temperatures in weather station shelters of about 27°C/42°C in July). Two of the cowpea accessions screened had the ability to produce flowers and set pods in these hot field conditions, and growth chamber studies showed they have heat tolerance during reproductive development (Warrag and Hall, 1983). Screening studies with tomato and chickpea also uncovered only a small number of accessions with heat tolerance.

We crossed the heat-tolerant cowpea accessions with current California cowpea cultivars, and used a pedigree breeding program to develop cowpea lines with heat tolerance during reproductive development and desirable agronomic traits. The procedure involved selecting in two types of environments: selection for heat tolerance in field or greenhouse environments with stressfully high night temperatures, and parallel selection for desirable agronomic traits in field environments in the target production zone with a range of cool to hot temperatures (Hall, 1992, 1993a).

Subsequently, six pairs of cowpea lines, with each pair either having or not having heat tolerance during reproductive development but similar genetic backgrounds, were developed and compared in eight subtropical field environments in California with contrasting temperatures (Ismail and Hall, 1998). The heat-susceptible lines, including a currently grown cultivar, showed a 13.6% decrease in grain yield per 1°C increase in minimum night temperature above a threshold of 16°C. The reduction in grain yield was mainly due to decreases in the number of pods per peduncle and harvest index. In the three environments with the highest night temperatures, the heat-resistant lines had 54% higher grain yields than the heat-susceptible lines mainly due to greater pod set and harvest index. One of the heat-tolerant lines has been released as a cultivar for use in hot subtropical zones of California (Ehlers et al., 2000).

We have evaluated whether the heat-tolerant cowpea lines bred in California are effective in hot parts of Africa. The six pairs of cowpea lines, with each pair either having or not having heat tolerance during reproductive development but similar genetic backgrounds, also were compared in six tropical environments in the Sahelian and Savanna zones of West Africa (Figures 10.5 and 10.8). In all of the tropical environments, there was no difference in grain yield between the heat-tolerant and heat-susceptible lines, even though minimum night temperatures exceeded 20°C during early flowering. A possible explanation for these contrasting results comes from controlled-environment studies (Ehlers and Hall, 1998). The heat-tolerant California lines exhibited high pod set in both the hot long-day conditions occurring in subtropical zones and the hot short-day conditions occurring in tropical zones, whereas the heat-susceptible California lines only exhibited very low pod set in the hot long-day conditions. High night temperatures were much less damaging to reproductive development of the California lines on short days than on long days. Responses to red light during long nights, far-red light at the end of long days, and far-red then red light at the end of long days had indicated that the greater sensitivity of cowpeas to high night temperature under long days could be a phytochrome-mediated effect (Mutters et al., 1989b). Among cowpea cultivars developed by empirical selection for grain yield in Africa, some of those developed in the very hot Sahelian zone (Figures 10.5 and 10.8) had high grain yields in the hot short-day controlled environment, whereas cultivars developed in cooler tropical zones (Figure 10.10) had lower grain yields in this environment (Ehlers and Hall, 1998). The heat-tolerant African cultivars were effective only under hot short-day conditions, whereas the heat-tolerant lines developed in California were effective in both hot long-day and hot short-day conditions. Progress in breeding cowpeas with heat tolerance for use in tropical zones in Africa has been reviewed by Hall (2011).

It is important to ask why natural selection has not favored tolerance to high night temperature during reproductive development. Observations with cowpea provide a clue. The higher pod set associated with heat tolerance during reproductive development also was associated with substantial dwarfing (Ismail and Hall, 1998, 1999), which would confer a disadvantage in natural plant communities subjected to intense competitive. Farmers can compensate for the dwarfing present in heat-tolerant cowpeas by growing them on narrower-row systems (Ismail and Hall, 2000).

Heat stress can have another effect on reproductive development. Under very hot long-day field conditions, many cowpea genotypes produced floral buds but they did not produce any flowers (Ehlers and Hall, 1996). Suppression of floral bud development was greatest under a combination of hot nights and very hot days (Dow El-Madina and Hall, 1986). Responses to red and far-red light indicated that the suppression of floral buds under long days and high temperatures also may be a phytochrome-mediated effect (Mutters et al., 1989b). Transfer and heat-pulse experiments demonstrated that plants did not have a particular stage of development where they were sensitive to high temperature, but that the duration of heat experience may be critical for the suppression of floral bud development (Ahmed and Hall, 1993). A period of two weeks or more of consecutive or interrupted hot nights during the first four weeks after germination caused complete suppression of the development of the first five floral buds on the main stem. This heat-suppression effect is different from the classical short-day effect on day-length-sensitive cowpea cultivars, which determines whether floral buds are initiated and also is phytochrome mediated (Mutters et al., 1989b). Multiple forms of phytochrome are present in plants that have different physiological roles (Smith, 1995). Some cultivars of common bean have been shown to exhibit floral bud suppression under hot long days. Pima cotton also has been shown to not produce flowers under very hot conditions.

The studies with cowpea provide a model system for breeding for heat tolerance. The complex detrimental effects of high night temperature on grain yield have been shown to be caused by the heat sensitivity of a linear sequence of simpler processes for which major genes confer heat tolerance. Floral buds can be suppressed by the combination of high night temperatures and long days so that few flowers are produced, and heat tolerance at this stage is conferred by a single recessive gene (Hall, 1993a). Anther development is damaged by high night temperature so that few pods are set, and heat tolerance at this stage is conferred by a single dominant gene (Marfo and Hall, 1992). Embryo development is damaged by high night temperature such that fewer seeds are produced per pod, and two accessions have been shown to have tolerance to this effect (Ehlers and Hall, 1998). These effects are incorporated into the following model for cowpea grain yield under high temperatures (Y_h):

$$Y_h = (\#\,\text{flowers/m}^2) \times (\#\,\text{pods/flower}) \times (\#\,\text{seeds/pod}) \times (\text{g/seed}) \tag{5.1}$$

For the case where an environmental stress is reducing the extent of the reproductive sink much more than the activity of the photosynthetic source tissue, as is the case with effects of high night temperature on cowpea (Ismail and Hall, 1998), grain yield may be enhanced by selecting all of the various yield components in Equation 5.1 and increasing them to optimal levels. Conventional plant breeding using this model along with emphasizing selection for enhanced fruit set during hot weather has been effective in developing heat-resistant cultivars of cowpea (Ehlers et al., 2000), common bean, tomato, and both Pima cotton (*Gossypium barbadense*) and Upland cotton (*G. hirsutum*) (Hall, 1992, 1993a; Porch and Hall, 2013).

Selecting species with greater heat resistance also provides a method for overcoming hot conditions. Okra exhibits substantial fruit production in the hottest crop production environments such as in Sudan, where okra is a favorite vegetable, and during the summer in the Imperial Valley of California, where relatively few crops effectively produce fruits. Heat-resistant varieties of snap cowpea have been bred that are much more productive than either current snap cowpea varieties or all varieties of snap beans under hot conditions (Patel and Hall, 1986). Tepary bean, a native species from the Sonoran Desert of Mexico and southwestern United States, has much greater heat resistance than common bean (Porch and Hall, 2013).

Table 5.2 provides estimates of the temperature ranges of adaptation of cool-season and warm-season annual crop plants. These estimates are very approximate, since these two groups contain many different species (Table 5.1) and varieties, and extreme temperature limits depend on the duration of exposure and extent of hardening. Reproductive development of warm-season crop plants is sensitive to both high night temperatures and chilling temperatures. Reproductive development of cool-season crop plants is sensitive to both high temperatures and freezing temperatures. Optimum day temperatures are determined by the temperature sensitivity of photosynthesis.

Breeding new cultivars can take many years; consequently, breeders need to breed for future rather than present environments. The type of cultivars that should

TABLE 5.2
Temperature Ranges of Adaptation of Cool-Season and Warm-Season Crop Plants

	Night Temperatures (°C)		Day Temperatures (°C)	
	Minimum	**Maximum**	**Optimum**	**Maximum**
	Freezing	*Heat stress*		*Heat stress*
Cool-season annuals	–30 to –1	>16 to 24	18 to 28	>28 to 40
	Chilling	*Heat stress*		*Heat stress*
Warm-season annuals	0 to 18	>20 to 30	26 to 36	>30 to 50

be bred for future environments was evaluated by Hall and Allen (1993), Hall and Ziska (2000), and Hall (2011). In addition to considering breeding for heat tolerance the authors also considered breeding to take advantage of the increase in atmospheric [CO_2] that is occurring.

Prior to 1900, atmospheric [CO_2] fluctuated between 180 and 290 parts per million for about 220,000 years. During this long period, it is likely that the overall photosynthetic system of C_3 plants became adapted to these low [CO_2]s. During the industrial period, atmospheric [CO_2] has rapidly increased and is now 400 parts per million and is projected to reach about 700 parts per million by the end of the twenty-first century. Considering the balance between the extents of photosynthetic tissues and carbohydrate-sink tissues, it is possible that the overall photosynthetic system of current C_3 plants is not well adapted to [CO_2]s of 400–700 parts per million.

Doubling of [CO_2] using controlled environment chambers has increased grain yield of cereals by 32% and grain legumes by 54% at intermediate temperatures. More recent studies with free-air [CO_2] enrichment (FACE) experiments under field conditions, however, resulted in grain yield responses to elevated [CO_2] that were about 50% lower than those obtained using enclosures (Leakey et al., 2009). FACE experiments provide crop responses that are most likely to reflect those that will be experienced on farms because they were obtained with crops grown under natural open-air field conditions. But these yield increases often were less than the increases in photosynthesis that occurred with short-term doubling of [CO_2]. A major factor explaining this difference in responses is the down-regulation of photosynthetic capacity under long-term exposure to elevated [CO_2] that has been observed in some experiments and may be due to inadequate capacity of sink tissues.

The influence of elevated atmospheric [CO_2] is examined for several crops whose reproductive development is sensitive to high temperatures under the current atmospheric [CO_2]. If the sensitivity to heat stress of reproductive development is associated with reduced supplies of photosynthate then plants should be less sensitive to heat under elevated [CO_2], where the rate of photosynthesis is enhanced. For soybean growing under controlled-environment field conditions, harvest index (the ratio of grain yield to aboveground biomass) progressively decreased with increases in temperature under either 330 or 660 parts per million [CO_2] (Baker et al., 1989).

In addition, harvest index was lower with 660 parts per million [CO_2], indicating a more severe imbalance between the reproductive sink and the photosynthetic supply under elevated [CO_2]. Pima cotton can be sensitive to heat during flowering and unable to produce flowers or bolls under high night temperatures at 350 parts per million [CO_2]. It also did not produce flowers or bolls under high night temperatures at 700 parts per million [CO_2] (Reddy et al., 1997). For rice, elevated [CO_2] even resulted in high temperature having greater detrimental effects on viability of pollen grains and causing greater reductions in seed set and grain yield (Matsui et al., 1997). It is clear that the effects of high night temperature on reproductive development are not caused by heat-induced reductions in the supply of photosynthates and will not be overcome by elevated atmospheric [CO_2]. Also heat-induced increases in floret sterility may have been responsible for the downregulation of photosynthesis observed in rice under high temperatures and elevated [CO_2] through indirect effects associated with reductions in reproductive sink strength (Lin et al., 1997).

A unique insight into the interactive effects of high night temperature and elevated [CO_2] was obtained from growth chamber studies with contrasting cowpea genotypes that are either heat tolerant or heat sensitive during reproductive development. Three types of cowpea genotypes were evaluated in four growth chamber environments with either high or optimal night temperatures and either 350 or 700 parts per million [CO_2] (Ahmed et al., 1993a). Under high night temperatures, a heat-sensitive cowpea genotype did not produce any flowers under either 350 or 700 parts per million [CO_2]. Under high night temperatures, a partially heat-tolerant genotype produced flowers but did not set any pods under either 350 or 700 parts per million [CO_2]. Interestingly, a completely heat-tolerant genotype had greater flower and pod production under 700 parts per million [CO_2] at both high and optimal night temperatures than a heat-sensitive genotype. Note that the heat-tolerant genotype is genetically similar to the heat-sensitive genotype except for the few heat-tolerance genes that were incorporated by back-crossing.

These studies demonstrate that elevated atmospheric [CO_2] will not overcome heat-stress effects on reproductive development. This supports the argument that these effects are direct and are not mediated by heat-stress effects on the photosynthetic source. The cowpea studies indicate that incorporating heat tolerance during reproductive development of the many crops that are sensitive may also enhance their yield responses to elevated atmospheric [CO_2] over a range of temperatures (Hall and Ziska, 2000). Hall and Allen (1993) hypothesized that cultivars with heat tolerance during reproductive development, higher harvest index, higher photosynthetic capacity per unit leaf area, small leaves, and low leaf area per unit ground area will be most responsive to elevated atmospheric [CO_2] under a range of temperatures.

In summary, it is likely that current crop plants with either the C_3 or the C_4 photosynthetic system will not be able to take full advantage of the high [CO_2]s likely to occur in the twenty-first century. Cultivars need to be bred that have greater investment in sinks to accommodate a greater flow of photosynthetic products and less investment in the overall size of the photosynthetic system but with investment in a greater photosynthetic capacity per unit leaf area.

5.5 CLIMATIC ZONE DEFINITION BASED ON TEMPERATURE

I am providing partial definitions of climatic zones based only on crop responses to temperature. In Chapter 10, I extend these definitions to include consideration of rainfall and evaporative demand (the aridity of the climate).

5.5.1 Tropical Zones

Tropical zones are where all monthly mean air temperatures are >18°C, and there is no frost and minimal chilling (e.g., the locations in Figures 5.3, 10.11, and 10.13). The overall zone occurs at low elevations between the Tropic of Cancer (23.5°N latitude) and the Tropic of Capricorn (23.5°S latitude) passing through the equator. This is the main zone for the production of chilling-sensitive evergreen perennials such as mango, cacao, rubber, and banana. Warm-season annuals (e.g., maize, cowpea, and cotton) are widely grown in the tropics, whereas cool-season annuals (e.g., pea or lettuce) are not very successful in this zone. Crops that require chilling (apple, peach, or winter wheat) or benefit from chilling (oranges for fresh fruit) are not very successful in this zone, as will be discussed in Chapter 6.

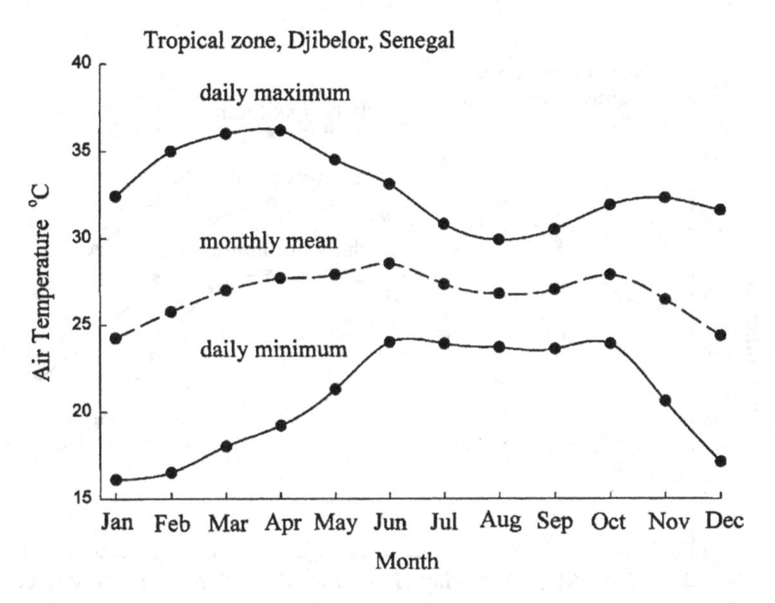

FIGURE 5.3 Monthly mean daily minimum and maximum air temperatures (1972–1980) measured in a weather station shelter in the tropical zone at Djibelor, Senegal (12°33′N, 16°16′W, elevation 23 meters). The dashed line provides the means of these values.

5.5.2 Subtropical Zones

Subtropical zones are where the coldest month has a mean air temperature <18°C, there is a long period (8–12 months) when plants can actively grow (mean monthly air temperatures >10°C), and only occasional frosts occur. This zone occurs at low elevations in latitudes between the 20s and the 30s. Florida (Figures 5.4 and 10.12), low-elevation valleys in central and southern California (Figures 5.2, 10.3, 10.4, and 10.6), low-elevation areas around the Mediterranean Sea, and high elevations near to the equator are subtropical. This is a zone where a broad range of crop species can be grown: evergreen perennials such as coffee and those that benefit from some chilling but are damaged by frosts (e.g., orange for fresh fruit), deciduous tree crops that require high heat units in summer but low to moderate chilling in winter (e.g., peaches and apricot), warm-season annuals in the summer, and cool-season annuals in the winter.

5.5.3 Temperate Zones

Temperate zones are where there are only four to seven months when temperatures are high enough for plants to actively grow (mean monthly air temperatures >10°C), and there is a long, cold winter. Temperate zones occur at high latitudes or at very high elevations in more equatorial latitudes. Some warm-season annuals, such as maize, can be successful in the warmer of the temperate zones (Figure 5.5),

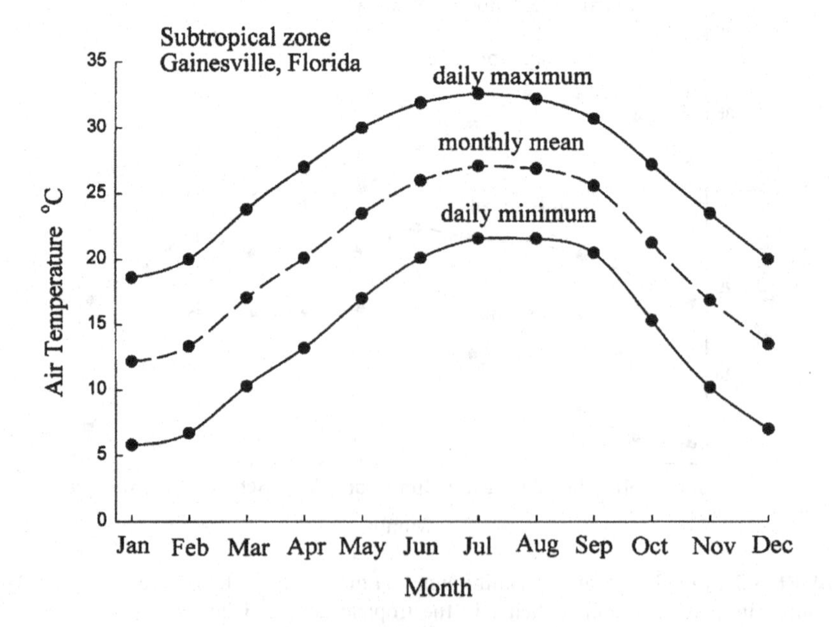

FIGURE 5.4 Monthly mean daily minimum and maximum air temperatures (1961–1990) measured in a weather station shelter in the subtropical zone at Gainesville, Florida, U.S.A. (29°41′N, 82°15′W, elevation 42 meters). The dashed line provides the means of these values.

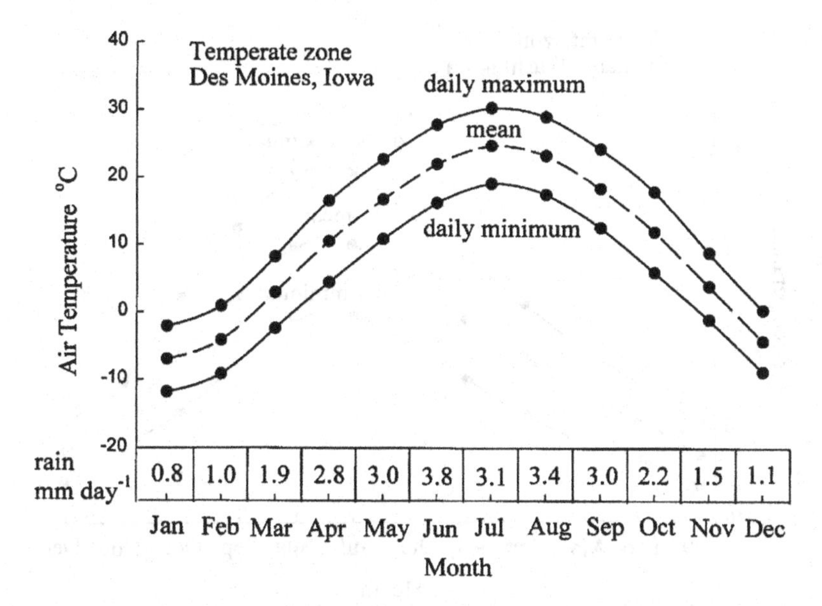

FIGURE 5.5 Monthly mean daily minimum and maximum air temperatures and precipitation (1961–1990) measured in a weather station shelter at Des Moines, Iowa, U.S.A. (41°32′N, 93°39′W, elevation 294 meters), subhumid temperate zone with average annual precipitation of 841 nanometers. The dashed line provides the means of these values.

providing their growth cycle is not too long. Cool-season annuals can be successful in many temperate zones, with spring wheat being grown in the colder temperate zones and winter wheat being grown in the warmer temperate zones. Deciduous trees that require substantial chilling (such as most apple varieties) can be successful commercially in this zone (Figure 5.6). Frost-hardy evergreen perennials, such as conifers, are successfully grown in this zone, whereas chilling-sensitive (e.g., mango) and frost-sensitive (e.g., orange and avocado) evergreen perennials would be killed in most locations and years in this zone.

5.5.4 Boreal Zones

Boreal zones, also known as subalpine zones, are where there are only one to three months when temperatures are high enough for plants to actively grow (mean monthly air temperatures >10°C). In general, little cultivation of crops is possible in field conditions. Cold-hardy short-cycle barley varieties might produce significant quantities of grain. A warmer boreal zone is described (Figure 5.7) where some short-cycle cool-season annuals can be grown in fields in summer.

The subtropical and temperate zones located at high elevations near the equator have different seasonal day lengths from the subtropical and temperate zones that occur at higher latitudes. Consequently, there are some differences in plant adaptation to equatorial and higher-latitude subtropical or temperate zones. For example, pyrethrum (*Chrysanthemum cinerariaefolium*), a crop grown

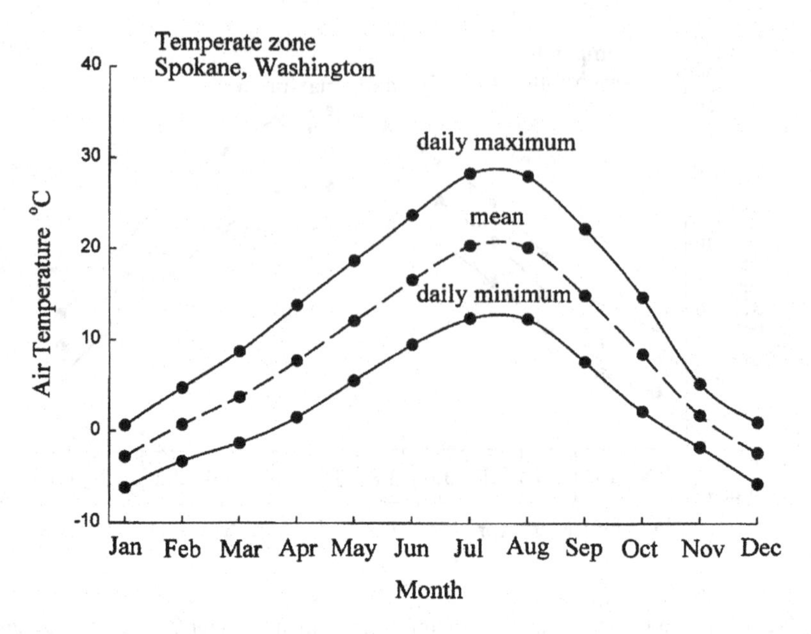

FIGURE 5.6 Monthly mean daily minimum and maximum air temperatures (1961–1990) measured in a weather station shelter in the temperate zone in the Yakima Valley at Spokane, Washington, U.S.A. (47°38′N, 117°32′W, elevation 23 meters). The dashed line provides the means of these values.

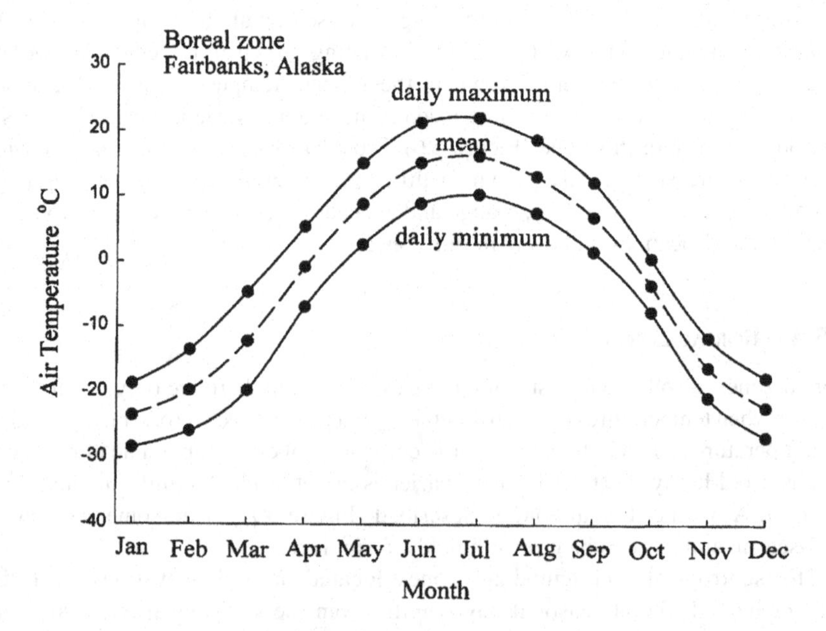

FIGURE 5.7 Monthly mean daily minimum and maximum air temperatures (1929–1993) measured in a weather station shelter in the boreal zone at Fairbanks, Alaska, U.S.A. (64°50′N, 147°52′W, elevation 145 meters). The dashed line provides the means of these values.

to produce the pyrethrins present in the flowers, is more productive near the equator at very high elevations (2,000–3,000 meters) in the temperate zone of Kenya than it is in temperate zones at higher latitudes and lower elevations (Purseglove, 1968).

Smaller-scale descriptions of climatic zones are available that are useful for determining where different crop species or even cultivars can be commercially successful or effective in home gardens. The *Sunset Western Garden Book* (the 1995 fortieth anniversary edition published by Sunset Books, and the newest edition published by Oxmoor House in 2012) provides maps that subdivide the subtropical and temperate zones of the western United States into 24 different climatic zones with respect to their suitability for different plant species and varieties. Temperature data from weather stations at many locations in California may be found at the website of the Integrated Pest Management Project of the University of California (www.ipm.ucdavis.edu). Monthly mean daily maximum and daily minimum air temperatures (and rainfall) for many cities in the world may be found at www.worldclimate.com.

5.6 COMPARISON METHOD FOR DETERMINING WHERE CROPS CAN BE GROWN

Knowing the climatic zone of a location can be useful for estimating which crop species can be grown at the location. More detailed analysis of the thermal regime must be conducted to determine whether the crop may be commercially successful. For example, cotton requires at least five consecutive months with monthly mean temperatures in the 20s to be successful. The hot season at the subtropical location at Riverside, California (Figure 10.4), is only marginally long enough for growing cotton, whereas hotter subtropical locations in the southern San Joaquin Valley of California between Bakersfield (Figure 10.3) and Fresno (Figure 5.2) and the Imperial Valley of California (Figure 10.2) have suitable long hot seasons. Tropical zones (Figures 5.3, 10.10, and 10.11) have sufficient warmth for growing cotton in all months of the year.

The comparison method can provide more precise estimates of whether the thermal environment is suitable for the commercial production of a specific crop species. Applying the comparison method involves a series of procedures as follows:

1. Select regions where a specific crop species is presently grown successfully on a commercial basis.
2. Obtain weather station data for average daily minimum and maximum air temperatures, over several years, during the season when the crop is grown for each region.
3. Analyze the results. For example, Kimball et al. (1967) analyzed the temperatures for regions in the western United States where head lettuce was being grown commercially. They discovered that monthly mean values of daily minimum temperatures were between 3°C and 12°C, while daily maximum values were between 17°C and 28°C.

4. Make predictions. Kimball et al. (1967) predicted that regions with three consecutive months having daily maximum temperatures between 17°C and 28°C and daily minimum temperatures between 3°C and 12°C had suitable thermal regimes for growing head lettuce. For the locations presented in the figures, the Imperial Valley of California (Figure 10.2) and Gainesville, Florida (Figure 5.4), are predicted to have suitable thermal regimes for growing head lettuce from late fall through winter; while the Yakima Valley of Washington (Figure 5.6) and Fairbanks, Alaska (Figure 5.7), are predicted to have suitable thermal regimes during the late spring and summer. The model predicts that lettuce could be grown at any time of year in the Salinas Valley of California, which has a subtropical climate that is not too hot and exhibits little diurnal or seasonal variation due to the moderating effects of the Pacific Ocean.
5. Consider other factors. Lettuce needs a frequent supply of rain (refer to Chapter 9) or irrigation water with low salt content (refer to Chapter 11), wind speeds that are not too high, and enough workers who are willing to do the hand labor at a price that is acceptable to the grower.

More complex analyses of the thermal regime often are required. For grain and fruit crops, the minimum and maximum temperatures during reproductive development may have substantial effects on yield and quality, and often there is a narrower window of acceptable temperatures during this stage of development than the difference between the maximum and minimum temperatures for vegetative growth. Temperatures also influence rates of plant development, and Chapter 6 describes heat-unit approaches for modeling these effects.

For some perennial crops, the probability of damaging freezes may be critical for commercial success. In this case, more site-specific temperature data may be needed due to the strong effects on winter minimum temperatures of local microclimates in areas where radiation frosts occur (Chapter 7). For long-lived crops it is prudent to consider possible climate changes that might occur such as reductions in chilling units during winter as they influence adaptation of tree crops requiring chilling (Chapter 6).

Prior to planting large fields of an annual crop species that are new to an area, it is advisable to plant small test plots to see if there are any unanticipated problems. In the 1980s, thousands of hectares of cowpeas were sown for the first time on a specific ranch on the west side of the San Joaquin Valley in California. Cowpea had been successfully grown on other ranches in the same area. This planting, however, was a total failure, because the soil had boron levels that were far above the toxic threshold for this crop (refer to Chapter 11), due to high levels in the well water that had been used for irrigation. This very expensive mistake could have been avoided if a more cautious approach had been used, beginning with tests of the well water and soil, then followed by sowing only small plots during the first year and conducting plant tissue analysis and making careful observations to look for unanticipated pest, disease and other problems.

ADDITIONAL READING

Ahmed, F. E. and A. E. Hall. 1993. Heat injury during early floral bud development in cowpea. *Crop Sci.* 33: 764–767.

Ahmed, F. E., A. E. Hall, and D. A. DeMason. 1992. Heat injury during floral development in cowpea (*Vigna unguiculata*, Fabaceae). *Amer. J. Bot.* 79: 784–791.

Ahmed, F. E., A. E. Hall, and M. A. Madore. 1993a. Interactive effects of high temperature and elevated carbon dioxide concentration on cowpea [*Vigna unguiculata* (L.) Walp.]. *Plant Cell Environ.* 16: 835–842.

Baldocchi, D. and S. Wong. 2008. Accumulated winter chill is decreasing in the fruit growing regions of California. *Clim. Change* 87 (Suppl.): 153–166.

Bedi, S. and A. S. Basra. 1993. Chilling injury in germinating seeds: Basic mechanisms and agricultural implications. *Seed Sci. Res.* 3: 219–229.

Board, J. E. and M. L. Peterson. 1980. Management decisions can reduce blanking in rice. *Calif. Agric.* 34(11/12): 5–7.

Cornish, K., J. W. Radin, E. L. Turcotte, Z. Lu, and E. Zeiger. 1991. Enhanced photosynthesis and stomatal conductance of pima cotton (*Gossypium barbadense* L.) bred for increased yield. *Plant Physiol.* 97: 484–489.

Hall, A. E. 2011. Breeding cowpea for future climates. In S. S. Yadav, R. J. Redden, J. L. Hatfield, H. Lotze-Campen, and A. E. Hall (Eds.), *Crop Adaptation to Climate Change.* Wiley, Chichester, UK, pp. 340–355.

Hall, A. E. and L. H. Ziska. 2000. Crop breeding strategies for the 21st century. In K. R. Reddy and H. F. Hodges (Eds.), *Climate Change and Global Crop Productivity.* CABI Publishing, New York, pp. 407–423.

Ismail, A. M. and A. E. Hall. 1998. Positive and potential negative effects of heat-tolerance genes in cowpea. *Crop Sci.* 38: 381–390.

Ismail, A. M., A. E. Hall, and T. J. Close. 1997. Chilling tolerance during emergence of cowpea associated with a dehydrin and slow electrolyte leakage. *Crop Sci.* 37: 1270–1277.

Ismail, A. M., A. E. Hall, and T. J. Close. 1999. Allelic variation of a dehydrin gene cosegregates with chilling tolerance during seedling emergence. *Proc. Natl. Acad. Sci. USA* 23: 13569–13573.

Kimball, M. H., W. L. Sims, and J. E. Welch. 1967. Climatographs for head lettuce in western producing areas. *Calif. Agric.* 21(4): 3–4.

Leakey, A. D. B., E. A. Ainsworth, C. J. Bernacchi, A. Rogers, S. P. Long, and D. R. Ort. 2009. Elevated CO_2 effects on plant carbon, nitrogen and water relations: Six important lessons from FACE. *J. Exp. Bot.* 60: 2859–2876.

Mayer, A. M. and A. Poljakoff-Mayber. 1989. *The Germination of Seeds*, 4th ed. Pergamon Press, Oxford, UK, p. 270.

Mutters, R. G. and A. E. Hall. 1992. Reproductive responses of cowpea to high temperature during different night periods. *Crop Sci.* 32: 202–206.

Nielsen, C. L. and A. E. Hall. 1985b. Responses of cowpea (*Vigna unguiculata* [L.] Walp.) in the field to high night temperature during flowering. II. Plant responses. *Field Crop Res.* 10: 181–196.

Reynolds, M. P., M. Balota, M. I. B. Delgado, I. Amani and R. A. Fischer. 1994b. Physiological and morphological traits associated with spring wheat yield under hot, irrigated conditions. *Austral. J. Plant Physiol.* 21: 717–730.

Roberts, E. H. 1988. Temperature and seed germination. In S. P. Long and F. I. Woodward (Eds.), *Plants and Temperature, Symposium of the Society for Experimental Biology, Number XXXXII.* Company of Biologists, Cambridge, UK, pp. 109–132.

ADDITIONAL READING

6 Crop Developmental Responses to Temperature, Photoperiod, and Light Quality

The development of plants can be influenced by temperature through influences on the rate of production of nodes and on the triggering of flowering, which can be determined by either accumulated heat units or photoperiod depending on the species or variety. Development also can be influenced by light quality. *Development* should be distinguished from *growth*, which is different in that it involves increase in mass, length, volume, or area of a plant or organ.

The life cycle of adapted plants is synchronized with the seasonal changes in average weather (climate) through effects of photoperiod and/or temperature on development that result in an optimal phenology (Bunting, 1975). Phenology is the sequence of developmental events during the plants' life cycle as they are determined by environmental conditions.

Annual crop plants are sown when the environment permits effective germination, emergence, and establishment. Adapted cultivars grow vigorously and then flower on an optimal date, enabling them to produce fruit and/or seed during a period of the year when average environmental conditions permit the plant to produce many fruit and/or seed of good quality.

Bunting (1975) provided examples from England and West Africa of how plant breeders and agronomists have developed cultivars and management methods that enable annual crops to have an optimal phenology with respect to the timing and duration of their period of reproductive development such that it fits the available season and results in maximum productivity. Clearly, when breeding improved cultivars, it is important to be able to predict the dates of flowering and harvest of these cultivars. Methods have been developed for making these predictions.

6.1 HEAT-UNIT SYSTEMS FOR PREDICTING PLANT DEVELOPMENT

Heat-unit systems are empirical approaches used to predict the rate of development of plants (or insects) where the process and genotype are insensitive to photoperiod. They may be used to predict dates of flowering or harvest or the locations and dates where a particular fruit quality can be achieved, such as with citrus or grapes. The following is an example of the development of a heat-unit system for predicting date of flowering of an annual plant that includes an interaction with genotypic heat tolerance:

1. Sow the cultivar at different dates and in different locations choosing ones that are effective for commercial production and provide different thermal regimes during the vegetative period.
2. Record dates of sowing and first flowering.
3. Measure daily (24 hours) maximum (T_{max}) and minimum (T_{min}) air temperatures in a standard weather station shelter at each location.
4. Analyze the data by plotting average rate of development $1/D$, where D is the number of days from sowing to flowering, versus average daily temperature = $[\Sigma(T_{max} + T_{min})/2]/D$ over the period from sowing to flowering (Figure 6.1). A linear curve with an acceptable correlation coefficient indicates that rate of development can be modeled as being proportional to degrees average air temperature above a base temperature (T_{base}).

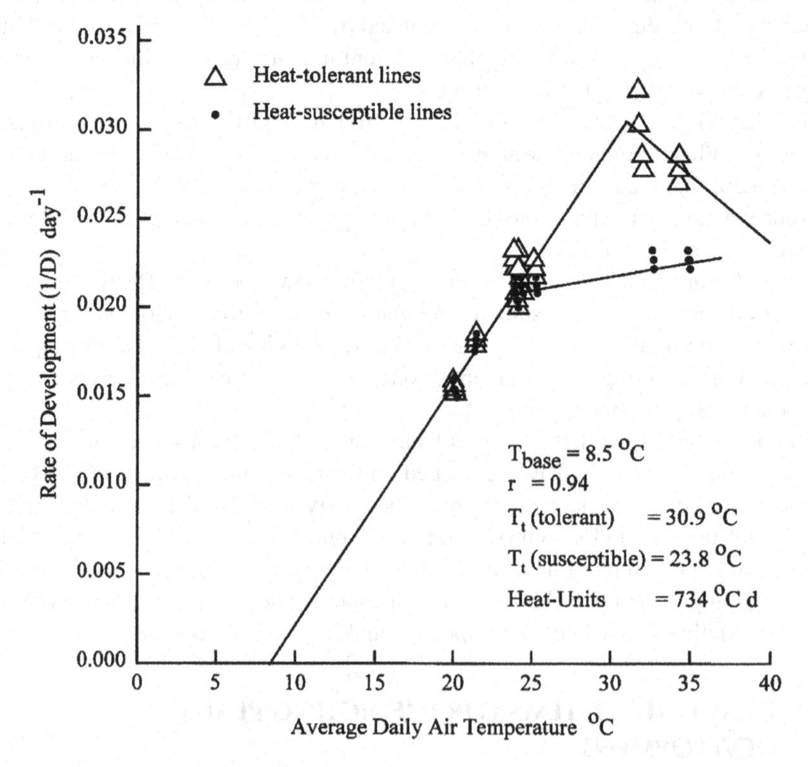

FIGURE 6.1 Rate of development ($1/D$) of heat-tolerant and heat-susceptible cowpea lines, where D is the period between sowing and first flowering as a function of average air temperature. The heat-unit and r values are based on the linear part of the curve. (From Ismail, A. M., and A. E. Hall, *Crop Sci.*, 38, 381–390, 1998.)

The projected intercept on the x-axis is T_{base} and, for the cowpea data in Figure 6.1, it is 8.5°C. The inverse of the slope of the line is the total heat unit (degree-days) for the process as described by Equation 6.1:

$$\text{Heat units} = \sum_{i=D}^{i=1} \frac{T_{max} + T_{min}}{2} - T_{base} \quad (6.1)$$

where T_{max} and T_{min} are daily data from sowing to first flowering, and T_{base} is a constant for a particular cultivar and developmental process. In Figure 6.1 the total heat unit is 734°C day.

The heat-unit value should be effective for predicting the number of days from sowing to flowering of the specific cultivar in the range of environments over which the model was developed, as long as average daily air temperatures do not exceed threshold values (T_t). Where average daily air temperatures (T) exceed threshold values, the model may be modified by including an algorithm stating that for $T > T_t$, $T = T_t$. This modification assumes that $1/D$ becomes constant once T exceeds T_t, which is approximately consistent with the data in Figure 6.1 for the heat-susceptible lines. For the example presented in Figure 6.1, however, it would be necessary to have two different threshold values for the different types of cultivars. The heat-susceptible cowpea lines exhibited a lower T_t (23.8°C) because their floral bud development was suppressed by high temperatures, whereas the floral bud development of the heat-tolerant lines was not affected by high temperatures until a higher T_t (30.9°C) was reached. Note that the experiments of Figure 6.1 were conducted under long-day conditions. Under the short-day conditions that prevail in many environments near the equator, the floral bud development of cowpea is not suppressed by high temperatures (Dow El-Madina and Hall, 1986). Other effects of photoperiod on photoperiod-sensitive plants are described later in this chapter.

Heat-unit values can differ between different cultivars and developmental processes, but the T_{base} values tend to be similar for different cultivars and developmental processes but different for different types of species. For example, cool-season species such as pea have T_{base} values of about 4°C, whereas warm-season species such as sweet corn and cowpea have T_{base} values of 8°C–10°C.

Heat-unit systems are used to guide the commercial production of annual crops that are suitable for market only over a few days, that is, sweet corn or peas for market or for freezing. If the objective with sweet corn is to supply a major market, such as July 4 in the United States, the cultivar used and sowing date must be chosen such that the crop is ready for market a few days before July 4. By using average daily air temperature data and working backward in time, it is possible to predict the combination of cultivar (heat-unit requirement and T_{base}) and sowing date that will produce a crop that is ready for harvest a few days before July 4. (Note that the actual temperatures during the cropping season probably will differ from the climatic average and that the soil must be warm enough at sowing to permit emergence as discussed in Chapter 5.) For producing packets of frozen peas and other crops, it is necessary to achieve a continuous flow of fresh produce from farms to the processing factory. This can be done by choosing cultivars with different heat-unit requirements from sowing to the date when crop quality is optimal, and sowing on different dates and in different regions with different thermal regimes.

The heat-unit system described by Equation 6.1 has been used to define climatic zones for growing different varieties of grapes for various end uses in California. In this case, a base temperature of 10°C is used, and the heat units are obtained by summing average degree-days over the growing season from April 1 through October 31. Cooler zones with a specific small number of degree-days are useful for producing good dry wines; hotter zones with more heat units can produce sweeter wines; and the hottest zones, with even more heat units, are effective for producing table grapes or raisins.

Different citrus scions have different heat-unit requirements for producing high-quality fruit. For example, most grapefruit cultivars require more heat units than oranges, except two new grapefruit cultivars—Oroblanco and Melogold—that have smaller heat-unit requirements for fruit development, similar to those of navel oranges.

The overall objective of a heat-unit system is to provide an empirical method for effectively predicting a specific developmental process in a particular commercial production region using available weather station data. Different types of heat-unit systems have been used. In some cases, it is assumed that the diurnal curve of air temperature is a sinewave, and average temperature is determined based on this assumption. Some different heat-unit models are described on the website www.ipm.ucdavis.edu.

6.2 CHILLING REQUIREMENTS OF PLANTS

Optimal development of some plants requires that they experience chilling. For example, deciduous trees can require sufficient chilling during winter to overcome bud dormancy. Insufficient chilling results in delayed foliation and some of the following symptoms: A tree may produce small tufts of leaves near the tips of the stems and be devoid of leaves for 30–50 centimeters below the tips. Lower buds may break eventually, and substantial suckering can occur from lower parts of the tree. Bloom can be delayed and the flowers can be abnormal, resulting in reduced fruit set and yield. To avoid these problems, deciduous tree species and cultivars must be chosen whose chilling requirements will be met every year in the environment where they are to be grown. The physiologically effective temperatures are between 1°C and 12°C but, in earlier years, chilling requirements of different varieties were defined on the basis of the number of hours experienced with air temperatures <7°C. More effective models have been developed, such as one based on the hours when air temperatures are between 0°C and 7°C or the more complex Utah Model, which includes negative effects on chilling units of high temperatures. Average chilling units for different locations in California can be calculated using different models at www.ipm.ucdavis.edu.

Average chilling requirements of different species rank as follows: apples > cherries > peaches. But new cultivars have been developed that have much lower chilling requirements than the traditional cultivars of a species. For example, Fuji and Gala cultivars of apple are commercially successful in warm subtropical zones such as near Bakersfield, California (Figure 10.3), whereas traditional apple cultivars require more chilling and are successful in temperate zones such as the foothills of the Sierra Nevada in California, the Yakima Valley of Washington (Figure 5.6), and in England (Figure 10.14).

Global warming has caused accumulated winter chilling hours to decrease across fruit and nut growing regions of the world. Baldocchi and Wong (2008) computed chilling hours between 0°C and 7.2°C for about 30 sites in California where deciduous fruit and nut trees are grown and observed that decreases were occurring of between 50 and 260 chilling hours per decade. They predicted that this reduction in winter chilling will have deleterious economic impact on fruit and nut production in California by the end of the twenty-first century, when orchards are expected to experience less than 500 chilling hours in the winter. On a global scale, many regions where fruit and nut trees that have chilling requirements are grown will be subjected to detrimental decreases in winter chilling (Luedeling et al., 2011). As yields decrease growers probably will need to remove orchards and plant them with tree crops that require less chilling or grow other crops that do not require chilling.

Winter-type cereals (e.g., winter wheat) are sown in the fall in temperate zones because they require several weeks of chilling (this process has been called vernalization) if they are to produce flowers in the spring, whereas spring wheat does not require any chilling. It should be noted that, in some cases, there is confusion in the labeling of wheat cultivars in that some labeled as being *winter wheats* do not have a chilling requirement.

Some biennials, such as sugar beet, carrots, and celery, have a chilling requirement for flowering. Sugar beet is sown as seed in the first year and, in appropriate commercial production environments, it produces a storage root that is harvested and sent to factories for the extraction of sucrose. If the root is left in the ground and experiences chilling during the winter, it will produce a flowering stem as the days become longer in the following spring. Sometimes, there is a problem in fields where sugar beet is being grown to produce roots for sucrose if plants experience too much chilling during the seedling stage. In this case, some plants become annuals and produce both a storage root and a flowering stem (bolting) during the first year. This is a problem because the hard, woody, flowering stem interferes with the operation of machinery used to harvest the storage roots. A solution to this problem is to breed sugar beet cultivars that have a higher chilling requirement such that they do not bolt during the first season when they are producing a storage root. In temperate zones (e.g., Figure 10.14), sugar beet cultivars with resistance to bolting can be sown earlier and thus reach full canopy development sooner in the season and have a more optimal phenology (Bunting, 1975).

Chilling during the winter season can be beneficial for commercial orange production because it results in fruit having more carotenoids and less chlorophylls so that they have the bright orange color that is favored on the fresh market. In contrast, oranges produced in warmer, more tropical environments can have a yellow/green appearance. Those oranges not sold on local markets are mainly sold on commercial juice markets, which have much lower prices than fresh-fruit markets. Sometimes, mature Valencia oranges are kept on trees in California into the summer while waiting for a suitable market. During this period, chlorophyll returns to the rind and carotenoid content decreases, changing fruit color from a bright orange back to a more yellowish green color. This re-greening reduces the appeal of the fruit to consumers. Another advantage of chilling during the winter season for commercial citrus production is that it induces a quasi-dormancy, which results in a concentrated spring bloom and thus a short period when there are many mature fruit on the trees, which facilitates picking. In contrast, in more tropical environments, oranges tend to be ever-bearing, producing small amounts of fruit during different periods of the year. This ever-bearing is good for trees such as lemons used for home consumption but is not suited to commercial production, because it increases the amount of labor required for harvesting.

6.3 PLANT DEVELOPMENTAL RESPONSES TO PHOTOPERIOD

Some developmental responses of the cultivars of some species are influenced by photoperiod. This usually is considered as a response to day length, although the mechanism involves the length of the night. Where photoperiod determines the date of flowering, it also determines the node at which flowering first occurs. Photoperiod-sensitive cultivars can begin flowering at different nodes, depending on their date of sowing and the prevailing photoperiods. In contrast, the development of cultivars that are insensitive to photoperiod (day-neutral cultivars) is determined by heat units. For all cultivars, the rate at which nodes appear is positively correlated with the average temperature (above a baseline temperature), and day-neutral cultivars produce their first flowers at the same node even with different temperatures and photoperiods.

Effective day length is defined as the period between sunrise and sunset plus twilight of about 26 minutes. Effective day length changes with season and latitude. More extreme latitudes have longer days in summer and shorter days in winter (Figure 6.2).

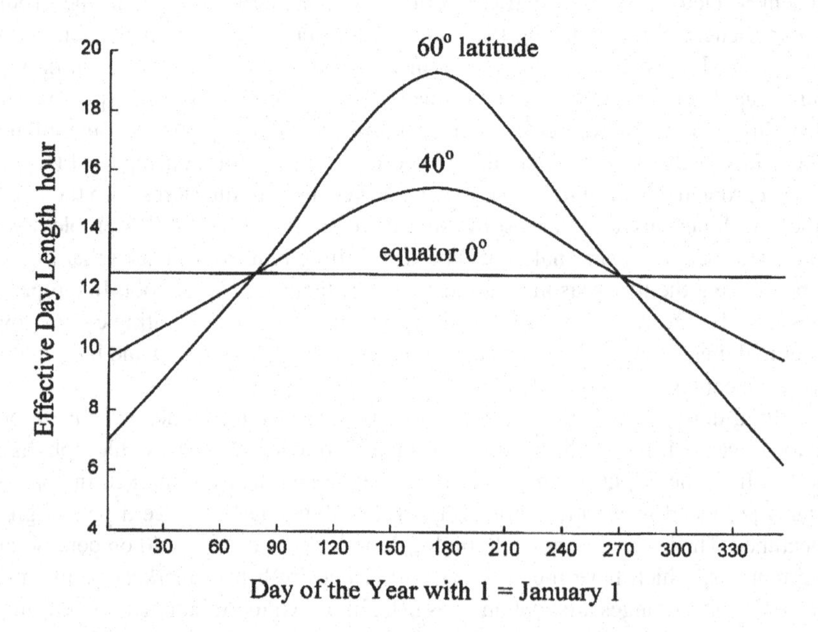

FIGURE 6.2 Day length in hours between sunrise and sunset plus 26 minutes of twilight for the equator and two locations with different latitudes in the northern hemisphere at different days of the year.

In the northern hemisphere, the longest day occurs at the summer solstice on about June 22 (day 173), and the shortest day occurs on the winter solstice on about December 22 (day 356). Effective day length is about 12.5 hours on the equinoxes, about March 21 (day 80) and about September 24 [day 267]) and year-round at the equator. Generally, there are larger seasonal changes in both photoperiod and temperature at more extreme latitudes, which led some scientists to believe that plant responses to photoperiod may be more prevalent at extreme latitudes. However, adaptively significant responses to photoperiod occur with some cultivars of sorghum at latitudes not far from the equator (e.g., 10°), where the changes in day length are small (Bunting, 1975).

The aspect of photoperiod that determines the first flowering of different types of crops may be understood by considering the annual cycle of cropping in a subtropical zone at 34°N latitude (Figure 6.3). A photoperiod-sensitive cool-season annual that is sown in early winter would be induced to flower during the lengthening days of spring. In contrast, a photoperiod-sensitive warm-season annual that is sown in late spring would be induced to flower during the shortening days of the late summer. This leads to the following definitions:

- *Cool-season annuals* that are sensitive to photoperiod begin flowering when photoperiods become longer than a critical value. They are called LD (longer day) plants and include some wheat cultivars.
- *Warm-season annuals* that are sensitive to photoperiod begin flowering when photoperiods become shorter than a critical value. They are called SD (shorter day) plants and include most soybean cultivars and some tropical cowpea, sorghum, and rice cultivars.

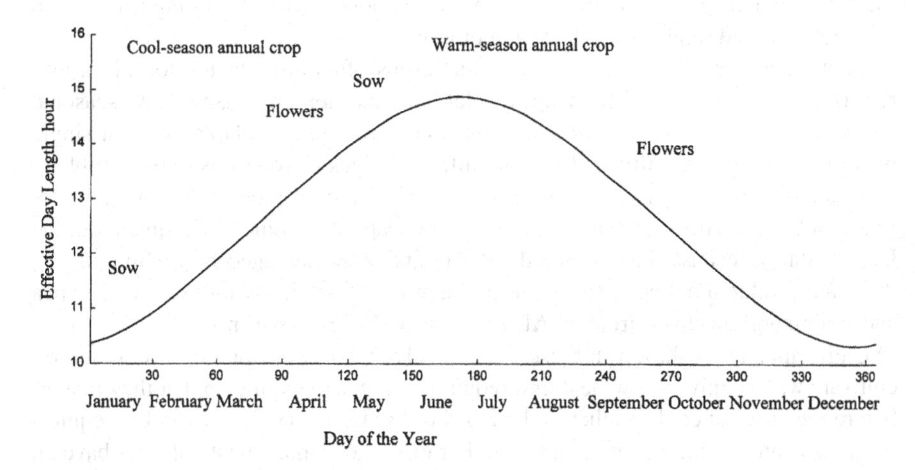

FIGURE 6.3 Effective day length throughout the year at Riverside, California, U.S.A. (location 33°58′N, 117°21′W). Dates of sowing and first flowering are indicated for a cool-season annual crop and a warm-season annual crop.

The strength of the photoperiod effect varies among cultivars. For some cultivars, the photoperiod effect is *obligate*, and the plant will remain vegetative until it receives the appropriate photoperiod. For most photoperiod-sensitive cultivars, the photoperiod effect is *quantitative* in that the plant will flower under non-inductive photoperiods, but it begins flowering later. In this case, the photoperiod trigger causes flowering to occur earlier.

In addition to flowering, several other developmental processes are influenced by photoperiod, including tuber formation, bulb formation, runner formation, and the formation of dormant buds in deciduous trees that occurs in the fall.

Day-neutral annuals exhibit no effect of photoperiod on the time at which flowering occurs. They begin flowering at a fixed nodal position, and their time of flowering is determined by heat units. This group includes many important agricultural crops: virtually all cultivars of cucumber, sweet corn, tomato, and pea; modern cultivars of cotton, maize, and spring wheat; and most cultivars of rice, sorghum, and cowpea that are grown in subtropical zones.

The photoperiod-induced developmental responses have some adaptive significance. For cool-season annuals in a Mediterranean climate (Figures 10.4 and 10.6), the LD trigger results in seed maturation prior to the hot and dry weather of the early summer. In contrast, for warm-season annuals, such as soybean, in temperate (Figure 5.5) or subtropical (Figure 5.4) climatic zones, the SD trigger results in seed maturation prior to cold weather in the fall. For warm-season annuals in tropical semiarid zones (Figure 10.8), the SD trigger can result in seed maturation prior to the extreme droughts of the dry season. These photoperiod responses are less important in agriculture than in natural conditions. When using day-neutral cultivars, farmers can control the date of flowering simply by either varying the sowing date or choosing a cultivar with a different heat-unit requirement. For some deciduous trees, such as those being grown in temperate zones (Figures 5.6 and 10.14), an SD response in the fall can result in the formation of dormant buds that have substantial resistance to the cold weather of the winter. Some types of outdoor artificial lighting can disturb this response and reduce the adaptation of trees.

Photoperiod-sensitive cultivars of annual crops often are adapted to only a narrow range of latitudes. With changes in latitude, day lengths change, and seasonal temperatures change in a manner that does not favor broad adaptation of a single photoperiod-sensitive cultivar. There are different types of solutions to this problem. Traditional cultivars of maize were SD types that were adapted to the low latitudes found in Central America. They were not well adapted to some environments in the United States because they flowered too late and were damaged by chilling during the grain production stage. The same problem occurs with SD cultivars of cowpea and grain sorghum from tropical Africa when they are grown in the Central Valley of California. The solution in these cases involved the development of day-neutral cultivars with relatively low heat-unit requirements for flowering so that they mature before the onset of cool weather in the fall. Cool-season annuals with an LD requirement also can exhibit narrow adaptation. For example, some wheat cultivars have an LD requirement for flowering that confers adaptation to climatic zones around the Mediterranean Sea. At high elevations near the equator, temperature conditions are suitable for growing wheat, but cultivars with an obligate LD requirement do not flower. In this case also, the solution involved the development of day-neutral cultivars.

In contrast, modern soybean cultivars have an SD requirement for flowering. Individual cultivars flower at similar dates when they are sown at different dates, and they are adapted to a narrow range of latitudes. At more extreme latitudes, they tend to flower too late and become damaged by chilling temperatures during grain development. When grown near the equator, they can be triggered to flower too early by the short days, and the plants are too small when entering the reproductive stage and produce small grain yields. The solution taken to solve this problem involved the development of many different types of soybean cultivars with different photoperiod responses for commercial production in regions with different latitudes. The problem with this solution is that it has required a substantial investment by society in soybean breeding programs. Large investments are now being required as genetic engineering is being used to incorporate genes to confer traits, such as herbicide resistance and resistance to insect pests, because it must be done with many different types of soybean cultivars. A potential advantage of the large numbers of cultivars is that the soybean crop may be less likely to suffer large-scale damaging effects from disease epidemics than crop species where a few cultivars with similar genetic background are grown over very large areas. For example, the maize crop in the United States suffered from an epidemic in 1970 caused by a *new* strain of the fungi that causes corn leaf blight, which reduced grain yield by about 50% in some southern states and 15% nationwide. At that time, most of the cultivars of maize grown in the United States had the same *Texas* cytoplasm that had been used to confer the male sterility that facilitated the breeding of hybrid cultivars. All cultivars with the *Texas* cytoplasm were susceptible to the *new* fungal strain that causes corn leaf blight.

Multiple seasons of cropping are possible with warm-season annuals in the tropics because of the year-round warm to hot temperatures. Traditional tropical cultivars of SD-type rice were only grown in the rainy season. With the availability of irrigation it became possible to grow three crops of rice per year on the same field. In some cases, this required the development of day-neutral cultivars of rice for growing in the dry season. Growing the same crop species season after season or year after year on the same field usually is not sustainable because of the buildup of soil pests, diseases, and specific weeds. For rice, continuous cropping may be effective (Dobermann et al., 2000) and may be needed in those areas where the density of the human population is too high in relation to the area of arable land that is available for producing food.

There are cases where flowering causes agronomic problems and must be prevented. One such case is where annual legumes are grown as green manure or forage crops. Cultivars of cowpea are being developed for use as green manure crops in the United States. They have extreme sensitivity to photoperiod (they are SD types) and, when sown in June, they grow actively, without flowering, until September, when they are incorporated to enrich the fertility of the soil. The advantage of this phenology is that the plants abundantly fix atmospheric nitrogen for a longer period compared with cowpea cultivars that initiate reproductive activity early, because nitrogen fixation rate decreases when plants produce many pods. Commercial supplies of seed of these SD cowpea cultivars can be produced by sowing them in mid-August in a location with a hot fall season, such as the Coachella Valley in California, which has a similar climate as the location in Figure 10.2. In this valley, day lengths become shorter than the critical photoperiod in late September and trigger flowering, and the temperatures are warm enough until mid-November to enable the crop to produce adequate yields of seed.

Another case where flowering can be detrimental is where sugar cane is being grown for commercial sucrose production. Some SD-type cultivars of sugar cane tend to begin flowering if grown for two years in certain locations, such as Hawaii. This is undesirable because flowering causes decreases in sucrose concentration in the cane. The first method used to prevent flowering was to install large lights in the field to lengthen the day. This was effective but used too much electricity. The next method was based on the idea that the length of the night, not the length of the day, determines the photoperiod response. This second method consisted of turning the lights on for a period in the middle of the night to effectively shorten the night. This also was effective and used less electricity. Subsequently, effective results were obtained using sprays of an herbicide that destroys the apical meristem and prevents flowering and its detrimental effects. The most effective solution is to breed sugar cane cultivars that do not flower in the particular commercial production region.

Floriculturists, plant breeders, and scientists often wish to achieve year-round production of flowers by annual plants growing in greenhouses. For day-neutral plants, it is simply necessary to maintain optimal temperatures in the greenhouses. For photoperiod-sensitive plants, special procedures must be used. I will provide an example of procedures that can be effective with an SD annual plant. First, it is necessary to achieve a vegetative period of adequate duration. If the day lengths are longer than the critical value, this will happen naturally. If the day lengths are shorter than the critical value, the plants may begin flowering too soon when they are very small. In this case, supplemental lighting must be used to either lengthen the day or, more practically, to shorten the night with night-break lighting. This consists of providing 2–5 hours of low intensity light (5–50 watts [input electricity] per m^2 of bench). It also is necessary to induce flowering at appropriate times. If the day length is shorter than the critical value, this will occur naturally. If the day length is longer than the critical value, it will be necessary to shorten the day, which can be done by enclosing the plants with opaque curtains for a few hours or minutes prior to sunset.

Phenological responses such as flowering can be complex, depending on both photoperiod and temperature, and they vary among cultivars as well as species. When working with a crop species, it is important to learn the specific developmental responses to photoperiod and temperature of the major cultivars within the species and of the cultivars and accessions you are working with.

6.4 LIGHT QUALITY EFFECTS ON PLANT DEVELOPMENT

Photoperiod-sensitive plants detect and respond to variations in night length through a specific system involving the pigment phytochrome and circadian regulation. Several phytochromes are present in plants that enable them to detect changes in the quality of light and have different functions. The aspect of light quality that phytochromes respond to is the red (660 ± 5 nanometers)/far-red (730 ± 5 nanometers) quantum flux ratio. Some examples of developmental responses to variation in light quality obtained from the review of Smith (1995) are presented.

Forest soils commonly contain large amounts of dormant seed of pioneer species, with much of the seed present at the soil surface. When a gap of sufficiently large size appears in the canopy, many seeds germinate and produce seedlings, some of which are successful and colonize the gap. The dormancy of the seed was due to the low R/FR ratio of the light beneath the canopy of the forest. Sunlight has an R/FR of about 1.2, but leaves absorb much red light and transmit much of the far-red light so that the R/FR beneath the canopy is less than 1.2. When a gap occurs, the R/FR ratio of the light on the soil surface increases, approaching 1.2, and the dormancy of the seeds is broken.

The ability of plants to detect and respond to changes in R/FR also enables them to detect and respond to shading by leaves and potential competing plants. For example, the recumbent weed and pot herb purslane appears to be able to detect FR that is reflected from neighboring vegetation and respond by growing away from it. Similarly, cucumber seedlings, which are day-neutral with respect to flowering responses, exhibit this negative phototropism to neighboring vegetation and actively project new leaves into light gaps present in patchy canopy environments. The overall effect involves both proximity detection and shade-avoidance reactions through changes in development. It should be noted that competition in the aerial environment can be very strong in that, when one leaf shades another, the *PFD* falling on the shaded leaf can be reduced as much as 90% (Chapter 7), and the effect can be amplified. The plant receiving more *PFD* has greater P_n and produces more carbohydrate and more leaves, thus shading the other plant to an even greater extent.

Plant responses to light quality can explain some of the effects of crop management practices on plants. Soybean and many other species grown on narrower row spacing at higher densities tend to be taller, with longer internodes and fewer branches. Kasperbauer (1987) hypothesized that this effect could be explained by the fact that stem regions of plants grown at higher densities are subjected to lower R/FR ratios than for plants grown on wider rows. He also observed that plants sown on north-south rows had slightly longer internodes, initiated fewer branches, and received light with slightly lower R/FR than plants sown on rows oriented east-west. With controlled environments, he demonstrated that plants subjected to low R/FR light at the end of the day had longer internodes and more biomass in stems and petioles, but less biomass in roots than plants subjected to high R/FR light at the end of the day. He hypothesized that since the R/FR ratio influences development, it also should influence partitioning of photosynthate, with high R/FR benefiting root crops and low R/FR benefiting crops that produce shoot products. This hypothesis was tested by comparing crops grown under field conditions with either a black plastic mulch or a red plastic mulch that reflects more R than the black mulch but much more FR, resulting in low R/FR of the light reflected from the red mulch. Fruit yields of tomato (Kasperbauer and Hunt, 1998) and strawberry (Kasperbauer, 2000) were greater with the red plastic than the black plastic mulch. (Refer to Chapter 7 for a discussion of the merits of the clear plastic mulches commonly used in California compared with the black plastic mulches widely used elsewhere.)

It has been hypothesized that self-organization can occur in plant populations (López Pereira et al., 2017). The authors reported that sunflower plants cultivated in high-density stands perceive light signals from their immediate neighbors, adopt alternate positions of their single stems along the crop row, and collectively increase production per unit land area. This resulted from neighboring sunflower plants in a row avoiding each other by growing toward a more favorable light environment as determined by their detection of R/FR ratios.

Artifacts can occur in controlled-environment studies due to the use of fluorescent lamp systems that have a much higher R/FR ratio than sunlight. Floral bud development of cowpea was inhibited by high night temperatures and long days in field conditions with sunlight having R/FR of 1.2, and in growth chambers with artificial light having R/FR of 1.3–1.6, but not in growth chambers with only fluorescent lamps that had a R/FR of 1.9 (Ahmed et al., 1993b). Many growth chambers used in the past mainly had fluorescent lamps and a very high R/FR and thus the potential to generate artifactual plant responses. Note that, in field conditions, the R/FR deep in the canopy is less than 1.2, due to the enhanced transmission of FR than R by leaves. Consequently, the extent of heat-induced suppression of floral buds may be influenced by factors that change the R/FR experienced by the buds, such as the row width and plant density.

Several aspects of plant morphology can vary among plants growing in different growth chambers that have different lighting systems. These effects are mainly caused by differences in R/FR ratio, and it usually is advisable to use lighting systems in growth chambers that provide a R/FR ratio of 1.2–1.6. When fluorescent lamps were used to provide most of the *PFD*, a sufficient number and wattage of tungsten lamps also should have been used to increase the FR component of the light to achieve the desired R/FR ratio. Note that the R/FR ratio of the lighting system will tend to change with time, primarily due to differential aging and changes in lamp output of the different types of lamps. Artifactual plant responses that have resulted from growing plants in controlled environments often have been caused by differences in light quality, level of *PFD*, or rooting volume compared with plants growing in field environments. New LED lighting systems are now available for growing plants, and their effects on plant morphology and development should be evaluated.

ADDITIONAL READING

Ahmed, F. E., R. G. Mutters, and A. E. Hall. 1993b. Interactive effects of high temperature and light quality on floral bud development in cowpea. *Austral. J. Plant Physiol.* 20: 661–667.

Baldocchi, D. and S. Wong. 2008. Accumulated winter chill is decreasing in the fruit growing regions of California. *Clim. Change* 87 (Suppl.): 153–166.

Bunting, A. H. 1975. Time, phenology and the yields of crops. *Weather* 30: 312–325.

Kasperbauer, M. J. 1987. Far-red light reflection from green leaves and effects on phytochrome-mediated assimilate partitioning under field conditions. *Plant Physiol.* 85: 350–354.

Kasperbauer, M. J. 2000. Strawberry yield over red versus black plastic mulch. *Crop Sci.* 40: 171–174.

Luedeling, E., E. H. Girvetz, M. A. Semenov, and P. H. Brown. 2011. Climate change affects winter chill for temperate fruit and nut trees. *PLos ONE* 6(5): e20155.

Smith, H. 1995. Physiological and ecological function within the phytochrome family. *Ann. Rev. Plant Physiol. Plant Mol. Biol.* 46: 289–315.

7 Radiation and Energy Balances and Predicting Crop Water Use and Temperature

7.1 SOLAR RADIATION AT THE SURFACE OF THE EARTH

In quantifying the amount of solar radiation incident on crops, two systems of measurement are useful. The use of photon flux density for wavelengths between 400 and 700 nanometers (*PFD* in units of mol photon area^{-1} time^{-1}) for predicting photosynthesis, biomass production, and grain yield was described in Chapter 4. Solar irradiance for wavelengths between 300 and 3,000 nanometers (R_s in units of energy area^{-1} time^{-1}) also can be used to predict productivity, and is particularly useful for predicting the short-wave radiant energy load, crop water use, and temperature. An approximate reference level for solar irradiance is the full sun value for clear skies with the sun overhead at sea level where R_s = 1,000 joule m^{-2} s^{-1}. Note that sunlight provides about 2 micromoles of photosynthetically active photons (*PFD*) per joule of irradiance (but this value does vary with changes in spectral quality of the light), and the full sun value for *PFD* is 2,000 µmol photon m^{-2} s^{-1}. On clear, long, summer days in the San Joaquin and Sacramento valleys of California, R_s is about 25 M joule m^{-2} day^{-1}. Values of R_s for many locations in California may be found at the www.ipm.ucdavis.edu website. On clear days in subtropical (Figure 7.1) and temperate climatic zones, daily R_s is much greater in the early summer than in the late fall and early winter. Factors responsible for the seasonal variation in daily R_s are discussed in this chapter. Note that seasonal variations in daily R_s are responsible for the seasonal variations in temperature at the Earth's surface that were discussed in Chapter 5.

There are two types of solar irradiance: direct radiation (R_{direct}) and diffuse radiation ($R_{diffuse}$). The latter results from the scattering of light by molecules and particles in the atmosphere and is responsible for the blue appearance of the sky.

$$R_s = R_{direct} + R_{diffuse} \tag{7.1}$$

On a clear day, as little as 10% of R_s is diffuse, whereas as much as 100% of R_s may be diffuse on a cloudy day.

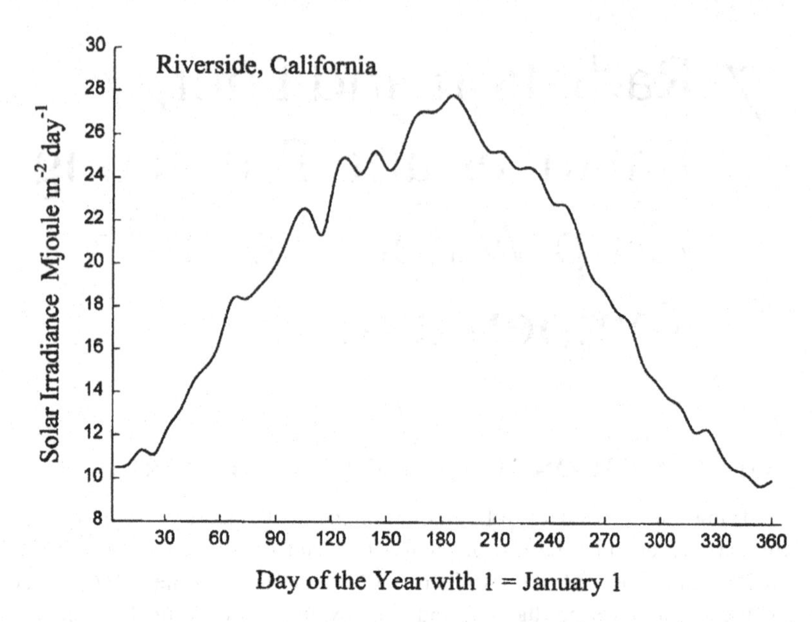

FIGURE 7.1 Seasonal variation in mean daily solar irradiance at Riverside, California, U.S.A. (location 33°58′N, 117°21′W, elevation 301 meters) for the period from 1935 to 1964.

The direct radiation on any surface depends on the orientation of the surface to the radiant beam. The amount of R_{direct} on a horizontal surface on the Earth depends on the zenith angle of the sun (where θ is the deviation of the direct beams of solar radiation from the perpendicular) and the cosine law.

$$R_{direct} = R_i \times \cos\theta \tag{7.2}$$

where R_i is the irradiance of the direct radiation when the sun is directly overhead of the sensor and the direct beams are perpendicular to the sensor. Note that R_{direct} approaches zero as θ approaches 90 degrees and equals R_i when θ is 0 degree (and cos of 0° = 1.0). A model is available to estimate the value of θ for different latitudes, dates, and times of day (Campbell, 1977). I will provide some benchmarks to illustrate how, at solar noon, θ varies with latitude and season for the northern hemisphere. Additional information may be found in Loomis and Connor (1992).

The *winter solstice* (December 22) is when the sun is overhead at the Tropic of Capricorn (latitude 23.5° S). In the northern hemisphere, θ = 23.5 + the latitude of the location. Since the sun is now at its lowest elevation in the sky (θ has its largest value), and this is the date of the shortest day in the year, R_s per day has its smallest clear-skies value on this day.

The *spring equinox* (March 21) is when the sun is overhead at the equator, and θ equals the latitude, and the time between sunset and sunrise is 12 hours for all places on Earth except for extreme polar locations.

The *summer solstice* (June 22) is when the sun is overhead at the Tropic of Cancer (latitude 23.5° N). In the northern hemisphere, θ equals latitude of the location–23.5

degrees. Since the sun is at its highest elevation in the sky, and this is the date of the longest day in the year, R_s per day has its highest clear-skies value on this day.

The *fall equinox* (September 24) is when the sun is again overhead at the equator, and θ equals the latitude and the time between sunset and sunrise is 12 hours for all places on Earth except for extreme polar locations.

The changes in positions of the sun (θ) and the length of the day largely determine the magnitudes of R_{direct}/day and therefore clear-skies values of R_s/day. The changes in R_s/day cause the progressions of the seasons from winter through summer and back to winter through their influences on the degree of warming and then cooling of the Earth's surface. In the United States, the magnitude of R_s/day is greater in summer than winter and greatest at intermediate latitudes of about 30°N, because the sun is overhead at about 23.5°N latitude and summer day length increases with latitude. Also, R_s/day is greater when there are no clouds, which often occurs in arid zones (defined in Chapter 10) and thus in some southwestern locations in the United States.

Solar irradiance at the Earth's surface is determined by several factors: θ, day length, cloudiness, turbidity, altitude (there is less depletion of solar radiation by the atmosphere at higher altitudes), and the slope of the land surface. The slope of the land determines the amount of R_{direct} intercepted per unit area of land following the cosine law (Equation 7.2). In the northern hemisphere, south-facing slopes intercept more R_s per unit area except on cloudy days, when R_s is mainly $R_{diffuse}$. Consequently, soils on south-facing slopes tend to warm up faster in the spring than soils on north-facing slopes. Farmers can take advantage of this effect by choosing fields with different slopes for different purposes. For example, a deciduous orchard on a north-facing slope could exhibit bud break later in the spring, because the soil in the root zone warms up slower, than an orchard on a south-facing slope. Later bud break can be advantageous because developing buds are very sensitive to cold weather. Later bud break enables sensitive developmental stages to escape damaging frosts that can occur in the early spring in subtropical and temperate zones.

Land can be made to slope at least on a microscale. In winter in the Coachella Valley of California, soil beds are sometimes made with surfaces that slope to the south, which is toward the sun. This practice is used when producing an early vegetable crop that benefits from warmer soil, or a prostrate but not an erect crop. Note that the amount of solar radiation intercepted by an erect crop is not influenced by microscale variation in the slope of beds it is planted on, because plants grow vertically irrespective of the slope of the land on which they are planted.

Large-scale variation in slope can influence the duration of daytime hours if it influences the time at which the sun rises and sets, such as can occur in the bottom of canyons. When estimating these effects using geometric drawings, it is important to draw the direct beams of solar radiation as being parallel.

The quality of solar radiation at the Earth's surface can vary with latitude and elevation, especially with respect to variation in amounts of ultraviolet (UV) radiation. Ozone in the stratosphere absorbs UV and prevents all of the UV-C (wavelengths of 250–280 nanometers) and most of the UV-B (wavelengths of 280–320 nanometers) from reaching the Earth's surface (Lambers et al., 1998). Most of the UV-A (wavelengths of 320–400 nanometers) in extraterrestrial solar radiation reaches the Earth's surface, but it is less damaging to biological systems because it has less energy per photon than UV-B, which in turn has less energy per photon than UV-C. Germicidal lamps that are used to kill bacteria produce UV-C radiation with short wavelengths (253 nanometers) and much energy per photon. Due to differences in the path length of the stratosphere and atmosphere through which the radiation passes, UV at the Earth's surface is greatest at high altitudes and low latitudes (e.g., the Andes Mountains). The effect of altitude on UV levels at the Earth's surface is much less than that of latitude, because most of the ozone in the stratosphere is above 15 kilometers. The latitude effect is due to the fact that when the zenith angle of the sun is larger (the sun is lower in the sky), the beams of direct radiation pass through a greater mass of both stratosphere and atmosphere (the path length is longer) before reaching the surface of the Earth.

Atmospheric pollutants such as chlorofluorocarbons (the Freon used in many refrigeration systems) and methyl bromide have been shown to deplete the stratospheric ozone layer, thereby causing the levels of UV at the Earth's surface to increase. Use of Freon has decreased in the United States. Also, in 1993, the U.S. Environmental Protection Agency enacted a phase-out of the use of methyl bromide, involving a 25% decrease in production in 1999 and a total ban by 2005, with exceptions to be granted for critical uses. Methyl bromide was extensively used in the 1990s as a fumigant for soils to control pests and diseases, in grain silos to control weevils, and in buildings to control termites. The ban was based on the assumption that much of the methyl bromide in the stratosphere was coming from this fumigation. It has now been shown that plants are capable of producing methyl bromide and may release large quantities to the atmosphere. Consequently, the ban on commercial production and use of methyl bromide may not be very effective in controlling the amount of methyl bromide reaching the stratosphere. This would be unfortunate because the depletion of the stratospheric ozone layer is a major environmental problem in that higher levels of UV can be very damaging to many organisms.

The most destructive effect of UV radiation on plants involves damage to DNA that results in mutations. Physiological effects also can occur, such as reductions in biomass production and non-stomatal aspects of photosynthesis (Teramura et al., 1991). The damage to DNA and photosynthesis caused by UV radiation can be partially repaired by plants during both the day and the night. Studies have been conducted with a specific cultivar of rice that is more UV-sensitive in that it exhibits greater growth inhibition and leaf browning than another cultivar of rice when subjected to UV-B radiation. The UV-sensitive cultivar was deficient in both photo-repair during the day and excision repair of DNA during the night compared with the UV-tolerant cultivar (Hidema et al., 1997). Morphology of plants also may be influenced by UV. Orange trees growing at very high altitudes in the Andes Mountains near Quito in Ecuador have a more branched appearance than the same scion growing at low elevations in a Mediterranean climatic zone at Riverside, California. The effect was eliminated by installing UV-absorbing screens above the orange trees, indicating the excessive branching may be due to an effect of the high UV in the Andes on the meristems of the branches. Protection against high UV may occur in plants by epidermal cells or leaf hairs that contain phenolic compounds (i.e., flavonoids and flavones). The adaptive significance of these compounds may be the selective absorption of UV, thereby preventing its penetration into tissue, but the phenolic compounds also may have other roles, such as in attracting pollinating bees.

7.2 TYPES OF RADIATION IN THE EARTH'S ENVIRONMENT AND OPTICAL QUALITIES OF PLANTS

In addition to solar radiation, which is defined as short-wave radiation, long-wave radiation (far infrared) with wavelengths of 3,000–100,000 nanometers is present in the Earth's environment. All mass emits radiation, and mass on the surface of the Earth and in the atmosphere is sufficiently cool that it emits radiation that has long wavelengths. The amount of radiation emitted is described by the Stefan–Boltzmann law, as shown in Equation 7.3:

$$R = \varepsilon \times \sigma \times T_K^4 \tag{7.3}$$

where R is the emitted radiation per unit area per unit time; ε is the emissivity, with values from 0 to 1.0, and is a property of the material and its temperature, with values of 0.90–0.98, for many plant parts and soils; σ is a constant that equals 5.67×10^{-8} joule m^{-2} s^{-1} $°K^{-4}$; and T_k is temperature in degrees Kelvin = °C + 273. Plants emit large quantities of long-wave radiation and energy. For example, a plant at a room temperature of 27°C with an emissivity of 0.98 emits 450 joule m^{-2} s^{-1}, which is large, since R_s of full sun is 1,000 joule m^{-2} s^{-1}. However, in most environments, plants do not quickly cool down because they also are receiving similar large amounts of long-wave radiation from adjacent plants, other masses on the surface of the Earth, and the clouds above them.

At night with low humidity, plants that are exposed to clear skies which emit little radiation (because there is little mass in the skies and the mass is cool) can cool down rapidly, because their net loss of long-wave radiation (and energy) to the atmosphere is large. This phenomenon is the physical basis for the occurrence of radiation frosts. A pane of glass placed over plants will reduce the net loss of radiation to the atmosphere during the night, because the glass absorbs long-wave radiation (such as that emitted by the plants), stays relatively warm, and emits substantial quantities of long-wave radiation back to the plants.

Instruments are available, called infrared thermometers, that can remotely sense the surface temperatures of plants, soils, or any other type of mass. When infrared thermometers are pointed at a surface, they measure the rate at which long-wave radiation is being emitted by the surface. The instruments contain a microprocessor that permits them to calculate the surface temperature using Equation 7.3 and a value for ε that is put in by the operator. Information on the surface temperature of crop plants can be used to determine whether a crop needs to be irrigated. With drought, stomata of plants partially close, causing a reduction in transpiration and evaporative cooling that results in an increase in leaf temperatures as compared with recently watered plants. Additional discussion of this method for determining whether crops need to be irrigated is presented in Chapter 10.

The reactions of materials to radiation are described by Equation 7.4:

$$\text{Reflectivity } (r) + \text{Absorptivity } (a) + \text{Transmissivity } (t) = 1.0 \quad (7.4)$$

where r, a, and t have values between 0 and 1.0, which depend on the wavelength of the radiation. Corn leaves absorb substantial quantities of photosynthetically active radiation ($a = 0.8$), reflect mainly green light ($r = 0.1$), and transmit very little white light ($t = 0.1$), which explains their strong ability to shade other leaves. In contrast, for near-infrared radiation (wavelengths of 700–3,000 nanometers), corn leaves absorb very little ($a = 0.05$), and reflect ($r = 0.4$) and transmit ($t = 0.55$) large quantities, thereby contributing to the prevention of overheating.

Stresses due to drought, salinity, diseases, and pests can cause a change in the reflectance of near-infrared radiation by leaves. The reflected near-infrared is light that is scattered by refractive index discontinuities with major effects at the interface of hydrated cell walls with intercellular air spaces. Various stresses can influence leaf morphology in a long-term manner; consequently, they also can influence near-infrared reflectance in a long-term day-to-day manner. In some cases, stressed plants can be detected by evaluating the reflection of near-infrared radiation from foliage using infrared photography. In the past film that responds to radiation having a wavelength of 900 nanometers was used; now sensors are available for digital cameras that operate in the near-infrared range. Prints from infrared film showed well-watered cowpea plants as being a light pink and severely droughted cowpea plants as being a dark red. Near-infrared photography was used to determine whether a center-pivot irrigation system was applying water in a uniform manner to alfalfa. This can be an important problem in that the nozzles on center-pivot systems must deliver different quantities of water, with the inner nozzle delivering the least, if they are to deliver the water uniformly to the plants. The infrared photographs showed concentric circles of pink (well-watered) and dark red (droughted) plants. The manufacturer had to change the nozzles on the center-pivot system to achieve a more uniform application of water. The method may not be sufficiently sensitive to detect whether an irrigation is needed in a manner that does not permit any drought-induced reduction in yield.

An attempt was made to use aerial near-infrared photography to detect genotypic differences in dehydration avoidance by sorghum lines growing in a field breeding nursery (Blum et al., 1978). I have the impression that the method may not have been sufficiently sensitive and consistent to be very effective in screening for genotypic differences which might be useful in breeding. However, aerial near-infrared photography has been useful for following the spread of diseases or pests over large areas of a crop in those cases where the disease or pest caused a distinct near-infrared signature.

Optical instruments have been used that measure reflectance at different wavelengths (spectral reflectance scanners). Empirical indexes have been developed for remotely quantifying the degree of stress experienced by crop canopies based on information obtained with scanners in aircraft or satellites. A widely used index is described by Equation 7.5:

$$NDVI = \frac{(R_{nir} - R_{red})}{(R_{nir} + R_{red})} \quad (7.5)$$

where *NDVI* is the normalized difference vegetation index, R_{nir} is the reflected near-infrared radiation at 900 nanometers wavelength, and R_{red} is the reflected red radiation at 680 nanometers wavelength. Stresses such as drought and salinity cause plant canopies to exhibit a decrease in R_{nir} and an increase in R_{red}, which causes substantial changes in *NDVI* from about 0.9 for healthy plants to 0.4 for plants subjected to extreme stress (Araus, 1996). The method does not appear sufficiently sensitive for use in managing the timing of irrigation or screening genotypes for differences in stress adaptation.

An advantage of remote sensing based on satellite information is that it can obtain information from large areas quickly. For example, one project used *NDVI* values from information provided by satellites to evaluate the state of crops in the Sahelian zone of Africa. The objective of this project was to predict the occurrence of low crop yields sufficiently early for food to be brought in to reduce the impacts of potential famines. A disadvantage of these types of remote sensing is that the near-infrared and red reflectance signatures of vegetation can be influenced by many stresses and other attributes of the canopies, such as the extent of ground coverage. Consequently, for most purposes, it is necessary to calibrate these indexes by comparing the reflectance signatures with the different conditions of the crop. An example of this calibration is provided by Aparicio et al. (2000), who evaluated reflectance signatures and crop characteristics of different durum wheat cultivars under well-watered and water-limited field conditions. They obtained some moderate correlations between *NDVI* and leaf area index (projected leaf area/ground surface area) and grain yield under water-limited field conditions for variation due to genotypic differences. The simple ratio R_{nir}/R_{red} was as effective as *NDVI*.

The reaction of plant materials to far-infrared radiation (long wavelengths of 3,000–100,000 nanometers) is substantially different from that to near-infrared radiation (short wavelengths of 700–3,000 nanometers). Plants absorb most of the far-infrared radiation ($a = 0.98$), since $a = \varepsilon$ according to Kirchhoff's law.

Leaves of some of the plants that are native to deserts appear white. What are their spectral properties? Generally, these leaves have higher reflectance of visible light (e.g., for wavelengths of 400–700 nanometers, $r = 0.3$ compared with 0.1 for green leaves), and they absorb less photosynthetically active radiation than green leaves. Consequently, for plants under optimal soil conditions, white leaves may be less likely to suffer from photoinhibition in intense sunlight and can be a little cooler but also less effective in photosynthesis than green leaves. The white leaf trait has both benefits and costs with respect to its influence on plant adaptation.

7.3 RADIATION AND ENERGY BALANCES

The radiant energy loading (daytime) or loss rate (nighttime) from crop canopies or soils is determined by the net radiation (R_n), which is the difference between the incoming and the outgoing radiation fluxes described by Equation 7.6:

$$R_n = (R_s + R_a) - (R_r + R_g) \qquad (7.6)$$

where the short-wave components are the solar irradiance (R_s) and the reflectance of solar radiation ($R_r = r \times R_s$, with r having a value of 0.20–0.25 for green crop canopies), and the long-wave components are the incoming radiation emitted by clouds and molecules in the atmosphere (R_a) and the outgoing radiation emitted by plants and soil (R_g). The long-wave components depend on the temperatures of the clouds (and atmosphere), plants, and soil.

The *greenhouse effect* of the Earth's atmosphere can be explained by considering radiation exchanges. The Earth's atmosphere transmits much of the solar radiation (R_s), causing the surface of the Earth to warm up. Some of the long-wave radiation emitted by the Earth's surface (R_g) is absorbed by the atmosphere and clouds, causing them to warm up. In turn, the atmosphere emits long-wave radiation in all directions, with some of it going back to the Earth's surface (R_a). This partial trapping of thermal radiation by the atmosphere increases the average temperature of the Earth's surface by about 33°C compared with what would occur in the absence of an atmosphere. The greenhouse effect is smaller with clear skies having low humidity because R_a is small. (This can be appreciated by looking up at the sky on nights with clear skies and low humidity. Your face will feel very cold, because it is emitting much long-wave radiation and receiving very little long-wave radiation from the sky.) The R_n is very negative in these conditions, and the cooling of the Earth's surface is fast. If surface temperatures become very low, a radiation frost can occur in the late night.

Elevated atmospheric [CO_2] and other greenhouse gases such as methane can enhance the greenhouse effect, because they make the atmosphere more effective in absorbing long-wave radiation. Future increases in atmospheric [CO_2] and other greenhouse gases have been predicted to cause global warming due to an enhanced greenhouse effect of the Earth's atmosphere (Schneider, 1989). Global average temperatures are predicted to increase by about 2°C–5°C by the end of this century and to have substantial negative effects on certain crop yields in some climatic zones (Singh et al., 2011).

Black plastic mulch is widely used for strawberry production in the southeastern United States, whereas clear plastic mulch mainly is used in California. Which of these mulches is most effective in warming the soil? Black plastic absorbs most of the daily R_s and becomes hot, but much of this heat is not transferred to the soil due to an air layer between the plastic and the soil surface that has low thermal conductivity. During the night, the black plastic cools considerably due to emission of long-wave radiation to space. In contrast, clear plastic transmits much of the R_s to the soil surface, which absorbs it, causing the soil to become warmer. During the night, the soil surface emits long-wave radiation, but it is absorbed by the plastic, which in turn emits long-wave radiation, some of which goes back to the soil surface. Consequently, soils warm up faster under clear plastic than under black plastic mulch. However, the transmission of R_s by the clear plastic favors the establishment of weeds. When clear plastic mulch is used, weed seeds and other propagules must be killed by procedures such as fumigation. Methyl bromide has been used as a fumigant, but replacements are being sought because, as was discussed earlier, it is one of the components responsible for the destruction of the stratospheric ozone layer. Using plastic mulches produces better-quality strawberries in that, without these mulches, the fruit can contact moist soil and develop pink rot. Black plastic can be less effective than clear plastic in enhancing fruit quality if it develops a hot enough surface that it burns the fruit.

Nevertheless, care must be taken in the extensive use of plastic mulches in agriculture because through general wear and tear the plastic sheets are broken down into small pieces. Contamination of marine and terrestrial ecosystems by microplastics (100–5 millimeters) is putting individual organisms at risk, including humans (Ng, 2017). A 2015 survey involving samples taken from 10 vegetable crops grown with plastic mulch in suburban greenhouses around Nanjing, China, found that phthalates were present in all of the crops (Ng, 2017). Levels of these toxic compounds were dangerously high in some samples. Pollution by small plastic particles is particularly worrisome. These particles are likely to be entering the human food supply. For example, microplastic particles have been found in salt. Birds and organisms in the ocean and in the soil are being harmed by nanoplastic (<100 nanometers) and microplastic particles.

The radiation balance is an important part of the overall energy balance shown in Equation 7.7. This equation is useful for predicting crop water use and temperatures. The energy balance equation is based on conservation of energy: that the steady-state fluxes of energy into a surface, in this case the crop and soil surface, must equal the steady-state fluxes of energy at and out of the surface.

$$R_n - LE - H - G - P_n = 0 \tag{7.7}$$

where P_n is the energy flow due to net photosynthesis, LE is the latent energy flux associated with the evaporation of water from soil and plant transpiration ($LE = L \times E$ where L is the latent heat of vaporization, which is 2,442 joule/g at 25°C, and E is the total evaporation rate, which includes both soil evaporation and plant transpiration), H is the heat flux between the canopy or soil surface and air (with the direction depending on the temperature gradient), and G is the heat flux between the canopy or soil surface and the lower regions of the soil (also with the direction depending on the temperature gradient).

Equation 7.7 can be simplified by considering the relative magnitudes of the various energy fluxes. P_n can be ignored because it is small, being less than 1.5% of R_s and less than 2% of R_n. G can be substantial during the day but small on a 24-hour basis, and significant on a monthly basis but negligible on an annual basis. The other terms, H, LE, and R_n, always are substantial except for non-vegetated deserts, where LE can be very small due to the lack of water for either transpiration or soil evaporation.

Three contrasting ecosystems will be considered to illustrate how the magnitudes of LE, H, and G might vary in relation to R_n. Consider the energy balance of a well-watered ecosystem with relatively uniform vegetation that is completely covering the ground over a very large area such that there are virtually no horizontal transfers of heat (Figure 7.2). During the day, the solar radiation energy input into the crop/soil surface provides a radiant energy load that mainly is dissipated by the evaporation of transpired water (LE is about 60% of R_n). For example, on a clear day with R_n of about 20 M joule m^{-2} day^{-1}, a transpiration rate of 0.5 cm/day could occur and result in an LE of 2,442 joule/g × 0.5 cm/day × 1 g/cm^3 (density of water) × 10^4 cm^2/m^2 × 10^{-6} joule/M joule = 12.2 M joule m^{-2} day^{-1}. About 28% of R_n could be dissipated in

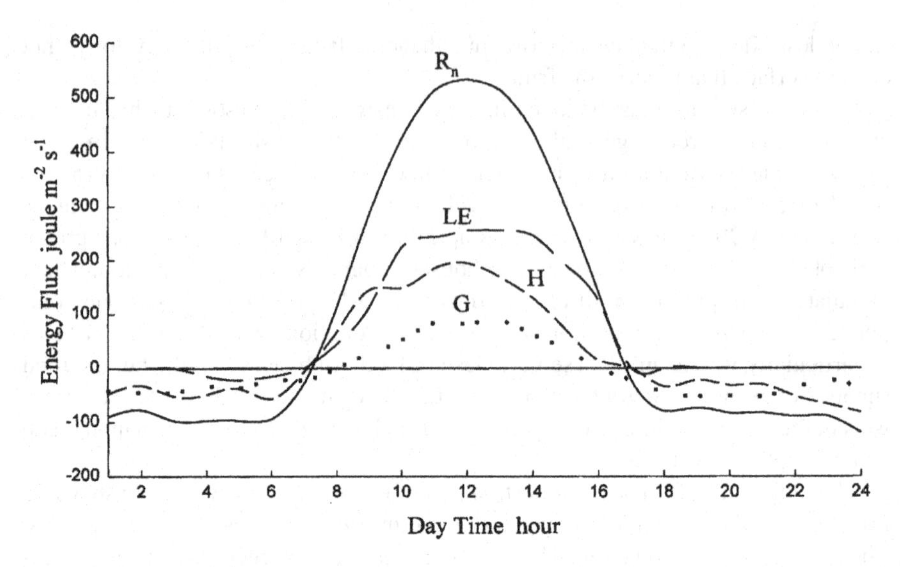

FIGURE 7.2 Energy balance components (Equation 7.7) of a well-watered herbaceous crop on a sunny day, surrounded by a large expanse of well-watered vegetation with virtually no horizontal transfers of heat. R_n is net radiation, LE is the latent energy flux associated with the evaporation of water, H is the heat flux between canopy and air, and G is the heat flux between the canopy and soil surface and the lower regions of the soil.

heating the air (H), while 12% is dissipated by heating the soil under the crop canopy (G). During the night, R_n is negative because there are no shortwave fluxes, and R_g is greater than R_a. LE is very small because the stomata are closed, and LE may become slightly negative just before dawn if the air temperature becomes sufficiently cold to cause dew formation. Consequently, the negative R_n during the night results in a cooling of the air and the soil beneath the canopy. H and G now flow toward the crop surface (i.e., in opposite directions from the ones occurring during the day), and most of the heat that flowed into the soil during the day flows out during the night.

For the next two examples, consider the energy balance of a well-irrigated, cropped area (Figure 7.3) surrounded by a large expanse of dry desert (Figure 7.4). In this case, horizontal transfers of heat are occurring between these ecosystems. This is called *advection* and must be considered when predicting crop water use in the oasis. During the day over the desert, LE will be very small if little water is available for evaporation from the soil or transpiration from the plants. Consequently, in the desert, most of the radiant energy load (R_n) is dissipated by a strong heating of the air (H) and some heating of the soil (G). As hot air from the desert blows across the irrigated area, it becomes cooler, which means that there are two types of energy loading on the irrigated oasis: radiant energy and a negative H. In these cases, crop transpiration rate can be very high, and LE can exceed R_n. Also note that the air blowing from the desert will be dry but pick up moisture as it flows over the cropped area, resulting in a horizontal transfer of water vapor. A practical consequence of hot dry air blowing in from deserts into large cropped areas is that not only will the average crop water use and LE be high relative to R_n, but there also will be gradients such

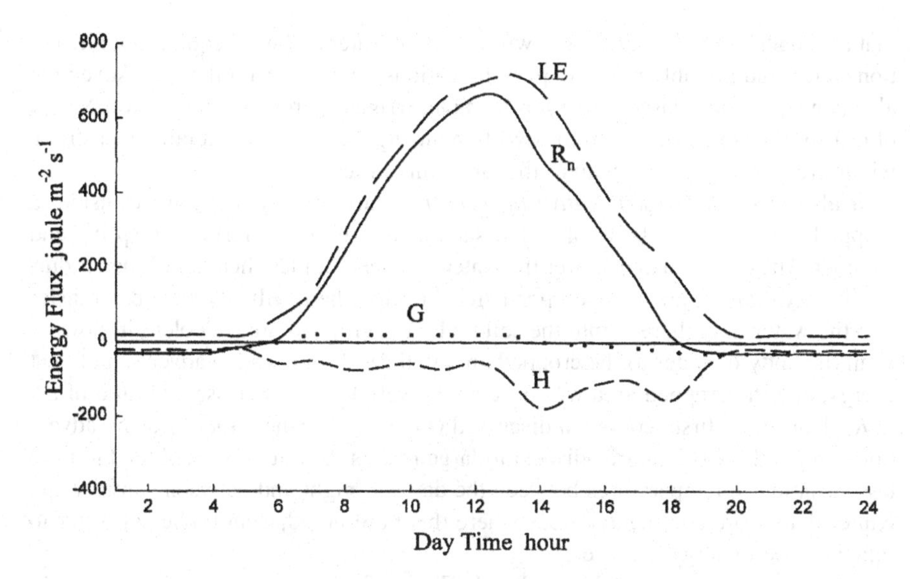

FIGURE 7.3 Energy balance components (Equation 7.7) of a well-irrigated herbaceous crop on a sunny day in an oasis that is surrounded by a large expanse of dry desert. R_n is net radiation, LE is the latent energy flux associated with the evaporation of water, H is the heat flux between canopy and air, and G is the heat flux between the canopy and soil surface and the lower regions of the soil.

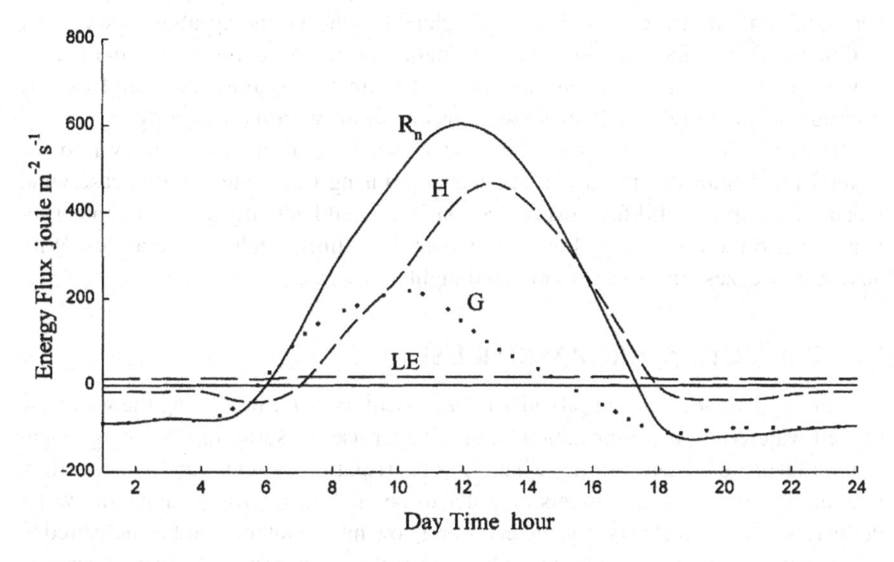

FIGURE 7.4 Energy balance components (Equation 7.7) of a dry desert on a sunny day, surrounded by a large expanse of dry desert. R_n is net radiation, LE is the latent energy flux associated with the evaporation of water, H is the heat flux between canopy and air, and G is the heat flux between the desert surface and the lower regions of the soil.

that the edges of the cropped areas will use more water and will require more irrigation water, and possibly more frequent irrigations, than the central areas. Advection also can occur on a large scale; for example, irrigated crops in the Central Valley of California can be subjected to advection during the summer and fall when strong winds are blowing from dry areas that are many miles away.

It also is possible for *negative advection* of heat energy to occur. Consider a cropped area next to a body of water, such as a large lake, during the spring and summer. The energy balance over the water will result in less heating of the air during the day than occurs in the cropped area, because there will be more heat transfer into the water than there is into the soil under the cropped area. Cooler air blowing from the body of water to the cropped area will result in negative advection of heat energy, and the cropped area will have a relatively lower water use and ratio of LE to R_n than in the first ecosystem discussed (Figure 7.2), where there was no advection. Cropped areas that are adjacent to large bodies of water also will tend to have less variation in temperatures between the day and night and between summer and winter than more continental areas where there are no adjacent bodies of water to influence their energy balances.

There are two types of cold weather. Radiation frosts can occur in winter at night when clear skies coupled with low humidity cause R_n to be very negative (Figure 7.4). Cooling of the Earth's surface causes heat to flow from the air to the canopy and soil surface (e.g., H is negative as in Figure 7.4). Throughout the night, air near the Earth's surface becomes progressively cooler, with greater effects if there is little wind and the air stays near the Earth's surface. Cold air is heavier than warm air and will tend to drain to lower elevations, causing some areas to have lower temperatures (frost pockets). By dawn, the air near the Earth's surface and crop canopy can be very cold, and an inversion will have developed such that the air above the canopy will be warmer. Wind machines that are operated when air temperatures in the canopy become low can provide an increment of protection against radiation frosts by mixing the warmer air from above with the cooler air within the canopy.

Advection freezes occur on a large scale when cold air masses blow into the lower United States from Alaska and Canada during the winter. In this case wind machines are not useful for frost protection, and would actually aggravate the problem because the air in and above the canopy has similar cold temperatures. With advection freezes, temperatures are cold night and day.

7.4 PREDICTING CROP WATER USE

The energy balance concept provides a theoretical basis for predicting the water use of well-watered crops. Information on crop water use is useful for managing irrigation on farms (e.g., determining what types of irrigation systems can be most effective and the timing and amounts of water to be applied to crops), managing water deliveries in water districts (e.g., determining how much water should be delivered to different areas in different seasons), planning regional water projects (e.g., determining the sizes of reservoirs and canals), and managing rain-fed farming (e.g., for dry areas, estimating the length of the growing season to permit choices of crop species and cultivars with optimal cycle lengths, and for wet areas, estimating the extent of deep drainage and leaching of nutrients and pesticides). Why should we try to predict crop water use? Why not measure it? Currently, measuring crop water use takes considerable time and effort and would have to be done at many different locations and seasons and for many different crop species and varieties. Also, predictions of crop water use can be very effective for irrigated crops that are intensively managed.

Consider predicting crop water use where supplies of water and nutrients are optimal, there are no significant pests and diseases, and temperatures are optimal (this is the same definition as was used in Chapter 4 for potential productivity), and define it as ET_m. Note that ET is plant transpiration + soil evaporation, and it has units of depth of water per unit time so that it can be compared with rainfall given in the same units in hydrologic budget analysis, as is shown in Chapter 10. Seasonal ET_m mainly is determined by the evaporative demand of the atmosphere (R_s, air temperature, advection, wind speed, and humidity) and crop factors such as percentage ground cover, stomatal characteristics, and length of growing season—but with only small effects of total leaf area and root or soil characteristics.

The following model has been used to predict seasonal crop water use where ET_0 is a measure of the evaporative demand and has been called either the *potential evapotranspiration* or the *reference crop evapotranspiration*. ET_0 is determined for a standard vegetative surface consisting of a short, well-managed grass completely covering the ground and actively growing. Surprisingly, extensive empirical research has shown that different grass species and cultivars have similar values of ET_0, providing they actively grow in the specific location and season, are completely covering the ground, and have a short canopy with a smooth surface.

$$\text{Seasonal crop water use} = \frac{\Sigma ET_m}{\text{day}} = \Sigma\left(\frac{K_{crop} ET_0}{\text{day}}\right) \tag{7.8}$$

where K_{crop} is a coefficient that varies with crop species and stages of growth. For annual crops, Equation 7.8 is summed on a daily basis from crop emergence to maturity.

Reference crop evapotranspiration (ET_0) has been predicted using different equations (Doorenbos and Pruitt, 1977). Radiation and temperature Equation 7.9 is simple and can be effective if properly calibrated:

$$ET_0 = \frac{b \times W\, R_s}{L} \tag{7.9}$$

where R_s is solar irradiance, W is a factor that depends on average daily air temperature and altitude (the latter adjusts for the effects on evaporation of variations in atmospheric pressure), L is the latent heat of vaporization, and b is a factor that must be obtained by calibrating the equation for different locations and seasons to account for effects on ET_0 of variations in advection, wind speed, and humidity. Note that temperature has a substantial effect on transpiration because the humidity inside the leaf increases exponentially with increases in temperature and increases in temperature also cause increases in stomatal opening over a large range of temperatures. In contrast drier air has limited effect on transpiration because it causes progressive stomatal closure. Refer to "Stomatal Responses to Environment" in Chapter 8 for additional discussion of this topic.

Calibrating equations for predicting ET_0, such as Equation 7.9, requires that measurements are made of the ET of a standard grass surface together with measurements of R_s and air temperature in a standard weather station shelter. Lysimeters can provide accurate and precise measurements of ET of either a standard grass surface or of crops. A weighing lysimeter consists of a tank that is sunk into the ground, contains the grass or the crop and the soil profile, and has a sensitive device to weigh the tank. Any change in weight of the tank is mainly due to the loss of water by ET from the system. Increase in weight of the tank due to crop growth is relatively small. For example, the fastest recorded crop growth rate (increase in CH_2O) is about 50 g m^{-2} day^{-1}. However, the loss of water due to ET would be much greater than this. For example an ET of about 5 mm day^{-1} is equivalent to a loss in weight of 5,000 g m^{-2} day^{-1}.

Instruments for measuring the concentrations of water vapor and carbon dioxide in the atmosphere, when combined with a sonic anemometer and appropriate computer hardware and software, make it possible to determine directly and accurately both the water vapor (ET) and CO_2 (P_n) fluxes between canopy and air with minimal disturbance to the system. In principle, all that is needed is to place the instruments in the air above the canopy and make the necessary calculations. The theory for this approach, the eddy covariance (or correlation) method (Rosenberg, 1974), has been available for many years, but instruments were not available to exploit the method—especially those needed for determining rapid changes in atmospheric [CO_2]; in recent years, however, these instruments have become available. For example, eddy covariance was used to measure the evapotranspiration and carbon exchange in an irrigated citrus orchard over three seasons (Maestrie-Valero et al., 2017).

An overall approach to predicting water use of intensively managed, well-watered crops has been described in detail by Doorenbos and Pruitt (1977). I will describe this approach for an annual crop. First, one should choose an appropriate crop species, cultivar, and sowing date for the specific agro-climatic zone. Next, one predicts the lengths of the initial, development, mid-season, and late-season growth stages of the crop (Figure 7.5) based on a knowledge of the overall cycle length of the cultivar and data in the literature such as in Doorenbos and Pruitt (1977). The value of K_{crop} for the initial stage is dependent on the frequency of soil wetting and soil properties, because there is little leaf area above the ground at this stage. Some values are presented in Doorenbos and Pruitt (1977). During the mid-season stage, most crops have developed 100% ground cover, and K_{crop} depends on stomatal characteristics and canopy roughness. An appropriate value for K_{crop} at the mid-season stage can be obtained from the literature (Doorenbos and Pruitt 1977) or, for scientists, by comparing the ET of the crop growing in a lysimeter with values of ET_0 predicted by a calibrated radiation and temperature equation (Equation 7.9). Note that most intensively managed annual crops have a K_{crop} of 1.0 ± 0.2 at mid-season. Some perennial crops have lower values, for example, K_{crop} is less than 0.8 for pineapple and citrus due to strong stomatal restrictions on water loss even for well-watered crops. The values of K_{crop} that can be used during the development stage are obtained by linear interpolation between the values

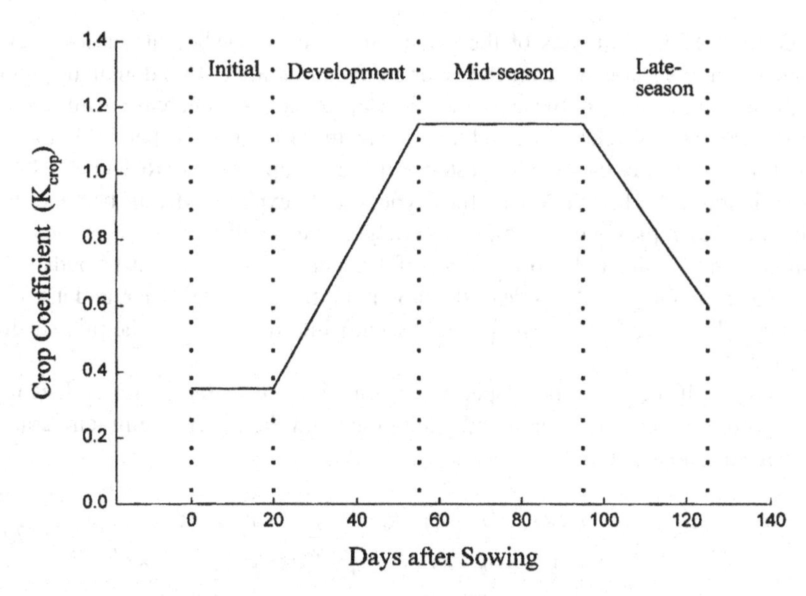

FIGURE 7.5 Approximate variation in crop coefficient (K_{crop}) during the growing season for a maize crop. Note that the value of K_{crop} during the initial stage depends on the frequency of soil wetting and soil properties. The lengths of development, mid-season, and late-season stages will depend on the cultivar and the environment as described in Chapter 6.

for the initial and mid-season stages (Figure 7.5). During the late-season stage of annual crops, K_{crop} decreases due to leaf senescence that causes stomatal closure and decreases in percentage ground cover. Note that irrigation is terminated during this period for many grain crops to promote senescence, which facilitates harvesting, and to promote grain drying. Values of K_{crop} for the late-season stage can be obtained from references such as Doorenbos and Pruitt (1977).

We now have values of K_{crop} for each day during the growing season. We next need to obtain data for ET_0 for each day, which can be done using Equation 7.9 on either a real-time (current weather) or an average weather (climate) basis. In rural areas of California, and many other irrigated areas of the world, radio stations announce ET_0 values, obtained from scientists, on a daily basis. Values of ET_0 for many locations in California may be obtained from the following website: www.ipm.ucdavis.edu. This information helps farmers to predict the water use of their crops and thereby manage irrigations, as will be discussed in Chapter 10. The last step is to sum values for daily $ET_0 \times K_{crop}$ to obtain seasonal crop water use using Equation 7.8.

7.5 PREDICTING TEMPERATURE DIFFERENCES BETWEEN CROP CANOPY AND AIR

Typically, only the air temperature is known, yet the canopy temperature determines plant function and can substantially differ from air temperature. For example, in one study, plants on a mountain at 2,000 meters elevation had the same leaf temperature (15°C) as plants of the same genotype in a valley at sea level, even though air temperatures were much cooler on the mountain (10°C) than in the valley (20°C) due to the effects of the lapse rate. The lapse rate is the decrease in air temperature with increase in elevation, and it varies from –4°C to –10°C per 1,000 meters, depending on whether the air is saturated with water vapor or dry (Rosenberg, 1974). This variation in the lapse rate with the dryness of air explains why air becomes hotter and drier as it passes over a mountain range. The initially moist air is forced up the mountain by wind and cools at a rate of 4°C per 1,000 meters until clouds form, and it may rain. On the other side of the mountain, the air is now drier, and it warms at a rate of 10°C per 1,000 meters as it descends into the valley in the rain shadow of the mountain.

Equation 7.10 has been developed from Equation 9.6 in Jones (1992). It is useful for predicting environmental and plant effects on the difference in temperature between leaf and air ($T_l - T_a$).

$$\left(T_l - T_a\right) = \frac{C_1 \times r_b' \times (r_b + r_l) \times R_{ni}}{C_2 \times [C_3 \times (r_b + r_l) + C_4 r_b']} - \frac{r_b' \times vpd}{C_5 \times (r_b + r_l) + C_6 \times r_b'} \tag{7.10}$$

where C_1 through C_6 are constants, r_b' is the boundary layer resistance to heat transfer, r_b and r_l are the resistances of the boundary layer and the leaf surface to water vapor transfer (refer to Chapter 8 for more discussion of these resistances), *vpd* is the vapor pressure deficit of the air ($e_{\text{sat at air temperature}} - e_a$), and R_{ni} is the isothermal net radiation as defined by Equation 7.11:

$$R_{ni} = \left(R_s + R_a\right) - \left(R_r + R_t + \varepsilon \times \sigma \times T_a^4\right) \tag{7.11}$$

where R_t is the transmitted solar radiation ($t \times R_s$), and T_a is temperature in Kelvins.

From Equation 7.10, it is apparent that when wind speeds are high and leaves are small, such that r_b' is very small (Equation 8.4), convective heat transfer (H) is very effective in minimizing differences in temperature such that ($T_l - T_a$) is small, day or night. For low wind speeds and large leaves, where r_b' is large, leaves can either be hotter or cooler than the air. For example, they can be much hotter (up to +10°C) if the radiant energy load (R_{ni}) is much larger than the evaporative cooling (LE), such as with sunny conditions, closed stomata (a large r_l), and cool humid air (a small *vpd*). Leaves can be cooler than air (as much as –8°C during the day) either during the day when evaporative cooling is greater (due to open stomata and a large humidity gradient between leaf and air) than the radiant energy load or during the night when net radiation is very negative due to clear skies accompanied by low humidity, resulting in little long-wave radiation coming from the atmosphere (R_a). Also note that canopies can be much cooler than air in environments subjected to substantial advection of hot air from surrounding dry areas.

ADDITIONAL READING

Aparicio, N., D. Villegas, J. Casadesus, J. L. Araus, and C. Royo. 2000. Spectral vegetation indices as nondestructive tools for determining durum wheat yield. *Agron. J.* 92: 83–91.

Araus, J. L. 1996. Integrative physiological criteria associated with yield potential. In M. P. Reynolds, S. Rajaram, and A. McNab (Eds.), *Increasing Yield Potential in Wheat: Breaking the Barriers*. CIMMYT, Mexico, D.F, pp. 150–164.

Campbell, G. S. 1977. *An Introduction to Environmental Biophysics*. Springer-Verlag, New York, p. 159.

Doorenbos, J. and W. O. Pruitt. 1977. *Crop water requirements*. FAO Irrigation and Drainage Paper No. 24 (revised), FAO, Rome, Italy, p. 144.

Jones, H. G. 1992. *Plants and Microclimate*, 2nd ed. Cambridge University Press, Cambridge, UK, p. 428.

Loomis, R. S. and D. J. Connor. 1992. *Crop Ecology*. Cambridge University Press, Cambridge, UK, p. 538.

8 Crop Transpiration and Water Relations

Plants require large quantities of water if they are to be productive. On a hot, sunny day, the roots of a well-watered C_3 or C_4 crop species whose canopy is completely covering the ground might take up 50–80 tons of water per hectare per day. What happens to the large amounts of water taken up by plant roots? This can be determined by evaluating the amount of water used in different processes per unit of carbohydrate produced by net photosynthesis (Table 8.1).

The amounts of water used in metabolism and storage are relatively small. Clearly, most of the water taken up by plant roots is transpired from leaves, but what is the role of transpiration? The large flow of water to and from leaves does not appear to be directly necessary for plant function in that a slow transpiration rate will suffice for the transport of nutrients and hormones from the roots to the shoot. Consider, for example, two plants with large differences in transpirational flow rate in the xylem but no other differences. If their roots deliver nutrients and hormones to the transpirational stream at the same rate, under steady-state conditions, these materials will arrive at the leaves at the same rate. The only differences will be that the plant with slower transpiration will take longer to achieve a steady state, and the nutrients and hormones in its xylem fluid will be more concentrated. Empirical evidence for the lack of an advantage of a fast transpirational stream is that plants usually grow more rapidly in humid environments, where transpiration is slower, than in dry environments with a high evaporative demand (Hall, 1982b). Evaporative cooling can be useful for some species in hot environments. For example, leaves of crops such as cotton and sunflower can be as much as 8°C cooler than the air when

TABLE 8.1
Amounts of Water Used in Different Plant Processes per Unit of Carbohydrate Produced in Net Photosynthesis

Process	Tons of Water Used/Ton (CH_2O) Produced
Metabolism	0.6
Stored inside cells	4
Transpired from leaves:	
C_3 plants	>400
C_4 plants	>200
CAM plants	>50

they are growing in irrigated desert environments subjected to advected heat, and the same is true for cattail (*Typha latifolia* L.) growing in water canals in Death Valley, California, during the summer.

But high rates of evaporative cooling are not essential for the adaptation of most plants in most environments. Instead, high rates of transpiration appear to have resulted from an evolutionary process that favored another aspect of plant function. Plants transpire large quantities of water as a consequence of the evolution of structures that favor photosynthesis, that is, large planar leaves for efficient interception of solar radiation that also result in heating providing the energy required to evaporate water, and stomata to permit the inward diffusion of CO_2, which also let water vapor out.

It is unlikely that cultivars of C_3 and C_4 plant species can be bred that are both productive and only require small quantities of water, because it would appear impossible to design a photosynthetic structure that permits daytime uptake of CO_2 while restricting the loss of water vapor. CAM plants partially overcome this problem when their CO_2 uptake and much of their water loss are restricted to nighttime conditions. At night, the evaporative demand (ET_o) is very small due to the lack of a radiant energy load (R_n is negative).

8.1 TRANSPIRATION

What determines the rate of transpiration at the plant level? Those factors that influence crop water use (Chapter 7) influence transpiration directly: the evaporative demand, percent ground cover, and leaf stomatal characteristics. In contrast, root and soil characteristics influence transpiration indirectly. Dry, cold, or saline soil results in reduced transpiration and therefore reduced water uptake by roots only if it causes either stomatal closure or reduced plant growth or leaf wilting such that the percentage ground cover is less, compared with plants in more optimal soil conditions. Damage to root systems by diseases or pests results in reduced transpiration only if it either causes stomatal closure or reduces shoot growth rate such that percent ground cover is less, compared with plants whose root systems are not damaged. Water flow in the soil–plant–atmosphere continuum is controlled at the top of the plant: the interface between leaves and the atmosphere.

The transpiration rate of leaves (T_r) can be modeled using Equation 8.1, which has a similar form as Equation 4.4 for CO_2 exchange and was developed from the general transport law for gases (Equation 3.4) and used to define leaf conductance to water vapor, g_w, in Equation 4.9.

$$T_r = g_w \times \left(H_i - H_a\right) \tag{8.1}$$

where H_i and H_a are the volumetric concentrations of water vapor in the air inside the leaf and in the air outside the leaf, respectively. Volumetric concentrations are equivalent to the ratio of partial pressure for water vapor (e) to atmospheric pressure ($P_{atmosphere}$), such that $H_a = e_a/P_{atmosphere}$ and, when e_a is 20 mbar, H_a is about 20 parts per thousand because atmospheric pressure is about one bar. The H_i strongly depends on leaf temperature, and it often is assumed that the air inside the leaf has a relative humidity of 100% and an e_i value equal to the saturated value at leaf temperature. For a more accurate method for determining the value of H_i, refer to the discussion of Equation 4.8 in Chapter 4. Note that when using Equation 8.1, the leaf conductance to water vapor, g_w, has the same units as T_r (flow × area^{-1} × time^{-1}, whereas resistances have the inverse of these units. Much of the older literature used a different definition of g_w that has units of length time^{-1} (values of the old and newer systems are compared in pages 55 and 56 in Jones [1992]). The problem with the older definition of g_w was that in addition to depending on plant properties, it depended on pressure and temperature. The same plants growing at different elevations would have different values of g_w because of the difference in atmospheric pressure at the different elevations, even when the plants had the same stomatal densities and stomatal apertures. The new definition for g_w (Equations 4.9 and 8.1) does not depend on pressure and has only a small dependence on temperature (Hall, 1982b). Consequently, when differences in g_w occur between plants with the new system, they indicate that differences are present between the plants, such as in stomatal properties.

The limitations of Equation 8.1 as a model for predicting T_r should be recognized. For example, if a change in solar radiation causes stomatal aperture to increase, g_w will increase and tend to cause T_r to increase, but this will tend to cool the leaf and cause H_i to decrease, which will tend to cause T_r to decrease. A more complete model would include Equation 8.1 with H_i as a function of leaf temperature and leaf temperature as a function of the energy balance components described in Equation 7.10.

The conductance to water vapor (g_w) of a single leaf surface can be related to resistances to water vapor transfer through the leaf surface (r_l) and leaf boundary layer (r_b) using Equation 8.2:

$$g_w = \frac{1}{(r_b + r_1)} \tag{8.2}$$

Resistance to water vapor transfer of a single surface of a leaf (r_1) can be related to the resistances to water vapor flow through the parallel pathways of the stomatal pores (r_s) and cuticular surface (r_c) using Equation 8.3:

$$\frac{1}{r_1} = \frac{1}{r_s} + \frac{1}{r_c} \tag{8.3}$$

The magnitude of r_b (and the boundary layer resistance to heat transfer r_b') increases with decreases in wind speed (u) and increases in the smallest dimension of the leaf surface (d), as described by Equation 8.4:

$$r_b = c \times \left(\frac{d}{u}\right)^{0.5} \tag{8.4}$$

where c is a constant whose value depends on the shape of the leaf (Jones, 1992).

The magnitude of r_s varies from being very small when stomata are fully open to approaching infinity when stomata are completely closed. The r_c tends to vary with species but not to vary with time and to be relatively large compared with either r_b or the r_s value for fully open stomata.

In most physiological studies, it is more effective to use conductances rather than resistances. Transpiration rate is more proportional to the conductance to water vapor than to the resistance to water vapor, and stomatal conductance (g_s) is proportional to the area of the stomatal aperture, whereas stomatal resistance has an inverse hyperbolic relationship (Jones, 1992). Even though resistance is simply the reciprocal of conductance, this issue is not trivial. Earlier use of resistance led to erroneous concepts such as threshold stomatal responses to solar photon flux density, even though stomatal apertures (and stomatal conductance) often respond linearly over a large range of solar photon flux densities (Figure 8.1). Note that the leaf resistance data appear to be approaching a minimum value at *PFD* of about 400 μmol m^{-2} s^{-1}, whereas the leaf conductance data appear to be achieving maximum values at much higher *PFD* of about 800 μmol m^{-2} s^{-1}. Thus, the use of leaf resistance data has resulted in errors in interpretation concerning stomatal function. Also, the use of resistances led to the use of instruments that did not provide good measurements of stomatal function (i.e., transit-time porometers that measure the time required for a leaf to increase humidity by a set increment in a closed system) in that they provide good measurements of leaf resistance but poor measurements of leaf conductance. In contrast, porometers that measure the steady-state flow of water vapor from leaves provide good measurements of leaf conductance (and stomatal aperture) but poor measurements of leaf resistance.

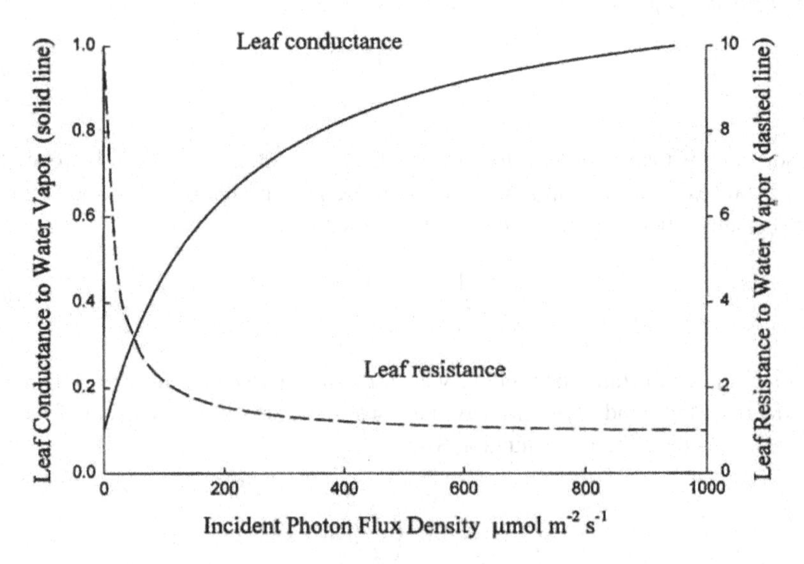

FIGURE 8.1 Leaf conductance to water vapor (solid line) and leaf resistance to water vapor (dashed line) in response to incident photon flux density (*PFD*). Leaf conductances are relative values, and leaf resistance values were calculated from 1/leaf conductance.

The conductance of a leaf surface (g_l) can be related to g_s and the cuticular conductance (g_c) using Equation 8.5:

$$g_l = g_s + g_c \tag{8.5}$$

Note that the regulation of water loss by the leaf surface (variation in g_l) is mainly due to variation in g_s. Variation in g_c, such as may be present between different cultivars, has substantial influence on g_l only on those occasions when stomata are fully closed and g_s approaches zero, such as with extreme drought stress or during the night for either well-watered or dry C_3 and C_4 plants. Effects of differences in g_s on canopy transpiration depend on canopy and atmospheric conditions (Jarvis and McNaughton, 1986, and Chapter 2). For a tall, isolated plant with small leaves subjected to a strong wind in an area with no other vegetation, transpiration rate would be proportional to stomatal conductance. In contrast, for extensive smooth and dense canopies of leaves subjected to low wind speed, substantial changes in g_s may have little influence on canopy transpiration due to the following reasons: With a change in stomatal opening and g_s, the change in canopy conductance would be small, because of the relatively large resistances to flow of the boundary layer of the leaves (r_b) and within the canopy compared with stomatal resistance (note that resistances can be useful when comparing the effects of processes oriented in a series along a pathway). In addition, with a change in transpiration rate, there would be counteracting effects due to the humidification of the canopy (increasing H_a) and cooling of the leaves that would reduce H_i and thereby reduce the gradient driving T_r, which is ($H_i - H_a$). This explains why reference crop evapotranspiration determined with well-watered grasses completely covering the ground surface and with a short uniform canopy varies little between different varieties and species of grasses.

8.2 STOMATAL RESPONSES TO ENVIRONMENT

Changes in stomatal aperture are mainly due to changes in turgor pressure (ψ_p) in the guard cells. Note that the components of water potential are defined later in this chapter. As ψ_p increases, stomatal aperture increases. The increases in ψ_p are mainly due to decreases in solute potential (ψ_s) of the guard cells due to active inward pumping of K^+ and other ions and metabolic conversion of starch to sugars. With more K^+ inside guard cells, ψ_s becomes lower and ψ_p becomes higher. In addition, some small changes in ψ_p of the guard cells may occur due to changes in water potential (Ψ) of the bulk leaf that result in changes in Ψ of the guard cells. Changes in ψ_p of adjacent subsidiary cells also may have a small influence on stomatal aperture (Jones, 1992).

Stomatal conductance exhibits a hyperbolic response to *PFD* (or solar irradiance) with a horizontal asymptotic maximum g_s occurring at *PFD* levels of 10%–100% of full sun values (Figure 8.1), depending on factors such as the species, age, and nutritional status of the leaf. Often, there is a positive correlation between maximal g_s and photosynthetic capacity as it varies due to leaf age and plant nutrition (Schulze and Hall, 1982), although the mechanism for this long-term (days) regulation is not known. Two types of mechanisms have been proposed for the short-term (seconds) stomatal response to *PFD*:

1. Direct responses of guard cells to *PFD* that result in ion pumping by the guard cell plasma membrane
2. Indirect responses through influences of *PFD* on P_n and thus the $[CO_2]$ inside leaves (C_i) that then influence the ion pumping of the guard cells

Stomatal response to *PFD* may be modeled as being due to independent and additive effects of the direct and indirect responses using Equation 8.6, which provides a method for measuring the magnitudes of the direct and indirect responses (Wong et al., 1978). This equation can be described more elegantly using partial differentials.

$$\Delta g_s = S_{pfd} \times \Delta_{pfd} + S_{ci} \times \Delta_{ci} \tag{8.6}$$

where Δg_s is the total change in stomatal conductance in response to a change in *PFD* (Δ_{pfd}), S_{pfd} is the sensitivity of the direct response of g_s to a change in *PFD*, and S_{ci} is the sensitivity of the indirect response of g_s to a change in C_i (Δ_{ci}). One component of Equation 8.6, S_{ci}, can be determined by measuring g_s response to variations in atmospheric $[CO_2]$ and thus C_i at constant *PFD* (which gives values of S_{ci} at different levels of *PFD*). Stomatal conductance exhibits decreases with increases in C_i.

Next, it is necessary to measure g_s responses to *PFD* at constant atmospheric $[CO_2]$ while measuring changes in C_i, which makes it possible to calculate S_{pfd}. Conclusions reached from this type of study are that stomatal responses to *PFD* are both direct and indirect for C_4 species, whereas stomatal responses to *PFD* mainly are direct for C_3 species and can be explained by the $S_{pfd} \times \Delta_{pfd}$ component (Schulze and Hall, 1982).

When plants are subjected to decreases in air humidity (H_a), stomata progressively close and g_s decreases (Hall et al., 1976; Schulze and Hall, 1982). The mechanism of this effect is not known, but it has been shown to be independent of bulk leaf Ψ and may depend on either T_r or the rate of water vapor loss from the outer surface of guard cells. Consequently, g_s often is plotted as a function of $(H_i - H_a)$ when evaluating the humidity effect. The adaptive significance of this response is that when plants are subjected to decreases in H_a, there may be little effect on T_r, because the increases in $(H_i - H_a)$ are counterbalanced by decreases in g_s and thus also in g_w. In contrast, when plants are subjected to increases in temperature, with $(H_i - H_a)$ kept constant, stomata progressively open and g_s increases over a broad range of leaf temperatures (Hall et al., 1976). It should be noted that to conduct this experiment it is necessary to increase H_a as leaf temperature is increased by increasing air humidity. The physiological significance of this response is that increases in temperature can cause large increases in T_r because of both the tendency for g_s to increase and the large increases in H_i with increases in leaf temperature. Early empirical studies found that crop water use was more strongly related to variations in air temperature than to variations in air humidity. Refer to Equation 7.9 and the discussion of this topic in Chapter 7.

When plants are progressively subjected to soil drought, in most field conditions, stomata do not open as much on a day-to-day basis. Often, there is a linear correlation between daily maximal g_s and predawn Ψ_{leaf} but no relation with Ψ_{leaf} at the time when $g_s$was measured (Schulze and Hall, 1982). An approximate mechanism for this response is that with decreases in Ψ_{root}, there is a decrease in ψ_p in the roots and an increase in synthesis of abscisic acid, which is transported to the leaves in the xylem fluid and causes the stomata to not open as much as they did on the previous day. With soil drought that develops very rapidly, such as for plants grown in very small pots, stomata may either exhibit a threshold closing response when the bulk leaf achieves zero turgor or periodic oscillations of stomatal closure and then opening and then closure, and so on. These stomatal responses to very rapid drought rarely occur in nature.

8.3 OPTIMAL STOMATAL FUNCTION

An elegant conceptual definition and a theory for optimal stomatal function have been developed by Cowan and Farquhar (1977). They argued that, at any time, further stomatal opening has a *cost* to adaptation in terms of increased T_r but a potential *benefit* in terms of increased P_n. Their conceptual definition states that on a specific day, in a specific environment, for a specific leaf in the canopy, a specific diurnal course of g_s would maximize daily P_n for a specific level of daily T_r, and this response of the stomata would thereby maximize daily transpiration efficiency ($\int P_n / \int T_r$). A way to visualize this process is to consider that, during the day, root growth and soil properties make available to the plant a specific quantity of water and that optimal stomatal function would enable the plant to achieve maximum daily P_n by the whole canopy while using this fixed amount of soil water. This would require complex stomatal responses in that the diurnal course of g_s which is optimal would be different for different leaves in the canopy that are experiencing different levels of *PFD* or that have different photosynthetic capacities due to differences in age and other factors.

Cowan and Farquhar (1977) also developed a rigorous mathematical theory for optimal stomatal function based on gain analysis. Remarkably, for the cases that have been evaluated, stomatal responses to humidity and temperature are consistent with this theory for optimal stomatal function (Farquhar, 1978; Hall and Schulze, 1980), and stomatal responses to *PFD* and long-term effects of leaf age and shading are qualitatively consistent with the theory. The fit is so good that an effective way to model stomatal function is to assume that stomata respond in a manner that is consistent with the theory of Cowan and Farquhar (1977) as was proposed by Hall (1982a). This would require combining equations for the theory of optimal stomatal function (Appendix of Cowan and Farquhar [1977], noting that the + in the numerator of their Equation 8.17 should be a –) with those for a model of leaf photosynthesis and respiration (e.g., Hall, 1979) and leaf energy balance (e.g., Equation 7.10). A model of this type could provide reasonable predictions of stomatal responses to *PFD*, leaf temperature, and air humidity for younger and older leaves varying in photosynthetic capacity.

It appears unlikely that current paradigms of plant function could provide mechanisms to explain how this complex functioning of stomata is achieved, since it requires sensing mechanisms that are not known to exist. A weak argument has been proposed that the various stomatal and photosynthetic responses to *PFD*, temperature, humidity, and leaf age evolved independently and fortuitously produced an integrated system that functions optimally.

8.4 ADAPTIVE SIGNIFICANCE OF PLANT DIFFERENCES IN THE LEVEL OF DAILY WATER USE

In well-watered, hot environments, a high level of daily water use would appear to be adaptive, especially where T_{leaf} is at or above the optimum for P_n. In recent varietal improvements for hot, irrigated environments, the yield of different cultivars has been positively correlated with g_s for Pima cotton (Lu et al., 1994) and for spring wheat (Reynolds et al., 1994b). Positive associations between stomatal conductance and yield also have been reported for different cultivars of several annual crops growing in a range of environments, including ones with optimal temperatures or moderate drought. Condon and Hall (1997) speculated that the evolution of these annual crops had resulted in conservative performance with respect to g_s, which is a tendency for stomata to be partially closed on many occasions. This could have happened if plant performance during very dry years, when conservative stomatal performance may be adaptive, had disproportionate influences on seed production and long-term evolutionary success over many years due to soil *seed banks* being much less effective after one year. This then provides an opportunity to increase the potential yield of these crops by selecting to increase g_s.

For water-limited environments, the level of daily water use that is adaptive depends on the rainfall pattern and the type of crop species. Consider a determinate annual growing on stored soil moisture from the middle to the end of the cropping season in an environment that is initially cool and then becomes very hot. This often occurs with wheat growing under rain-fed conditions in a Mediterranean climate (Figures 10.6 and 10.7). Two contrasting optimal water-use strategies have been hypothesized. Passioura (1982) hypothesized that cultivars are needed with more conservative water use during the vegetative stage so that more water remains in the soil during the reproductive period when grain yield may be more sensitive to drought. In contrast, it may be hypothesized that more open stomata may be adaptive during the vegetative stage because even though this results in greater water use, it also results in greater photosynthesis. In addition, during this cooler and more humid time of the year, the transpiration efficiency (ratio of photosynthesis to transpiration) would be greater than later in the growing season. For this hypothesis to result in enhanced grain yield, the excess carbohydrate produced during the vegetative stage would have to be stored in the stems and then translocated to the developing grain during the reproductive stage.

Indeterminate annuals such as cowpea may be expected to have more conservative water use during the reproductive period than determinate annuals. Indeterminate annuals have the capacity to produce more flowers and leaves once the drought is ended, so there is an advantage from surviving the drought. In contrast, determinate annuals have only one period of leaf production followed by one period of reproduction. Consequently, adaptation to reproductive-stage drought in determinate annuals is favored by maintenance of stomatal opening and the production and filling of as many seeds as possible during the drought.

Woody plants usually have smaller g_s and more conservative water use than herbaceous plants. Presumably, this arose because survival of the adult plant was critical for the evolution of woody plants, whereas evolution of herbaceous annuals was strongly influenced by the extent of seed production. Evergreen woody plants typically have smaller g_s and more conservative water use than deciduous woody plants. Considering the efficiency of photosynthate acquisition for water-limited environments, the evergreen *strategy* would appear to be adaptive in environments with a small amount of rain every month, whereas the deciduous *strategy* would fit environments with distinct wet and dry seasons. Leaves of deciduous plants also tend to have greater investments in photosynthetic components per unit leaf area and greater photosynthetic capacity than leaves of evergreen plants, which have greater investment in compounds that deter herbivores (Chapter 12).

8.5 ADAPTIVE SIGNIFICANCE OF PLANT DIFFERENCES IN TRANSPIRATION EFFICIENCY

Substantial differences in transpiration efficiency ($TE = P_n/T_r$) have been observed among species. For plants with different photosynthetic systems growing in a specific environment to which they are well adapted, approximate values of *TE* in units of grams of CH_2O produced per kilogram of water transpired are as follows:

CAM species 20 >> C_4 species 5 >> C_3 species 2.5

The high *TE* of CAM species arises from the very low transpiration that occurs when their stomata close in the day and only open at night, when it is much cooler and more humid than in the day. This *strategy* enables CAM plants to be well adapted to arid environments, especially those where a small amount of rain occurs every month. The mechanism for the difference in *TE* of C_4 and C_3 species may be seen by considering the factors that influence this efficiency for plants whose stomata open during the day.

$$TE = \frac{P_n}{T_r} = g_c \times \frac{C_a - C_i}{g_w \times (H_i - H_a)} \tag{8.7}$$

Since $g_c/g_w = D_c/D_w = 0.61$ (refer to the discussion of Equation 4.6), Equation 8.7 can be simplified to Equation 8.8:

$$TE = 0.61 \times C_a \times \frac{1 - \frac{C_i}{C_a}}{H_i - H_a} \tag{8.8}$$

This equation shows that for different plants growing in the same aerial environment, differences in their *TE* will mainly be determined by any differences occurring in C_i, because C_a and H_a, will be the same and H_i will be similar if leaf temperatures are similar. The C_4 species often have higher *TE* than C_3 species because of the initial fixation of CO_2 by the enzyme PEP carboxylase, which has a higher affinity for CO_2, causing C_i to be lower than where the initial fixation of CO_2 is by rubisco. Note that a C_3 plant having more active rubisco per unit leaf area than another C_3 plant also would have lower C_i and greater *TE*. Also, plants with more closed stomata (smaller g_s) have lower C_i and thus higher *TE*. Where more than one factor differs between plants, the overall effect on *TE* will depend on the balance of the effects on both photosynthetic capacity and stomatal conductance. The difference in *TE* between C_4 and C_3 species has some adaptive significance. For native species the abundance of C_4 dicots correlates with aridity, however the abundance of C_4 monocots correlates most strongly with growing season temperature, and any competitive advantage of the high *TE* of C_4 plants over C_3 plants has been difficult to document experimentally (Ehleringer and Monson, 1993).

Studies of the variation in *TE* within species and its adaptive significance were constrained by the difficulty of measuring seasonal *TE*. Gas exchange systems are available to obtain instantaneous measurements of P_n/T_r, but continuous measurements are needed to determine seasonal or even daily values. A scientific breakthrough occurred with the theory of Farquhar et al. (1982), which demonstrated that the extent of stable carbon isotope discrimination by C_3 plants (Δ) in photosynthesis is related to C_i and thus that the stable carbon isotope composition of plants can indirectly provide information concerning their time-integrated *TE*.

Atmospheric CO_2 is 99% $^{12}CO_2$ and 1% $^{13}CO_2$. There is discrimination against $^{13}CO_2$ when leaves of C_3 plants take up CO_2. There is slight discrimination in diffusion because $^{13}CO_2$ is slightly heavier than $^{12}CO_2$, while the enzyme rubisco strongly discriminates against $^{13}CO_2$. The large differences in discrimination against stable carbon isotopes by diffusion and CO_2 fixation by C_3 species causes overall discrimination to vary in different environments and among different genotypes, depending on the relative extents of the limitations to CO_2 uptake imposed by stomata and rubisco. If we define R as the ratio mol ^{13}C/mol ^{12}C, which can be measured with a sensitive ratio mass spectrometer, then Δ can be defined using Equation 8.9:

$$\Delta = \frac{(R_{air} - R_{plant})}{R_{plant}} \tag{8.9}$$

The theory of Farquhar et al. (1982) quantified the effects of diffusion, fixation of CO_2 by rubisco, and respiration on Δ of C_3 plants as shown in Equation 8.10:

$$\underset{(1)}{\Delta = -d} + \underset{(2)}{a \times (C_a - C_i)/C_a} + \underset{(3)}{b \times C_i/C_a} \tag{8.10}$$

where (1) describes the respiration effect, (2) describes the diffusion effect, and (3) describes the CO_2 fixation effect.

Equation 8.10 can be simplified to Equation 8.11:

$$\Delta = -d + a + (b - a) \times \frac{C_i}{C_a} \tag{8.11}$$

where a, b, and d are constants to account for effects of diffusion, fixation, and respiration, respectively, with values of 4 ppt for a, 30 ppt for b, and 3 ppt for d (ppt = parts per thousand, also called parts per mil in this application).

The balance between the effects of stomatal opening and photosynthetic capacity influences C_i and thus Δ in the following manner. When rubisco is very active and stomata are partially closed, there is the least opportunity for rubisco to discriminate against $^{13}CO_2$, C_i/C_a approaches 0.5, and Δ approaches 14 ppt. When stomata are fully open and rubisco is not very active, there is the greatest opportunity for rubisco to discriminate against $^{13}CO_2$, C_i approaches C_a, and Δ approaches 27 ppt.

Since TE and Δ both depend on C_i/C_a, the relation between them can be described by combining Equations 8.8 and 8.11 to give Equation 8.12:

$$TE = \frac{0.61 \times C_a(b - d - \Delta)}{(b - a)(H_i - H_a)} \tag{8.12}$$

When two different plants (1 and 2) are grown in the same aerial environment, they experience the same C_a and H_a and if their leaves have the same temperature, they have the same H_i. Consequently, their relative time-integrated TE can be related to the values of the carbon that was assimilated during this time period by Equation 8.13:

$$\frac{\int TE_1}{\int TE_2} = \frac{b - d - \Delta_1}{b - d - \Delta_2} \tag{8.13}$$

The theory of Farquhar et al. (1982) has been shown to be robust. Studies based on direct measurements of TE and carbon isotope composition of plant tissues have shown that Δ decreases and TE increases when plants are subjected to soil drought, salinity, and mechanically resistive soils in many (but not all) cases. In these instances, the increase in TE is caused by decreases in g_s. Also, substantial genotypic differences in Δ and TE have been reported in many C_3 crop species (Condon and Hall, 1997). In wheat and cowpea, much of the genotypic variation in Δ and TE is due to differences in g_s. For these species, selection studies indicate genotypes adapted to both well-watered and moderately droughted environments have high g_s, high P_n, high biomass production, high grain yield, high Δ, and low TE. Essentially, adaptation was associated with less conservative stomatal function. In peanut, much of the genotypic variation in Δ and TE is due to differences in photosynthetic capacity. In this case, genotypes adapted to water-limited conditions have high photosynthetic capacity per unit leaf area, similar T_r, high biomass production, low Δ, and high TE. The relevance of these findings to adaptation to water-limited environments is discussed in Chapter 9 in the section on water-use efficiency.

Other naturally occurring stable isotopes can provide information on plant adaptation that complements the information provided by carbon isotope discriminations. It has been hypothesized that the $^{18}O/^{16}O$ ratio is useful for determining whether differences between genotypes in Δ are caused by differences in photosynthetic capacity or stomatal conductance (Farquhar and Lloyd, 1993). The process of evaporation tends to cause enrichment of ^{18}O relative to ^{16}O in leaf water, because the vapor pressure of H_2O^{18} is less than that of H_2O^{16} and because H_2O^{18} diffuses more slowly from the leaf, and this process would be influenced by stomatal conductance. For genotypic variation in both wheat and cotton, the observance of strong correlations between stomatal conductance and the enrichment in ^{18}O of whole-leaf material above source water offers support for this hypothesis. The $^{18}O/^{16}O$ ratio of leaves, stomatal conductances, and grain yields were determined for eight cultivars of spring wheat growing under irrigation in field environments over three years (Barbour et al., 2000). Supporting the theory, the genotypic mean values of oxygen isotope ratio of leaves were negatively correlated with genotypic mean stomatal conductance values. Genotypic mean values for grain yield also were negatively correlated with the genotypic mean oxygen isotope ratio values and positively correlated with the genotypic mean stomatal conductance values.

Water in very deep soil layers can have a different stable hydrogen composition, with respect to ^2H (deuterium), than water in the upper soil layer due to fractionation of hydrogen isotopes during transport and phase transitions in the hydrologic cycle. The stable hydrogen isotope composition of water in xylem can provide a quantitative measure of the extent to which plant roots are accessing different sources of water providing the different sources of water have sufficiently different stable hydrogen isotope signatures (Dawson, 1993).

In principle, the ^{15}N composition of plants growing in natural field environments can provide an assessment of the extent to which the plants have depended on biological fixation of atmospheric nitrogen, compared with the assimilation of nitrates or ammonium ions from the soil solution. In practice, the natural abundance method for estimating the extent of nitrogen fixation is not very precise or accurate. In the future, it is likely that crop physiologists and physiological ecologists will be able to analyze plant parts for their composition of several stable isotopes and then, by applying various models, learn much about the functioning of the plants.

8.6 LIQUID WATER TRANSPORT FROM SOIL TO LEAVES

The rate at which liquid water moves from soil to plant is similar to the rate at which water vapor is transpired from leaves during the day. The flow of water to the plant can be slightly less than the water vapor loss rate during the morning, resulting in the development of water deficits in the plant. In contrast, the flow of liquid water to and within the plant can exceed the water vapor loss rate in the evening, and especially at night, when transpiration rate is slow, resulting in some recovery of plant water status. The flow of liquid water between the soil and leaf is determined by the difference in water potential (Ψ) between the soil and the leaf.

$$\Psi = \frac{(\mu_w - \mu_w^*)}{V_w} \tag{8.14}$$

where μ_w is the chemical potential of water in the system in free energy/mol μ_w^* is the chemical potential of pure water at the same temperature and pressure, and V_w is the partial molar volume of water in the system, which can be assumed to be

TABLE 8.2
Values of Water Potential in Soils and Plants

System Component	Water Potential (Ψ) MPa	Water Potential (Ψ) Bar
Pure water	0	0
Water in well-watered and aerated non-saline soil	−0.01 to −0.03	−0.1 to −0.3
Water in drier soil just prior to irrigation	−0.1	−1
Water in very dry soil with no water available for plants	−1.5	−15
Water in well-watered herbaceous plants at night	−0.1	−1
Water in well-watered woody plants at night	−0.3 to −0.5	−3 to −5
Range of day values for well-watered to droughted plants	−0.5 to −5.0	−5 to −50

18 mL/mol. Note that units of Ψ = (energy/mol)/(volume/mol) = energy/volume = force/area = pressure and that 1 megapascal = 10^6 pascal = 10^6 newton/m^2 = 10^7 dyne/cm^2 = 10 bar = 9.87 atmosphere.

Some representative values of water potential in soils and plants are presented in Table 8.2 using both the SI (MPa) and cgs (bar) metric systems. Many scientific journals require use of the SI system of units. The bar unit is very convenient, however, because leaf water potentials can be discussed as integer values. Also, it is easy to appreciate its magnitude, since one bar is approximately equal to one atmosphere of pressure, and bar units are often used in lectures by scientists. I use both systems in this book.

Note that Ψ is always less than 0 in natural systems and is lower (more negative) in plants than in soil, lower under drought than for well-watered systems, and lower by day than by night. Some midday leaf water potential values for walnuts, almonds, and prunes under different levels of water stress may be found in *data interpretation* on the website www.fruitsandnuts.ucdavis.edu.

Water tends to move from regions with high water potential (closer to 0) to regions with lower (more negative) water potential until equilibrium is reached where the water potentials are the same in both regions. An example of this occurs when dry seeds are placed in a bell jar over saturated salt solutions of known water potential. Water will evaporate from the saturated solution and be taken up by the seed until the water potential of the seed equals that of the saturated solution. If the relative humidity of the air over the saturated solution at equilibrium is known, the water potential of the seed can be calculated using Equations 3.2 and 3.3.

Methods for measuring the water potential of plants and soils are discussed by Boyer (1995). Some caution is warranted in interpreting water potential data obtained from excised tissue. In a comprehensive review and analysis, Bradford (1994) discusses several intriguing reports of apparently large differences in water potential between developing seed (and pods) and adjacent leaves. Adjacent tissues with even modest hydraulic connections should approach equilibrium and have similar values of water potential. He argues that the differences in water potential that have been reported are artifacts due to the use of excised tissue and long equilibration times when measuring the water potential of developing seed. He proposes that, during the long equilibration times, ion pumping occurred that regulated apoplastic solute content and changed the total water potential of the excised tissue. Bradford (1994) also provides a radical new model for phloem transport to developing seeds that includes modified cell walls acting as semipermeable apoplastic membranes which retain solutes in the unloading tissue while allowing the return of water to the parental xylem. Prior to Bradford (1994), it was generally assumed that, except for effects of charged surfaces, cell walls were freely permeable to solutes that were not too large. Semipermeable apoplastic membranes may not be that rare in vascular plant systems in that other tissues appear to contain them, such as the stems of sugar cane (Welbaum and Meinzer, 1990).

The rate of flow of liquid water depends on the gradient in water potential and the magnitude of resistances present in the flow pathway. A widely used but conceptually erroneous model for the steady-state flow of liquid water (F_h) in the soil-plant system is described by Equation 8.15:

$$F_h = \frac{(\Psi_{soil} - \Psi_{leaf})}{(R_{soil} + R_{plant})} \tag{8.15}$$

where R_{soil} and R_{plant} are resistance to liquid water flow from the soil to root surface and from the root surface to the leaves, respectively. This model is conceptually erroneous because F_h is not the dependent variable in Equation 8.15; Ψ_{leaf} is the dependent variable, which changes as F_h changes in response to changes in T_r. Since F_h may be approximated by ($T_r \times$ leaf area) and Equation 8.1, by rearranging Equation 8.15 and substituting with these independent variables, as was suggested by Elfving et al. (1972), we arrive at the more valid Equation 8.16:

$$\Psi_{leaf} = \Psi_{soil} - (g_w \times [H_i - H_a] \times \text{leaf area})(R_{soil} + R_{plant}) \tag{8.16}$$

This equation provides only an approximate description of soil–plant–water relations and mainly is useful in a qualitative sense in that it describes the directions to which factors will influence Ψ_{leaf}. The equation states that Ψ_{leaf} will approach the value of Ψ_{soil} when transpiration is very slow and will decrease as Ψ_{soil} decreases (e.g., due to drying soil or soil becoming more saline). The equation also states that factors causing transpiration to increase, such as more open stomata (higher g_w), lower air humidity (H_a), or higher leaf temperature (which causes H_i to increase), will cause Ψ_{leaf} to decrease (become more negative). The equation states that factors causing increases in soil resistance or plant resistance also will cause Ψ_{leaf} to decrease. Note that R_{soil} increases as soil becomes drier, because the water moves through thinner films around soil particles, and therefore the unsaturated hydraulic conductivity of the soil decreases. Also, R_{plant} increases if roots malfunction due to pests, diseases, or cool temperatures.

When studying tall trees, Equation 8.16 must be modified to include the effects of gravity on liquid water movement as shown in Equation 8.17:

$$\Psi_{leaf} = \Psi_{soil} + \Delta\psi_g - (g_w \times [H_i - H_a] \times \text{leaf area})(R_{soil} + R_{plant}) \tag{8.17}$$

where $\Delta\psi_g$ quantifies the decrease in Ψ_{leaf} needed to overcome gravity as described in Equation 8.18:

$$\Delta\psi_g = \rho_w \times g \times (h_{soil} - h_{leaf}) \tag{8.18}$$

where ρ_w is the density of water, g is the acceleration due to gravity, h_{soil} is the height of the soil where Ψ_{soil} was measured above some arbitrary reference point, and h_{leaf} is the height in the tree where Ψ_{leaf} was measured above the same reference point. Equation 8.18 predicts that for a very tall tree with $(h_{soil} - h_{leaf}) = -100$ meters, $\Delta\psi_g$ would be –1 megapascal, and even with a well-watered soil and no transpiration, Ψ_{leaf} will be more negative than –1 megapascal simply to provide the suction needed to maintain liquid water in the lumens of the xylem elements.

Equation 8.17 can be useful in a quantitative sense if it is calibrated by determining T_r (or F_h) and Ψ_{leaf} for a plant during the day under optimal soil conditions and then plotting Ψ_{leaf} as a function of T_r (Figure 8.2). This approach originally

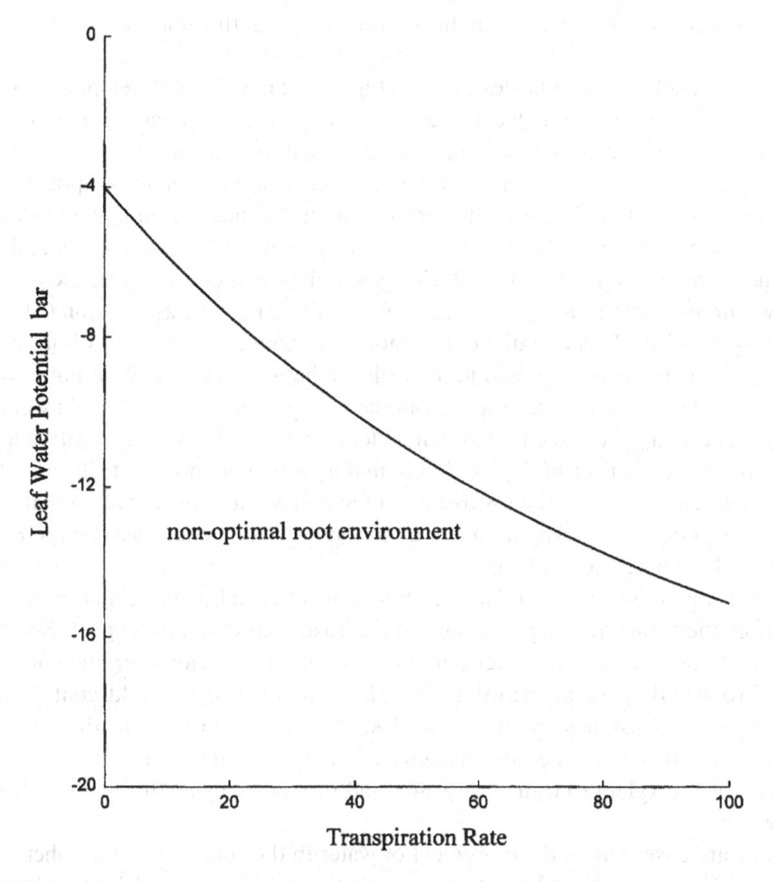

FIGURE 8.2 Leaf water potential as a function of transpiration rate per leaf area or per plant for a woody plant with an optimal root environment.

was developed to predict when the root environment was not optimal for the uptake of water by citrus (Elfving et al., 1972). For the model to be valid and the curve in Figure 8.2 to be useful, points determined during the morning should track on about the same curve as points taken during the afternoon. This curve describes the operation of the plant under optimal soil conditions. The intercept value of leaf water potential for transpiration rate approaching zero (–4 bar in the case of Figure 8.2) is the value that would be obtained just prior to dawn. If future measurements show that Ψ_{leaf} values fall below the curve, indicating lower values of Ψ_{leaf} at particular levels of T_p, then this indicates that soil conditions in the root zone are not optimal, that they may have become too dry or too cold, or that a disease or pest has attacked the root system. With soil drought, three changes occur to the response of leaf water potential to transpiration rate:

1. Transpiration rates are less than those of well-watered plants.
2. Leaf water potentials are lower (more negative) at particular levels of transpiration and predawn.
3. There is hysteresis in the curve in that at particular transpiration rates leaf water potentials are lower in the afternoon than in the morning.

Similar approaches to the one described in Figure 8.2 have been developed to predict when irrigation is needed in deciduous fruit and nut tree crops based solely on measurements of leaf water potential; they are discussed in Chapter 10.

The presence of what appears to be an emergent property may complicate relationships between liquid water transport within plants and the regulation of the loss of water vapor from plants by stomata. As sugar cane grows, maximum stomatal conductance can change in concert with changes in the hydraulic conductance to water flow within the plant ($1/R_{plant}$) such that they exhibit a positive association (Meinzer and Grantz, 1990). This coordination of stomatal conductances with hydraulic conductance, which could act to maintain a balance between the supply of liquid water to leaves and the loss of water vapor from them, appeared to be mediated by chemical signals coming from the root system (Meinzer et al., 1991). The coordination of long-term (days) changes in R_{plant} and maximal g_w would not invalidate the use of the models described by Equations 8.16 and 8.17 and may cause the curve for leaf water potential response to transpiration rate under optimal soil conditions (Figure 8.2) to be relatively constant as plants grow.

As the shoot system of an isolated plants grows, and it intercepts more solar radiation, the transpiration per plant will increase, and we might expect leaf water potential to decrease, even under optimal soil conditions. However, the coordination of root and shoot growth described in Chapter 2 also would contribute to the long-term constancy of the curve described by Figure 8.2 in that it would maintain a balance between the capacity of the root system to take up water, the capacity of the xylem to transport water, and the tendency of the shoot system to lose water.

There are cases where the movement of water in the soil–plant–atmosphere continuum does not approach steady-state conditions, and the models described by Equations 8.16 and 8.17 are not effective. For example, the water uptake per day can be much less than the T_r per day from leaves of conifers in cold soil in the winter or bottle trees (*Brachychiton australis*) during the season when the soil in the root zone is very dry. In these cases, a dynamic model is needed to describe the depletion of water from the body of the plant. Variables analogous to capacitance (C) have been used to describe this effect (Jones, 1992).

$$C = \frac{dW}{d\Psi} \tag{8.19}$$

where W is the water content.

Another case where the modeling of water movement is complex is for water flow into growing cells because one must consider the deformation of cell walls and metabolic effects on cellular constituents as they influence the components of Ψ. A model for water flow into living cells (Equation 8.27) will be discussed later in this chapter.

8.7 COMPONENTS OF TOTAL WATER POTENTIAL (Ψ)

Several factors influence Ψ in plant and soil systems and have been modeled as having independent additive effects as described in Equation 8.20.

$$\Psi = \psi_s + \psi_m + \psi_p \tag{8.20}$$

ψ_s is the *solute potential*. The presence of solutes decreases the free energy of water. The effect is always negative; for example, salts in soil water make the water less available to plants. The effect is dependent on the number (and not the type) of solute molecules or ions as described by Equation 8.21, which is a modified form of Equation 3.3, recognizing that $a_w = \gamma_w \times N_w$.

$$\psi_s = \frac{(R \times T \times \ln[\gamma_w \times N_w])}{V_w} \tag{8.21}$$

where R is the international gas constant, T is the absolute temperature, V_w is the partial molar volume of water, and N_w = mol of water/(mol of water + osmol of solute) where the osmol of a salt = mol × no. of particles formed after dissociation and no. of particles = (%dissociation × i + [100–%dissociation])/100 where i is the number of ions formed when a molecule dissociates. A salt such as NaCl may only dissociate about 80%, and 0.5 mol of NaCl has 0.9 osmol of solute. The term γ_w is the activity coefficient with values from 0 to 1.0 that approach 1.0 for dilute solutions.

A mathematical approximation of Equation 8.21 can be used to develop a formula, Equation 8.22, that is only approximately valid but can be more useful than Equation 8.21 in some situations.

$$\psi_s = -R \times T \times C_s \tag{8.22}$$

TABLE 8.3
Actual and Predicted Solute Potentials of Sucrose and Salt

	Molality	Molarity	Solute Potential Values (ψ_s) in bar		
				Predicted	
Substance	mol/L H_2O	mol/L solution	Actual	Using Equation 8.21	Using Equation 8.22
Sucrose	0.1	0.098	−2.62	−2.43	−2.39
@20°C	1.0	0.825	−26.99	−24.09	−20.10
NaC1	0.05	0.050	−2.30	−2.19	−2.19
@20°C	0.50	0.496	−22.41	−21.70	−21.80

where C_s = osmol of solute/total volume of the solution. This equation is only approximately valid, because its derivation assumes that there is no change in volume when solutes are added to water, which is reasonable for many salts but not for some cases with sugars. When making solutions of known ψ_s, it is better to use Equation 8.21 and develop molal and not molar solutions. But, even in this case, one should measure the ψ_s of the solutions. Values of ψ_s estimated with Equations 8.21 and 8.22 are compared with actual values in Table 8.3.

The calculations using Equations 8.21 and 8.22 assumed an 80% dissociation of NaCl into Na^+ and Cl^-, and $R \times T$ values of 24.37 liter bar mol^{-1} at 20°C, and for Equation 8.21, values of 0.01805 liter mol^{-1} for V_w at 20°C and 1.0 for γ_w.

The differences between the actual values and those predicted by Equation 8.21 arise because γ_w is less than 1.0. The differences between the values predicted for sucrose by Equations 8.21 and 8.22 arise because when sucrose is added to water, the volume of the solution increases and the molarity is less than the molality with a larger effect at the higher concentration of sucrose.

Equation 8.22 is useful for studying the water relations of cells. It predicts that with removal of water from within cells the relative change in ψ_s will be inversely proportional to the change in osmotic volume (V_o) of the cell, as shown in Equation 8.23. Osmotic volume is the volume within the plasma membranes, that is, the symplast.

$$\frac{\psi_s(\text{final})}{\psi_s(\text{initial})} = \frac{V_o(\text{initial})}{V_o(\text{final})} \tag{8.23}$$

ψ_m is the *matric potential* and accounts for the reduction in free energy of water by forces between water and solids, such as occurs in soils and cell walls but not in solutions such as those in the lumens of xylem elements.

ψ_p is the *pressure potential*—the extent to which the free energy of water is increased by pressure (e.g., turgor in living cells) or decreased by the tensions that can occur in macro systems such as the lumens of xylem elements.

Additional models have been developed for water movement in different parts of the soil–plant–atmosphere continuum and can provide useful insights into plant function. Steady-state water flow in the lumen of xylem elements (J_v) has been modeled using Poiseuille's law for the flow of water in tubes (Jones, 1992):

$$J_v = \frac{\Delta\psi_p \times r^2}{(8 \times \eta \times L)} \tag{8.24}$$

where $\Delta\psi_p$ is the difference in negative pressure (suction) along the length (L) of a tube, r is the radius of the tube, and η is the viscosity of the water. Jones (1992) used this equation to illustrate why the lumens of the xylem elements are much more efficient than cell walls for transport of water and consequently are likely to be the main factor in long-distance transport of water in plants. Assume that transpiring plants have a $\Delta\psi_p$ of –0.1 MPa m^{-1}. Xylem elements have radii of 20–100 micrometers which, according to Equation 8.24, would result in flow velocities of 5–125 mm s^{-1}. Actual maximum sap flow velocities have ranged from 0.3 to 0.8 mm s^{-1} in conifers, 0.2 to 12.1 mm s^{-1} in hardwood trees, and up to 28 mm s^{-1} in herbaceous plants. In contrast, the radius of the interstices (pores) along cell walls is about 5 nanometers and, according to Equation 8.24, would support a flow velocity of only 3×10^{-7} mm s^{-1}. Clearly, if this law is valid for liquid flow in both the lumens of xylem elements and cell walls, virtually all of the long-distance flow of water in plants occurs in the lumens of the xylem elements and not in the cell walls.

Equation 8.24 can be misused. A breeding program has been conducted to enhance the drought adaptation of wheat by selecting to reduce the diameter of the major xylem vessel in the seminal roots (Richards and Passioura, 1989). The objective of the program was to reduce transpiration during the vegetative stage so that more water is left for the plant during the reproductive stage, which had been shown to increase grain yield. Equation 8.24 implies that selecting to reduce r should reduce J_v, but this is incorrect in the context of the whole plant, as is Equation 8.15. Changes in plant resistance to water flow can only directly influence Ψ_{leaf} at a particular T_r (Equations 8.16 and 8.17) and do not directly influence T_r. The hypothesis of Richards and Passioura (1989) may be valid if the following emergent property is present: Plants with smaller xylem vessels may experience lower Ψ_{leaf}, and this in turn may in some way be linked to less leaf growth, and plant water use is reduced because there is less ground cover compared with plants that have larger xylem vessels.

The flow of liquid water in porous media, such as soils, has been modeled using Darcy's law which, in its simplest form, states that flux depends on a driving force times a conductivity. The important points to note are that the driving force for flow is the gradient in ψ_m in the soil, that the unsaturated conductivity of the soil decreases very rapidly as the soil becomes drier because water now moves in thinner films, and that water flow to roots is complex because it is three dimensional and dynamic (Hillel, 1971) and depends on the rate of root growth. Plants with more roots per unit volume of soil, that is, higher root length densities, may be more effective at extracting water because the average distance moved by a water molecule in the soil to the root surface is shorter. However, the uniformity of root distribution and degree of clumping of roots also could be important in that dry regions with low hydraulic conductivity may develop within the clumps (Petrie et al., 1992). The rates at which roots grow toward wet soil could have a major influence in facilitating water uptake by roots.

Capillarity provides a force whereby liquid water is held in soils and parts of plants. The potential upward distance (d) that water can be held up by capillarity can be determined using an equilibrium model (Jones, 1992).

$$d = \frac{2 \times \sigma \times \cos\alpha}{(r \times \rho \times g)} \tag{8.25}$$

where σ is the surface tension, α is the wetting angle (which is close to 0 degree for cellulose and water but close to 180 degrees for lipids and water), r is the radius of the tube, ρ is the density of water, and g is the acceleration due to gravity. For a cellulose/water system at 20°C, d in cm = 0.15/r in cm. Consequently, the extent of capillary rise in the lumen of xylem vessels having r of 1×10^{-2} to 0.2×10^{-2} centimeters is 15–75 centimeters, which is insufficient to take water to the top of trees. Note, however, that the xylem elements are filled with water as they are formed and that, in capillaries of non-uniform radius, the ability to hold a water column is determined by the region with the smallest radius (Zimmermann et al., 1994).

With respect to water movement, capillarity is only relevant to the movement of water over short distances in plants, because it is slow (refer to the discussion for Equation 8.24). Capillarity has importance for the force with which water is held in pores. The force needed to overcome capillarity effects can be calculated using Equation 8.26:

$$\text{Force} = d \times g \times \rho \tag{8.26}$$

which gives force (in MPa) = d (in m)/100.

If we now consider cellulose interstices in cell walls with radii of about 5×10^{-7} centimeters, according to Equation 8.25 they would have an effective d of 3,000 meters such that the water is held so strongly that a large force (a suction) of –30 megapascal would be required to remove it. Also note that Equations 8.25 and 8.26 predict that liquid water would not enter small pores in plant surfaces that are lipophilic with a wetting angle approaching 180 degrees. The equilibrium model of capillarity also tells us that when a tension is placed on a soil, the larger voids will drain first, followed by progressively smaller voids as the tension is increased.

8.8 FLOW OF WATER FROM ROOT TO SHOOT

The cohesion-tension hypothesis is widely believed to account for the movement of water in the xylem from roots to leaves (Kramer and Boyer, 1995). According to this hypothesis, water is pulled from the top of plants due to evaporation from within leaves. The evaporation of water from walls of parenchyma cells inside leaves causes the matric and total potentials of water in the cell walls to decrease, which results in water moving from the lumens of adjacent xylem elements to these cell walls. This creates a tension in the xylem lumens that is supported by the strong cohesive forces between water molecules and is transmitted down the xylem, lowering the water potential in the stele of the roots, which provides a water potential gradient to drive the movement of water from the soil into the stele of the roots. According to this hypothesis, large tensions must develop in the lumens of xylem elements when the water potentials of leaves are very negative because the xylem fluid often is very dilute with a solute potential close to zero, and the matric potential is close to zero because the radius of the lumens is relatively large. According to Equation 8.20, the pressure potential of the water in the lumens would be similar to the total water potential.

Zimmermann et al. (1994) have criticized the cohesion-tension hypothesis for various reasons, some of which are not consistent with the studies of Wei et al. (1999) that support the cohesion-tension hypothesis. The most compelling argument against the hypothesis, in my view, is the fact that certain insects are known to feed in xylem vessels; yet it appears unlikely that they can develop the necessary tensions to obtain the xylem fluid. For example, Andersen et al. (1992) reported that leaf-hoppers had no difficulty feeding on xylem that, according to their measurements and the cohesion hypothesis, had xylem tensions of about –18 bar. In contrast, other studies have estimated that these insects cannot suck with a force more negative than –3 bar. Andersen et al. (1992) also estimated that the energy content of the xylem fluid obtained by the insects, which is very dilute, was only slightly greater than the energy the insects would have had to expend to obtain the fluid against the large tensions proposed to be present by the cohesion hypothesis.

An alternative hypothesis for the ascent of sap has been proposed by Canny (1995), involving cohesion supported by tissue pressure from the living cells surrounding the xylem elements. According to this hypothesis, the tensions in the lumen could be small. However, Canny's theory does not explain how water would flow from soil to the stele. But water flow from the soil to the stele can be explained by an argument that also accounts for root pressure. When many plant species are supplied with adequate levels of soil water, and then their shoots are excised and plastic tubing is placed on the root stump, exudation of xylem fluid occurs from the stump and may develop a pressure as high as 6 bar (Kramer and Boyer, 1995). A possible mechanism for root pressure is that ions are being pumped into the stele across the endodermis, which acts as a semipermeable membrane, thereby lowering the solute potential in the stele but with a constant total water potential causing the pressure potential to increase (as it does in a physical osmometer). But what happens to the solutes flowing up with the xylem sap, since few solutes have been found in xylem sap higher up the plant in most studies? Possibly, some of the solutes are taken up by the living cells surrounding the xylem and other solutes are circulated back down to the root in the phloem elements.

The mechanism whereby sap ascends in the xylem lumen has not been rigorously established. Most scientists accept the cohesion-tension hypothesis, even though it has been effectively criticized. The alternative hypothesis of Canny (1995) with the modification that I described is complex and has not yet been shown to be valid. A more comprehensive model is needed that includes water flow in the xylem and interactions with sugar flow in the phloem (e.g., the Münch model), such as that described by Lampinen and Noponen (2003).

8.9 CROP WATER RELATIONS

The water relations of living cells can be strongly influenced by osmotic adjustment (accumulation of solutes). The growth of cells or tissues per unit volume, length, or area has been modeled by Equation 8.27:

$$dV/dt \text{ or } dL/dt \text{ or } dA/dt = K \times \left(\psi_p - \psi_{pthreshold}\right) \tag{8.27}$$

where K is an extensibility, and both K and $\psi_{pthreshold}$ depend on metabolism, synthesis, and cell wall loosening and therefore temperature. The ψ_p of actively growing cells is about 8 bar, which means that ψ_s inside the cell must be more negative than –8 bar. The low solute potential results from the continuous accumulation of solutes by either uptake of inorganic ions such as K^+ or Cl^- or the synthesis of small organic molecules such as sugars, organic acids, or amino acids. When the growth rate of plant tissues decreases due to drought or other factors, it should not be assumed that this is caused by decreases in ψ_p. Studies indicate that in many cases all of the variables in Equation 8.27 change, and in some cases ψ_p remains constant even though growth rate has decreased.

When a cell loses water, which occurs whenever its Ψ decreases, its ψ_p also decreases, as described by Equation 8.28 for the case where there is osmotic change due to concentration of solutes but no osmotic adjustment.

$$\frac{\Delta\psi_p}{\Delta\Psi} = \frac{\varepsilon_b}{(\varepsilon_b - \psi_{so})} \tag{8.28}$$

where ε_b is the bulk modulus of elasticity, that is, the change in pressure per unit change in relative volume $\varepsilon_b = dP/dV/V$, and ψ_{so} is the initial solute potential. With a rigid tissue having an ε_b of 62 bar (and a ψ_{so} of –15 bar), the change in pressure potential with a change in total potential is 0.80, whereas, with a softer tissue (ε_b = 42 bar), the cell shrinks more. Thus there is a greater concentration of solutes, and the change in pressure potential is only 0.74. If the decrease in Ψ is accompanied by osmotic adjustment, the ψ_p may not decrease or may only decrease slightly.

When species are subjected to long-term (days) drought, cells may experience an accumulation of solute. Measurement of relative water content (*RWC*) is useful when estimating the extent of osmotic adjustment.

$$RWC = 100 \times \frac{FW - DW}{TW - DW} \tag{8.29}$$

where *FW* is the fresh weight of the tissue, *TW* is the turgid weight after equilibration with pure water, and *DW* is the dry weight of the tissue after putting it into an oven to remove the water. The following procedures can be used to estimate whether osmotic adjustment occurs in a tissue: Estimate ψ_s values of the symplast of the tissue at full turgor prior to the drought and at full turgor after the drought. The extent to which ψ_s is more negative after the drought indicates the extent of osmotic adjustment, and the change in osmoles per unit volume can be estimated using Equation 8.22.

The overall procedure involves measuring *RWC* of the tissue, such as with one-half of a leaf and then measuring ψ_s of an extract from the other half of the leaf after it has been frozen and thawed. Next, one should correct for dilution of the extract by the apoplast solution using Equation 8.30.

$$\psi_{s(symplast)} = \psi_{s(extract)} \times \frac{RWC}{RWC - \text{apoplast}\%} \tag{8.30}$$

where apoplast% is the percentage of the water in the leaf at full turgor that is outside the plasma membranes, that is, in the cell walls and lumens of the xylem vessels. A value for the apoplast% can be either obtained from the literature or measured by using a pressure chamber to create a pressure-volume curve for the leaf (Boyer, 1995). The last step needed is to correct for changes in cell volume by normalizing the ψ_s value to full turgor using Equation 8.31:

$$\psi_{s(full\ turgor)} = \psi_{s(symplast)} \times \frac{RWC - \text{apoplast}\%}{100 - \text{apoplast}\%} \tag{8.31}$$

Osmotic adjustment in the vacuoles, which are the major part of the symplast, involves accumulation of inorganic ions, such as K^+, synthesis of organic acids, and the conversion of polysaccharides to simple sugars. Osmotic adjustment in the protoplasm includes accumulation of solutes that are compatible with the maintenance of the structure and function of macromolecules and membranes. Such compatible solutes include proline, glycine-betaine, mannitol, and sorbitol. Compatible solutes are highly soluble compounds that carry no net charge at physiological pH and are nontoxic at high concentrations. In addition to lowering the solute potential, compatible solutes act to stabilize proteins and membranes (McNeil et al., 1999).

The physiological significance of osmotic adjustment is that it occurs in all growing cells, such as cells in growing roots, and non-growing cells such as stomata as a mechanism that maintains the turgor pressure as a necessary force for pushing out the cell wall. It also occurs in some plant species when they are subjected to long-term drought or salinity and acts to maintain turgor pressure under these conditions. There may be physiological costs to osmotic adjustment in that some species, including ones that are well adapted to drought such as cowpea, exhibit very little osmotic adjustment. Possible costs are the need for strong cell walls, the capacity to de-osmotically adjust very rapidly, and the energy costs of ion pumping. For example, a creosote bush grown under dry desert conditions can osmotically adjust over several months without rain such that it develops ψ_s in its leaves of about –90 bar. On rare occasions, rain occurs at this time, and if the cells were to come into equilibrium with this pure water, they would develop a turgor pressure greater than 80 bar, which would cause them to burst. This probably does not happen, so I conclude that the cells must have the capacity to rapidly pump the solutes out of the symplast and into the apoplast, possibly in response to turgor pressure values exceeding a threshold value. The adaptive significance of osmotic adjustment will be discussed in Chapter 9.

ADDITIONAL READING

Bradford, K. J. 1994. Water stress and water relations of seed development: A critical review. *Crop Sci.* 34: 1–11.

Canny, M. J. 1995. A new theory for the ascent of sap-cohesion supported by tissue pressure. *Ann. Bot.* 75: 343–357.

Elfving, D. C., M. R. Kaufmann, and A. E. Hall. 1972. Interpreting leaf water potential measurements with a model of the soil–plant–atmosphere continuum. *Physiol. Plant.* 27: 161–168.

Hall, A. E., E.-D. Schulze, and O. L. Lange. 1976. Current perspectives of steady-state stomatal responses to environment. In O. L. Lange, L. Kappen, and E.-D. Schulze (Eds.), *Water and Plant Life Problems and Modern Approaches, Ecological Studies Vol. 19.* Springer-Verlag, New York, pp. 168–188.

Hillel, D. 1971. *Soil and Water Physical Principles and Processes.* Academic Press, New York, p. 288.

Jones, H. G. 1992. *Plants and Microclimate,* 2nd ed. Cambridge University Press, Cambridge, UK, p. 428.

Kramer, P. J. and J. S. Boyer. 1995. *Water Relations of Plants and Soils.* Academic Press, San Diego, CA, p. 495.

Wei, C., M. T. Tyree, and E. Steudle. 1999. Direct measurement of xylem pressure in leaves of intact maize plants. A test of the cohesion-tension theory taking hydraulic architecture into consideration. *Plant Physiol.* 121: 1191–1205.

Zimmermann, U., F. C. Meinzer, R. Benkert, J. J. Zhu, H. Schneider, G. Goldstein, E. Kuchenbrod, and A. Haase. 1994. Xylem water transport: Is the available evidence consistent with the cohesion theory? *Plant Cell Environ.* 17: 1169–1181.

9 Crop Adaptation to Water-Limited Environments

Subjecting crop plants to drought usually is considered detrimental. There are however, some cases when drought can be advantageous. For example, when coffee plantations growing in humid subtropical climates do not experience a significant drought during the year, they may exhibit sporadic flowering and sporadic berry production throughout the year (DaMatta, 2004). This makes harvesting inefficient and can reduce annual yields. Coffee plants subjected to a moderate drought on one occasion during the year exhibit substantial flowering after the drought is relieved and one large crop of berries. In irrigated coffee plantations, droughts can be managed to promote flowering of different blocks at different seasons to spread harvesting and processing of berries throughout the year.

Another example of a useful drought is where seed of an irrigated annual crop is sown into moisture and then subjected to a drought. This promotes deep root development such that the crop can better withstand subsequent droughts. For cowpea it has been shown that the first irrigation can be delayed until the first appearance of macroscopic floral buds without reducing grain yield (Ziska and Hall, 1983b; Ziska et al., 1985). This management practice reduces water-use costs and increases profits because the crop requires one less irrigation without effects on grain yield.

In considering the adaptation of crop plants to water-limited environments, it is useful to define *drought* and mechanisms of drought adaptation such as *drought escape* and *drought resistance*. Scientists and other people tend to use these terms in different ways, and the different definitions results in some confusion.

- *Drought* is defined as where a dry soil (due to lack of rain or delayed irrigation) or a hot dry wind (high ET_o) cause a substantial reduction in crop performance in terms of either plant survival or economic yield or crop quality.
- *Drought escape* is defined as where drought-sensitive stages of plant development (i.e., flowering and seed or fruit development) are completed during the part of the season when drought is not present.
- *Drought resistance* is defined as the ability of a cultivar to produce a greater economic yield (or to survive better) than another cultivar when it is subjected to soil or atmospheric drought, or the ability of a species to be more effective (e.g., profitable) than another species when they are subjected to soil or atmospheric drought.

For annual cropping in rain-fed areas where water is limited, well-adapted cultivars will have an optimal time of flowering and a cycle length from sowing to harvest that fits the patterns of rainfall and water availability. If the cycle is too long, drought during grain filling will reduce seed size and, in some cases, reduce seed viability, such as with some modern varieties of sorghum, and reduce grain yield. If flowering occurs too early, yield potential will be lower, resulting in smaller grain yields in wet years compared with cultivars that have a longer cycle length. Also, cultivars that flower too early may suffer from molds caused by rain during fruit or seed maturation.

Determining the time of flowering and cycle length that are optimal for a particular climatic zone is complex, due to the extreme variability in rainfall in semiarid zones. Hydrologic budget analysis, which is discussed in Chapter 10, can be used to determine the date of flowering and cycle length that are optimal in specific locations and soil conditions. The extent to which crop phenology can influence the extent of pest problems is discussed in Chapter 12.

9.1 CROP SPECIES DIFFERENCES IN DROUGHT RESISTANCE

Species differences in drought resistance depend on the type of economic product of the species. Species producing leafy vegetables, such as lettuce and spinach, have little drought resistance. The yield and quality of the leaves are reduced by even mild droughts. There is a need to maintain high plant turgor both by frequent irrigation or rain and by growing the crop in an environment with a low evaporative demand, that is, cool and humid, such as occurs year-round in the coastal Salinas Valley or in winter in the Imperial Valley of California (Figure 10.2).

Tuber crops, such as Irish potato, are more resistant to drought than leafy vegetables, but their yield and quality can be reduced by mild to moderate drought. They require frequent irrigation or rain when the tubers are expanding but can be grown in environments where the evaporative demand is high. For example, Irish potatoes are grown in either the fall or spring seasons near Hemet and Bakersfield in California (Figure 10.3), but when the evaporative demand is high, it may be necessary to irrigate them as frequently as every three days.

In contrast, hay crops such as alfalfa are even more drought resistant, and their yield is only reduced when drought becomes moderate. Hay crops can be productive in environments with high evaporative demands, providing they are adapted to the temperatures experienced in these zones. Irrigation cycles should be sufficiently frequent to prevent significant stomatal closure and reductions in photosynthesis. Alfalfa also is grown in the areas near Hemet and Bakersfield in California but, when the evaporative demand is high, it only requires irrigation about every 14 days, compared with the optimal interval of every three days for Irish potato. The greater drought resistance of alfalfa compared with Irish potato is partially due to the ability of alfalfa to extract more of the available water in the root zone without suffering reductions in hay yield, and partially due to it having deeper roots than potato and thus access to more soil water.

Where economic yield is a reproductive organ, resistance to drought depends on the stage of reproductive development, the type of economic product, and the determinacy of the plant. Plants often are more drought resistant during the vegetative stage than during early flowering or fruit development stages. Plants producing dry grain are more resistant to late-season drought than plants producing fleshy fruit, which require higher turgor. Indeterminate plants, such as cotton, cowpea, and tomatoes, can exhibit superior adaptation to mid-season droughts than determinate plants, such as maize, pearl millet, rice, sorghum, and wheat. This is because after the droughts, indeterminate plants can produce more leaves, fruits, and seeds, whereas determinate plants do not produce any more leaves or seeds on their main stems. Some cowpea cultivars are particularly plastic in that they have the ability to consistently produce a second flush of leaves flowers, pods, and grain, and this can be important when the first flush is destroyed by drought (Gwathmey and Hall, 1992) or other stresses such as insect pests. Also, indeterminate crops may have substantial resistance to vegetative-stage drought. Cowpea plants that were grown on a small amount of stored soil water and not irrigated until 43 days after sowing, but that received optimal irrigation after this, produced the same high grain yields as weekly irrigated plants (Turk et al., 1980). The study was conducted in a California summer environment with no rain and the drought was so severe during the vegetative stage that it reduced the leaf area of the cowpea plants by 74%, compared with the weekly irrigated plants, and would have killed most other annual crop species. Determinate crops can be well adapted to late droughts during grain filling because they can exhibit more complete remobilization of carbohydrate to their grain than indeterminate crops. For example, pre-anthesis reserves may contribute about 40% of the carbohydrate in the grain of barley and wheat subjected to late drought (a review of this topic is presented by Evans, 1993).

A model for the effects of drought on determinate annual crops can be used to quantify their susceptibility to drought at different stages, which is related to their overall drought resistance:

$$Y_d = Y_w\left(1 - D_v S_v\right)\left(1 - D_f S_f\right)\left(1 - D_s S_s\right) \tag{9.1}$$

where Y_d is the grain yield under drought; Y_w is the grain yield of a well-watered crop; D is the drought intensity during the vegetative (v), flowering (f), or seed-filling (s) stages (with $D = [1 - ET_d/ET_w]$, where ET is evapotranspiration of the droughted and well-watered crops during these stages); and S is the drought susceptibility of the crop during the vegetative, flowering, and seed-filling stages with values varying from 1.0 (highly susceptible) to 0 (highly tolerant). Values of S can be determined empirically by subjecting different experimental plots of plants to different droughts of different intensities imposed during different stages of development. ET would be measured to permit calculation of the D values. Y_w and Y_d would be measured. The S values would be obtained by solving a set of simultaneous equations. If the model is effective, the S values for specific stages would be relatively constant in different experiments. The model predicts that drought at any stage has a proportional effect on yield of determinate crops. Experimental studies indicate that S_f is often > S_s is often >S_v.

Maize is extremely sensitive to drought during the reproductive stage. A crop that experiences a drought of short duration at tasseling but is otherwise well irrigated may produce considerable biomass but cobs that have very few kernels. Two mechanisms may partially explain this sensitivity of maize to drought.

The first mechanism involves drought-induced delay in silk emergence but not tasseling so that the synchrony is disturbed and fewer silks are pollinated. For well-watered plants, silks, which are elongated styles with one silk per ovule, begin emerging from the ear leaf sheath at about the same time that tassels begin releasing pollen. Drought often delays silk emergence but not tasseling, and the pollen is viable, but most of it is shed prior to the emergence of the silks (Hall et al., 1982). Consequently, pollination is substantially decreased, and for every silk that is not pollinated, there would be one less kernel on the ear. A study by Herrero and Johnson (1981) provides insights concerning the mechanisms of these effects of drought on silk elongation. Silks of well-watered plants elongated rapidly at night and then slowed down to zero elongation during the day, when leaf water potential had more negative values. Drought resulted in slower silk elongation at night compared with well-watered plants and caused silks to shrink during the day, when leaf water potentials were even more negative than for the well-watered plants. Differences in silk turgor pressure may partially explain both the diurnal variation in silk elongation rate and the differences in silk elongation between well-watered and droughted plants. However, silk elongation rates were much slower for droughted plants that exhibited the same ear leaf water potentials as well watered plants. An additional factor, such as reduced extensibility (Equation 8.27), may have been responsible for the slow silk elongation rates of droughted plants. Drought-induced asynchrony in tasseling and silking would be more pronounced with populations of plants that are very uniform, such as with F_1 hybrids. Comparisons of maize cultivars under drought using typical small plots may not provide reliable predictions of performance in large fields in that the synchrony of tasseling and silking would be influenced by plants in neighboring plots, since maize is about 80% cross-pollinated by wind.

The second mechanism for the extreme sensitivity of maize to drought during reproductive development involves drought-induced embryo abortion. Maize was subjected to a few days drought at pollen shedding in a controlled environment where adequate pollen was provided to silks. Pollen germinated, pollen tubes grew within the silks, and the egg sack was fertilized, but embryo abortion occurred (Westgate and Boyer, 1986). The authors hypothesized that photosynthesis was inhibited in these plants and that a shortage of carbohydrates was responsible for the embryo abortion. Studies in which sucrose was supplied to droughted maize plants using a stem-infusion technique supported this hypothesis in that it prevented the embryo abortion (Boyle et al., 1991).

The drought resistance of tropical maize has been enhanced by selection based on the first mechanism, that is, by selection to reduce the anthesis-silking interval under drought. Recurrent selection of maize for three to eight cycles increased grain yield by 30%–50% with no change in total shoot biomass for plants grown in environments with drought during flowering (Edmeades et al., 1999). Grain yield under drought was strongly and negatively correlated with the anthesis-silking interval and was not associated with any morphological or physiological traits indicative of improved plant water status (Bolaños and Edmeades, 1996; Chapman and Edmeades, 1999). Bolaños and Edmeades (1996) recommended that when breeding maize to enhance drought resistance at flowering, two types of nurseries and selection programs should be used: selection for grain yield in a well-watered nursery to enhance general adaptation, alternating with selection for anthesis-silking interval, ears per plant, and grain yield in a nursery with severe water stress at flowering to enhance drought resistance.

Wheat also is sensitive to drought stress during an early stage of flowering, but the mechanisms are different from those with maize. Seed number per spike in wheat can be substantially reduced by drought stress occurring about seven days before anthesis (Fischer, 1980). The floret sterility is associated with pollen sterility (Jones, 1992) in contrast with maize, where pollen was not damaged by drought during flowering (Herrero and Johnson, 1981; Hall et al., 1982). Drought- and heat-induced male sterility in wheat may be caused by the high abscisic acid levels that can occur with these stresses (Blum, 1988).

A different model should be used for the effects of drought on grain yield of indeterminate annual crops:

$$Y_d = Y_w\left(1 - D_v S_v\right)\left[\left(N_{r1} - D_{r1}S_{r1}\right) + \left(N_{r2} - D_{r2}S_{r2}\right),\ \text{etc.}\right] \quad (9.2)$$

where subscripts r_1 and r_2 denote separate reproductive periods of the indeterminate crop; N is the proportion of grain yield attributed by a single flush of fruiting with the sum of N_{rl}, N_{r2}, etc., equal to 1.0; and D and S having the same definitions as were used for Equation 9.1.

The S_v of indeterminate crops usually is smaller than the S_v of determinate crops because the indeterminate crop can produce more leaves once the drought is ended. In the case of an indeterminate cowpea that survived a vegetative drought that would have killed most annual crop plants and then recovered sufficiently that it produced the same high grain yields as weekly irrigated plants, this indicates an S_v value of 0.0 (Turk et al., 1980). However, in another year, when the evaporative demand was high during the reproductive period, cowpea plants did not completely recover after the vegetative-stage drought and suffered 35% reductions in grain yield compared with weekly irrigated plants, and S_v was about 0.7 (Turk et al., 1980). The substantial variation in S_v indicates that there are conditions in which the model described by Equation 9.2 is not valid. Indeterminate crops, such as cowpea, have substantial adaptation to drought also because of their plasticity in being able to produce another flush of pods after the drought has ended (Gwathmey and Hall, 1992). According to Equation 9.2 the effects of drought on indeterminate crops tend to be independent and additive during the different reproductive periods. This model is only approximately valid in that, if the first flush of fruiting is substantially damaged by drought, the second flush may partially compensate by producing a flush of fruits that even is greater than the second flush produced by well-watered plants. It should be apparent that a model for indeterminate crops that closely simulated reality would be very complex and would have to include many emergent properties.

Effects of drought on seed size depend on the separate effects of drought on the relative sizes of the photosynthetic source and the reproductive sink (Fischer, 1980). Seed size has a hyperbolic relationship with the ratio of the photosynthetic source to the reproductive sink (Figure 9.1). Late drought tends to result in small seed, because it has little influence on the number of seed produced but strongly reduces the photosynthetic source by accelerating leaf senescence. In this case, the seed may have low vigor, as has been observed with some modern cultivars of sorghum. Drought-induced reductions in seed size tend to be more pronounced with determinate annuals, but they can occur with late drought in indeterminate crops such as cowpea (Turk et al., 1980). Mid-season drought can have little effect on seed size in both determinate annuals and indeterminate crops, if it reduces both the number of seed produced and the size of the photosynthetic source in a balanced manner (Figure 9.1).

Some differences in drought resistance among annual crop species include the following:

- *Warm-season cereals*: sorghum and pearl millet > maize and rice
- *Warm-season grain legumes*: cowpea > peanut > soybean > common bean
- *Cool-season crops*: barley > wheat > Irish potato

For perennial crops, such as trees, effects of drought can be complex in that drought in one year can affect yield in subsequent years. The simplest case is where young trees are subjected to drought, and it increases the number of years required for them

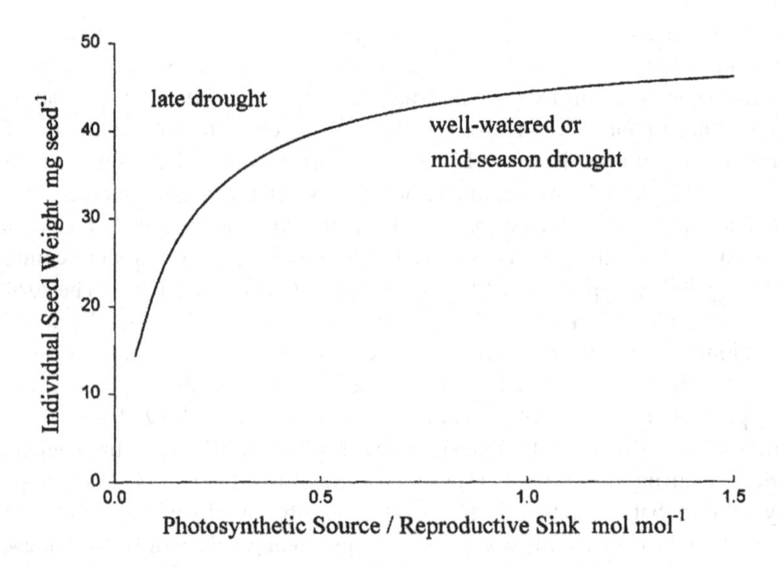

FIGURE 9.1 Relationship between individual seed weight and the ratio of photosynthetic source to reproductive sink for a determinate annual crop, such as wheat, subjected to late drought and drought at mid-season and when well-watered.

to gain near-complete interception of solar radiation and maximum yields. For some deciduous trees, such as apricot, flower buds are produced in the summer of the year preceding the fruit production year, and drought at this time can decrease the number of flowers and fruit produced in the following year. It should be noted, however, that trees producing many fruit often produce small fruit. Gross income per tree may depend more on the yield of large fruit, which gain a price premium, than on the total weight of fruit produced per tree.

The extent of carbohydrate reserves in roots may influence sensitivity to drought. Recently coppiced trees of *Eucalyptus camaldulensis* can be much more sensitive to drought, in terms of drought-induced leaf senescence, than seedlings, and it was shown that sensitivity to drought was associated with low levels of root carbohydrate reserves compared with levels of root carbohydrate reserves in seedlings (Hall, 1993b). Consequently, it is possible that trees experiencing biennial bearing may be more sensitive to drought, in terms of drought-induced leaf senescence, during and just after the *on year*, when they have produced many fruit and have low carbohydrate reserves in their roots, than during and after the *off year*, when the fruit load was very low and they have more carbohydrate reserves in their roots.

Some differences in drought resistance among evergreen perennial crop species include the following:

jojoba and olive > citrus > avocado

Drought due to high evaporative demands is a special case, in that high evaporative demands can be accompanied by high winds and sand blasting. Crop species that are adapted to these conditions include date palm, jojoba, pineapple, and sisal, which have very tough leaves. In contrast, some crops are not well adapted to environments with high evaporative demands, including leafy vegetables, tea, tobacco, and banana (because it can blow over in a strong wind).

9.2 MECHANISMS OF DROUGHT RESISTANCE

Drought resistance can be considered as depending on the extent of dehydration avoidance, feedforward responses, dehydration tolerance, and water-use efficiency.

9.2.1 Dehydration Avoidance

This refers to the extent to which relative water content (*RWC*) is maintained under drought compared with other plants. *RWC* is defined by Equation 8.29 in Chapter 8.

Plants can avoid dehydration by maintaining higher Ψ_{leaf} (closer to zero) when subjected to drought if they have deeper roots to access moisture present deep in the soil, slower growth of leaf area, and earlier drought-induced stomatal closure compared with other cultivars. Note, however, that adaptation requires a balanced response because all of these mechanisms of dehydration avoidance have costs in terms of either greater use or less acquisition of carbohydrate. Among plants with substantial resistance to drought, some exhibit little change in Ψ_{leaf} (it does not go below –2 megapascal) when subjected to drought (e.g., cacti, cowpea, [Petrie and

Hall, 1992] and siratro), whereas others exhibit large decreases in Ψ_{leaf} down to –5 megapascal (e.g., pearl millet [Petrie and Hall, 1992] and sorghum), and a few species can develop very low Ψ_{leaf} under extreme drought (i.e., creosote bush has exhibited Ψ_{leaf} values as low as –9 megapascal).

Plants that develop low values of Ψ_{leaf} can still partially maintain *RWC* if they also adjust osmotically, and experience decreases in ψ_s in cells. Plant breeders have developed varieties of wheat and sorghum with greater drought resistance by selecting for greater drought-induced osmotic adjustment in leaves. However, the adaptive mechanism may involve osmotic adjustment in roots that result in the maintenance of ψ_p in root cells and maintenance of root growth, enabling the improved cultivars to access more soil water than the older cultivars (Ludlow, 1993). It should be noted that this mechanism of adaptation would be effective only where the deeper roots access substantially more soil water. Some plant species with excellent adaptation to drought exhibit very little drought-induced osmotic adjustment, for example, cowpea (Petrie and Hall, 1992). However, osmotic adjustment would appear essential in the adaptation of halophytes to very saline conditions. Attempts are being made to enhance the salt tolerance of crops by incorporating the ability to produce compatible solutes through genetic engineering. An example of this involves incorporating the pathway for synthesizing glycine-betaine into species that lack this pathway (McNeil et al., 1999). Refer to Chapter 11 for a discussion of salt tolerance.

9.2.2 Feedforward Responses

There is now increasing evidence that roots sense difficult conditions in the soil and send signals to the shoot that cause partial stomatal closure and slow down leaf expansion before the supplies of water or nutrients are affected (Passioura and Stirzaker, 1993). Such behavior is known in control theory as *feedforward*, which contrasts with *feedback* in that it provides advance warning of change. If stomata partially closed or leaf area expansion rate decreased as a feedback response to a decrease in water status, the plant might suffer damage from the low plant water status. In contrast, feedforward mechanisms enable the plant to avoid extreme dehydration. Also, feedback systems can be undesirable in that they have a tendency to exhibit oscillations that represent an inefficient use of resources. Feedforward responses appear to be a component of the sophisticated system that maintains a balance between root activity and shoot activity.

The bonsai effect is an extreme example of the regulation of shoot growth to match root growth. Plants grown in small containers usually are much smaller than those grown in large containers, even when seemingly adequate supplies of water and nutrients are provided (Passioura and Stirzaker, 1993). The small bonsai plants do not manifest any obvious symptoms of water-deficit or nutrient stress. Crop plants can exhibit the bonsai effect in field conditions if the soil does not permit the plants to have significant root growth.

When the soil profile dries, many changes take place within it. It not only holds water more strongly, it also gets harder, it transmits water and solutes less readily, and salts become more concentrated. Studies with split root systems, where one part of the root system was kept well-watered while the other was permitted to dry, have provided evidence for root signals initiating feedforward responses causing stomata to partially close and leaf expansion to slow down (Passioura and Stirzaker, 1993). By using a pot pressurization technique to prevent the water potential of the leaf from changing, it was shown that the feedforward system responds to both the drying and the hardening of the soil (Passioura and Stirzaker, 1993). Plant responses to hard soils that can restrict rooting are discussed in Chapter 11.

Stomatal responses to low humidity (atmospheric drought) also may involve a feedforward response (Farquhar, 1978) if the stomata are responding to increases in cuticular water loss from the epidermis. This type of feedforward response would act to prevent very rapid rates of transpiration and large decreases in bulk leaf water potential. Note, however, that if stomatal response to humidity really is a stomatal response to transpiration rate, then the mechanism of the effect would be one of feedback (Farquhar, 1978). At this time, the extent to which stomatal response to humidity involves a feedforward system or a feedback system is not known.

The significance of feedforward responses to crop production is that they may be too strong and favor plant survival too much over plant productivity and thus be too conservative for annual crop plants in most target production environments. For perennial crops, strong conservative feedforward systems may be desirable. In this case, choice of root stocks may provide opportunities for modifying the root component of the feedforward system in a beneficial manner.

9.2.3 Dehydration Tolerance

This refers to the extent to which plant function is maintained when *RWC* decreases. The mechanisms of leaf dehydration tolerance are poorly understood. One hypothesis is that there is a critical *RWC* at which processes stop or start, such as leaf death. But higher plants may not respond to or be damaged by dehydration per se. The *RWC* of plant tissue provides a measure of relative symplast (i.e., cell) volume, and plants may respond to changes in volume or more likely to changes in turgor pressure but not to the level of dehydration. Ludlow and Muchow (1990) reviewed studies of lethal levels of low water status in plants. They commented that for pigeon pea subjected to different rates of soil drying, leaf death occurred at a specific critical *RWC* (32%), irrespective of substantial differences in level of osmotic adjustment and leaf water potential. For pigeon pea, the *RWC* at which zero turgor occurs is about 80% (Flower and Ludlow, 1986). Thus, an *RWC* of 32% could represent a leaf that has suffered a catastrophic irreversible inward collapse of cell walls (cytorrhysis) resulting from a critical level of negative turgor pressure.

An alternative hypothesis is that drought-induced leaf death is really a programmed leaf senescence caused by changes in hormonal signals coming from roots that are being subjected to day-to-day decreases in soil moisture content (Hall, 1993b). In this hypothesis, plasma membranes in leaf cells respond to the changes in hormonal signals by permitting osmotica to leave the cells, which would result in an adaptive recycling of nutrients from senescing leaves. The reduction in osmotic pressure (increase in solute potential) within the cells would result in a loss of water, negative turgor, and a collapse of the cells.

Tolerance to dehydration and reduction in activity of water appear to be important in seeds that become very dry at maturity. Certain compounds accumulate during seed development and are thought to play a role in preventing damage to the desiccating embryo. These compounds include sugars (Koster and Leopold, 1988) and proteins, among which the LEA D-11 family of proteins (dehydrins) have been suggested to play a role in desiccation tolerance (Close, 1996).

9.2.4 Water-Use Efficiency (*W*)

This is the ratio of biomass production to transpiration. Species differences in *W* have been observed. When grown in the same optimal warm to hot environment, warm-season C_3 species have lower *W* than C_4 species, mainly because the C_4 species produce more biomass per day. Refer to Chapter 8 and the discussion of transpiration efficiency, which is similar to *W* (see Equation 9.4).

The possibility that cultivars within a species also may vary in *W* has been investigated with several C_3 species in the hope that it may provide an opportunity for breeding varieties with greater drought resistance (Hall et al., 1994; Condon and Hall, 1997). This research was stimulated and facilitated by theory indicating that it may be possible to detect genotypic differences in *W* within C_3 species by indirect selection for differences in stable carbon isotope composition (refer to the discussion of transpiration efficiency in Chapter 8). This approach to breeding is based on the following model for grain yield (*Y*) under water-limited conditions:

$$Y = \sum_{i=n}^{i=1} E_i \times \left(\frac{T_i}{E_i}\right) \times W_i \times CP_i \tag{9.3}$$

where the total water use per day E_i, which includes evaporation from the soil and transpiration, the ratio of transpiration to total water use per day (T_i/E_i), the W_i per day, and the partitioning of dry matter to grain per day (CP_i) are summed over the period (*n*) when most photosynthesis occurs that provides dry matter for grain filling. For cowpea, this is the period from first formation of floral buds to physiological maturity (Hall, 1999). The water-use efficiency, *W*, can be related to the time-integrated transpiration efficiency ($\int P_n/\int T_r$) using Equation 9.4, developed by Hubick and Farquhar (1989):

$$W = \left(\frac{\int P_n}{\int T_r}\right)\left(\frac{1-\theta_c}{1+\theta_w}\right) \tag{9.4}$$

where θ_c is the influence of respiration expressed as a fraction of the carbon fixed, and θ_w is the influence of the water lost from plant surfaces that is independent of CO_2 uptake, such as the cuticular loss from shoots at night, expressed as a fraction of the transpiration.

A model of the type described by Equation 9.3 would be very useful for guiding breeding if the various components were either relatively independent or positively associated such that increases in any component would increase grain yield. Unfortunately, this appears not to be the case (Condon and Hall, 1997). For example, cultivars with deeper roots that are accessing more water would have greater E_i and T_i/E_i if they have greater ground cover, but smaller W_i because their stomata would be more open, which would benefit T_r more than P_n. Also, negative correlations have been reported between W and CP for several species (Condon and Hall, 1997). The physiological basis of this negative correlation is not known. It should be noted that enhanced W will lead to greater drought resistance only if it also is associated with near-complete use of available soil water and effective partitioning of carbohydrate to grain, and it is not clear at this time whether this suite of traits can be combined.

Progress has been made in Australia, however, in breeding wheat varieties with increased W by using selection for carbon isotope discrimination. These varieties had enhanced yield under specific water-limited conditions (Richards et al., 2002). In other cases, it is not absolutely clear at this time whether breeders should select to either increase or decrease W for different water-limited and well-watered environments.

Genotypic differences in W_i could arise from two causes:

1. Cultivars with greater photosynthetic capacity, due to higher levels of rubisco per unit leaf area, would have greater W_i and greater biomass production, as has been observed in peanut. Peanut genotypes have been selected with greater W_i, and they had greater photosynthetic capacity and produced more biomass under water-limited conditions. Unfortunately, the selected genotypes did not produce any more peanuts due to reductions in partitioning of carbohydrate to grain (Condon and Hall, 1997).
2. Cultivars with smaller stomatal conductance would have greater W_i but smaller biomass production, as has been observed in wheat. Some experiments have been conducted where segregating populations of wheat and cowpea were selected for higher and lower W (based on stable carbon isotope composition), and they indicate that low W may be adaptive in a range of well-watered to moderately droughted environments (Condon and Hall, 1997). For well-watered, hot environments, it may be argued that selecting for lower W may select cultivars with more open stomata, cooler leaves, greater photosynthesis, greater biomass production, and greater grain yield. It is more difficult to explain how lower W could be associated with greater grain yield in water-limited environments. A possibility is that the more open stomata are linked to the presence of more effective rooting systems with enhanced ability to access water deep in the soil. In this case, the better-adapted cultivar is less efficient in the use of water but accesses much more soil water than the poorly adapted cultivar such that it produces more biomass and more grain.

Crop plant adaptation to drought depends on the additive effects of drought escape and drought resistance. Drought resistance, however, does not depend on the additive effects of dehydration avoidance, feedforward responses, dehydration tolerance, and water-use efficiency. Plants with different mechanisms may be equally well adapted to the same semiarid environment. For example, cereals subjected to soil drought develop low (very negative) leaf water potentials but avoid dehydration by exhibiting osmotic adjustment. Some grain legumes growing in the same environment avoid dehydration by partial stomatal closure and paraheliotropic leaf movements that reduce transpiration (Shackel and Hall, 1979) such that leaf water potential decreases only slightly with soil drought. The possible benefits of the cereal strategy are that by keeping stomata more open, they may have more photosynthesis, and by developing lower leaf water potentials, they may access slightly more water from the soil. Possible costs of the strategy are associated with osmotic adjustment, which may have significant energy costs in terms of transporting osmotica and the construction of tough leaves that can withstand occasional high turgor pressures if de-osmotic adjustment is not sufficiently rapid following rain or sprinkler irrigation. (Refer to Chapter 8 for a more complete discussion.) In general, adaptation to drought is conferred by optimal levels of drought escape, dehydration avoidance, feedforward responses, and water-use efficiency in relation to the current and expected types of drought in the specific target production environment.

Some insights into plant adaptation to drought may be obtained from studies of native plants adapted to arid environments (Ehleringer, 1993). In these studies, it was found that different types of plants had different internal CO_2 concentrations as indicated by measurements of carbon isotope discrimination (Δ). The Δ values were considered to provide the set point for plant gas exchange activity, including both photosynthesis and transpiration. Most of the native plants growing in arid environments are C_3 plants. Annual plants had the highest Δ values, indicating high gas exchange activity, high growth rates, and little resistance to drought. Among the perennials, deciduous shrubs had higher Δ and higher growth rates than evergreen shrubs. The most conservative species were long-lived evergreen shrubs, which had the lowest Δ values irrespective of the age of the plant when the measurements were made. Adaptation to drought of different native species and crop plants involves different *strategies*, with different suites of plant traits associated with each *strategy*.

ADDITIONAL READING

Bolaños, J. and G. O. Edmeades. 1996. The importance of the anthesis-silking interval in breeding for drought tolerance in tropical maize. *Field Crops Res.* 48: 65–80.

Boyle, M. G., J. S. Boyer, and P. W. Morgan. 1991. Stem infusion of liquid culture medium prevents reproductive failure of maize at low water potential. *Crop Sci.* 31: 1246–1252.

Close, T. J. 1996. Dehydrins: Emergence of a biochemical role of a family of plant dehydration proteins. *Physiol. Plant* 97: 795–803.

Condon, A. G. and A. E. Hall. 1997. Adaptation to diverse environments: Variation in water-use efficiency within crop species. In L. E. Jackson (Ed.), *Ecology in Agriculture.* Academic Press, San Diego, CA, pp. 79–116.

Ehleringer, J. R. 1993. Carbon and water relations in desert plants: An isotopic perspective. In J. R. Ehleringer, A. E. Hall, and G. D. Farquhar (Eds.), *Stable Isotopes and Plant Carbon–Water Relations.* Academic Press, San Diego, CA, pp. 155–172.

Farquhar, G. D., M. H. O'Leary, and J. A. Berry, 1982. On the relationship between carbon isotope discrimination and the intercellular carbon dioxide concentration in leaves. *Austral. J. Plant Physiol.* 9: 121–137.

Fischer, R. A. 1980. Influence of water stress on crop yield in semiarid regions. In N. C. Turner and P. J. Kramer (Eds.), *Adaptation of Plants to Water and High Temperature Stress*. Wiley, New York, pp. 323–339.

Flower, D. J. and M. M. Ludlow. 1986. Contribution of osmotic adjustment to the dehydration tolerance of water-stressed pigeonpea (*Cajanus cajan* (L.) millsp.). *Plant Cell Environ.* 9: 33–40.

Gwathmey, C. O. and A. E. Hall. 1992. Adaptation to midseason drought of cowpea genotypes with contrasting senescence traits. *Crop Sci.* 32: 773–778.

Hall, A. E. 1993b. Is dehydration tolerance relevant to genotypic differences in leaf senescence and crop adaptation to dry environments? In T. J. Close and E. A. Bray (Eds.), *Plant Responses to Cellular Dehydration during Environmental Stress*, Current Topics in Plant Physiology, Vol. 10. American Society of Plant Physiologists, Rockville, MD, pp. 1–10.

Hall, A. E., R. A. Richards, A. G. Condon, G. C. Wright, and G. D. Farquhar. 1994. Carbon isotope discrimination and plant breeding. *Plant Breeding Rev.* 12: 81–113.

Hall, A. J., F. Vilella, N. Trapani, and C. Chimenti. 1982. The effects of water stress and genotype on the dynamics of pollen-shedding and silking in maize. *Field Crops Res.* 5: 349–363.

Herrero, M. P. and R. R. Johnson. 1981. Drought stress and its effects on maize reproductive systems. *Crop Sci.* 21: 105–110.

Koster, K. L. and A. C. Leopold. 1988. Sugars and desiccation tolerance in seeds. *Plant Physiol.* 88: 829–832.

Ludlow, M. M. and R. C. Muchow. 1990. Critical evaluation of traits for improving crop yields in water-limited environments. *Adv. Agron.* 43: 107–153.

McNeil, S. D., M. L. Nuccio, and A. D. Hanson. 1999. Betaines and related osmoprotectants. Targets for metabolic engineering of stress resistance. *Plant Physiol.* 120: 945–949.

Passioura, J. B. and R. J. Stirzaker. 1993. Feedforward responses of plants to physically inhospitable soil. In D. R. Buxton et al. (Eds.), *International Crop Science I*. Crop Sciences Society of America, Madison, WI, pp. 715–719.

Petrie, C. L. and A. E. Hall. 1992. Water relations in cowpea and pearl millet under soil water deficits: I. Contrasting leaf water relations. *Austral. J. Plant Physiol.* 19: 577–589.

Turk, K. J., A. E. Hall, and C. W. Asbell. 1980. Drought adaptation of cowpea. I. Influence of drought on seed yield. *Agron. J.* 72: 413–420.

Westgate, M. E. and J. S. Boyer. 1986. Reproduction at low silk and pollen water potentials in maize. *Crop Sci.* 26: 951–956.

10 Hydrologic Budget of Cropping Systems, Irrigation, and Climatic Zones

Hydrologic budget analyses can be used to estimate the amount of water in the root zone and extent of deep drainage on a day-to-day or weekly basis, providing water is applied uniformly to the land surface by rainfall, sprinkler irrigation, basin irrigation, or furrow irrigation. The change in moisture storage in the root zone (ΔS) on a daily or weekly basis can be estimated using Equation 10.1:

$$\Delta S = (R + I) - (E + T + RO + D) \tag{10.1}$$

where R is rainfall, I is amount of water applied in irrigation, E is soil evaporation, T is transpiration, RO is surface runoff, and D is the extent of drainage below the root zone. Note that in some cases there is significant upward movement of water from a water table, and in this case D is negative and provides water to the root zone.

10.1 IRRIGATION MANAGEMENT

During a period of several days with no R or I, D often is small and ΔS is approximately equal to $E + T$, which is equal to ET_m for intensively managed crops and can be estimated from weather station data and crop parameters as described in Chapter 7. Consequently, in this case, soil water depletion in the root zone (ΔS) can be estimated from the sum of daily values of ET_m.

During a period with a large amount of R or I, the deep drainage (D) can be estimated, provided RO is very small. In agriculture, soil surfaces typically are managed so that RO is small to prevent soil erosion. In this case, the deep drainage equals the extent to which $R + I$ on a day exceeds the difference between the current soil moisture content of the root zone and the maximum amount of water that can be stored in the root zone (S_{max}):

$$S_{max} = \text{effective depth of root zone} \times \frac{P_{v(FC)}}{100} \tag{10.2}$$

where, for the purpose of irrigation management, $P_{v(FC)}$ is the percentage volumetric water holding capacity of the soil as measured one to three days after an irrigation

TABLE 10.1
Maximum Effective Rooting Depths of Different Types of Crop Species

Types of Crops	Maximum Effective Rooting Depth[a] (centimeter)
Shallow-rooted crops (e.g., lettuce, onions, and Irish potatoes)	30–50
Medium-rooted crops (e.g., cereals, grain legumes, and many fruit trees)	75–150
Deep-rooted (e.g., some perennials that have grown for at least a few months)	150–300
Very deep-rooted (e.g., some species of old trees)	300–3,000

[a] Depth of roots responsible for 95% of water uptake by the crop.

or rain of sufficient quantity to fully charge the profile. If the soil has layers with different $P_{v(FC)}$ values, S_{max} should be calculated separately for each layer and then summed. The effective depth of the root zone is that depth responsible for about 95% of ΔS due to water uptake by roots. Maximum effective rooting depths are provided for different types of crops in Table 10.1. If the depth of the soil is less than the maximum effective rooting depth, use the depth of the soil instead as an indicator of the actual rooting depth.

The maximum amount of the soil water that is available to crops (S_{avail}) can be estimated using Equation 10.3:

$$S_{avail} = S_{max} - S_{min} = \text{effective depth of root zone} \times \frac{P_{v(FC)} - P_{v(LL)}}{100} \quad (10.3)$$

where $P_{v(LL)}$ is the lower limit of soil water availability, which can be estimated as the volumetric water content of the soil after plant extraction until death or permanent wilting or at a soil ψ_m of –1.5 MPa. Hypothetical examples of S_{avail} in soils of different textural classes are provided in Table 10.2. Actual values of $P_{v(FC)}$ and $P_{v(LL)}$ depend on soil structure as well as soil texture and must be determined experimentally for each type of soil in the rooting zone.

TABLE 10.2
Examples of Water Retention Characteristics of Different Textural Classes of Soil

Textural Class	$P_{v(FC)}$ (%)	$P_{v(LL)}$ (%)	S_{avail} in a 100 Centimeters Deep Root Zone (centimeter)
Sandy soil	10	5	5
Sandy loam soil	20	5	15
Loam soil	30	10	20
Clay soil	50	30	20

Soils with more clay particles that are small have many small voids. They hold more water but hold it with more force, such that a substantial portion of it is not available to plants, compared with soils having more sand particles that are larger and have larger voids between them.

A widely used approach to irrigation management involves irrigating when a certain percentage of S_{avail} has been depleted, as estimated from daily values of ET_m, in a specific depth of the root zone. For example, drought-susceptible crops such as lettuce and Irish potatoes might be irrigated when 30% of S_{avail} in the top 30 centimeters of soil has been depleted. For some crops, such as cotton, tomato, and cowpea, the permissible depletion percentage can vary with the growth stage. For example, 100% depletion in the top 100 centimeters of soil during the vegetative stage of cowpea may not reduce grain yield whereas, once floral buds are apparent, it is necessary to irrigate when 40%–80% of S_{avail} has been depleted in the top 100 centimeters, depending on the soil type (Ziska and Hall, 1983a, 1983b; Ziska et al., 1985). In a soil with high bulk density that restricted rooting, it was advisable to irrigate with 40% depletion, whereas with a soil that was more favorable for root growth, irrigations could be delayed until 80% of S_{avail} was depleted. Basing the decision of when to irrigate on percentage depletion in a specific depth of the root zone is more effective than basing the decision on the percentage depletion in the actual total root zone. Farmers and scientists usually do not know what the depth of the root zone is, and it is difficult to quantify.

An approach to conventional low-frequency irrigation management of annual crops based on estimating the hydrologic budget could involve the following steps:

1. Pre-irrigate before sowing to bring the full effective root zone to $P_{v(FC)}$, except where the water supply is limited or expensive and some rain may fall in the next few weeks, in which case a smaller amount of water should be applied. Advantages of pre-irrigation are that it provides an opportunity to leach any salts present out of the seed bed zone. Also, the seed bed will warm up more quickly, any hard surface crusts and weeds can be removed at sowing, and subsequent weed management will be simplified. In contrast, a crop that is sown into dry soil and then irrigated will be subjected to waterlogging when attempting to leach salts, cool soil due to evaporation, surface crusts (in some cases) if sprinklers are used, and moist surface soil that will cause weed seeds to germinate.
2. Provide subsequent irrigations when hydrologic budget estimates of ΔS indicate that the permissible amount of water has been depleted.
3. In individual irrigations, provide enough water to bring the effective root zone back to $P_{v(FC)}$, plus extra water to account for the inefficiency of the irrigation system and any drainage needed to leach undesirable salts.
4. Fine-tune the system by making occasional measurements of P_v in the soil profile using an instrument such as a neutron probe because errors in the hydrologic budget estimates can accumulate and result in progressively larger errors in estimating the amount of water in the root zone, or by making observations on plant traits toward the end of the drying cycle, as is discussed next.

It is possible to determine when crops should be irrigated based on plant or soil measurements. The simplest is visual assessment of leaf wilting, which can be effective for determining when to irrigate gardens that have a diverse set of plant species. Irrigation is provided when the most sensitive species wilts. This approach usually is not effective with fields of crop plants. Often, plant wilting either occurs too late after yield has been lost due to drought stress or does not occur at all. With many cereals, other grasses, and grain legumes, the leaves become more vertical with drought. Also, on hot days with a high evaporative demand, leaves of sugar beet or red beet may wilt, even though the soil is well watered, and applying irrigation would not be useful.

There are some effective visual symptoms of the need to irrigate specific crop plants. In the afternoon, the canopy of cowpea appears light green when the crop is well supplied with water but becomes a dark green-blue when irrigation is needed. Upland cotton should be irrigated when the growth of the shoot tip has slowed such that the tender green tip is shorter than 10 centimeters. Irrigation of pineapple can be based on maintaining a specific minimal depth of the layer of water-storage cells in the leaf, which can be determined visually after cutting the leaf with a sharp knife. When using visual symptoms as a guide to when irrigation is needed, plant conditions in the small areas of a field where symptoms develop sooner, such as where the soil has more sand, can provide an early warning. Deciding when to irrigate, however, should be based on the average condition of most of the plants in the field.

Pressure chamber measurements of leaf water potential (Boyer, 1995) can be used to determine when crops should be irrigated. Predawn measurements can be effective because this is when the plant water status approaches equilibrium with that in the soil (Equation 8.17 and Figure 8.2). Unfortunately, predawn is not a convenient time and environment for making measurements in fields, especially when poisonous snakes may be present. Approaches based on afternoon measurements have been developed for upland cotton in California. For the San Joaquin Valley, upland cotton should be irrigated when pressure chamber measurements between noon and 3 p.m. on clear days, using recently matured leaves, exhibit leaf water potential values that are more negative than –1.8 MPa. Specifying the location, time of day, and weather conditions represents an attempt to normalize to a specific level of transpiration because leaf water potential can vary with transpiration rate (Figure 8.2).

Detailed procedures for optimal irrigation of tree crops, such as prune, walnut and almond, and grape vines that are based on pressure chamber measurements have been developed by K. A. Shackel and colleagues and shared on the website www.fruitsandnuts.ucdavis.edu. The authors recommend putting a bag around the leaf for about two hours to cause it to come into equilibrium with the stem prior to excising the leaf and making the pressure chamber measurement. This bagging reduces variability in the water potential data. McCutchan and Shackel (1992) provide a table of target values for midday stem water potentials of prune for different weeks during the growing season in the Central Valley of California (target values vary between –0.6 and –1.5 MPa). When these target values are reached, the prunes should be irrigated. They also provides a table of baseline stem water potentials exhibited by well irrigated prunes under different air temperatures and relative humidities, which could be used to guide irrigation in other parts of the world. The different air temperatures and relative humidities represent an attempt to normalize the data and adjust for the variation in leaf water potential that occur with variation in transpiration rate (Figure 8.2). Another approach for normalizing the responses would be to develop a table of threshold stem water potentials in relation to values of reference crop evapotranspiration as are described in Chapter 7. Shackel et al. (1997) provide evidence that the approach to fine-tuning irrigation management developed with prune could be effective with several deciduous fruit and nut tree species. Pressure chamber measurements of leaf water potential provide more effective indicators of irrigation need with crops that exhibit large drought-induced changes in leaf water potential, such as woody ones, than with some herbaceous crops that exhibit only small changes in leaf water potential, such as cowpea.

For crops, such as cowpea, that exhibit sensitive drought-induced changes in stomatal aperture, the temperature difference between canopy and air can provide a sensitive indicator of the need to irrigate (Ziska et al., 1985). An infrared thermometer is used to measure canopy temperature because it does so remotely and quickly and provides a value averaged over many plants. The temperature difference is normalized against the vapor pressure deficit of the air, measured under standard conditions of solar radiation (i.e., clear skies), and must be calibrated for the specific crop species and climatic zone (level of advection). An example developed for cowpea growing in the San Joaquin Valley of California (Ziska et al., 1985) is presented in Figure 10.1. In this environment, cowpea should be irrigated when measurements of the difference between canopy and air temperature fall above the line for the well-irrigated crop. The dashed line indicates the sensitivity of the system in that it shows how far the deviation in temperature difference has to be at each irrigation during the reproductive period to reduce grain yield by 50%.

A simple device for measuring soil matric potential, the tensiometer, has been used to schedule the irrigation of drought-sensitive vegetable crops. Values must be established for the depth of placement of the tensiometer and the tension reading when irrigation is needed. For example the tensiometer can be placed at 30 centimeters depth in crops with shallow roots, such as lettuce, and irrigations could be provided when the tensiometer reading is as negative as –20 centibars. For a more drought-resistant crop, such as cowpea during its most drought-sensitive stages (early flowering and pod development), irrigation is needed when a tensiometer placed at 45 centimeters depth indicates –40 to –60 centibars, depending on how favorable the soil is for root development (Ziska and Hall, 1983b). Tensiometers are not very effective with drought-resistant stages and crops because they stop working due to air entry into the instrument at tensions more negative than –80 centibars.

Managing high-frequency irrigation systems (such as center-pivot, linear-move, and drip or micro-sprinkler systems) simply requires estimation of the amount of water to apply at each irrigation. Where there has been no rain, the amount to apply is the sum of daily ET_m since the last irrigation plus extra water to account for inefficiencies in the irrigation system. Correcting for any rain that occurs may be difficult if the irrigation system does not completely cover the ground surface (such as with drip or micro-sprinkler systems) since it may be difficult to estimate the extent of

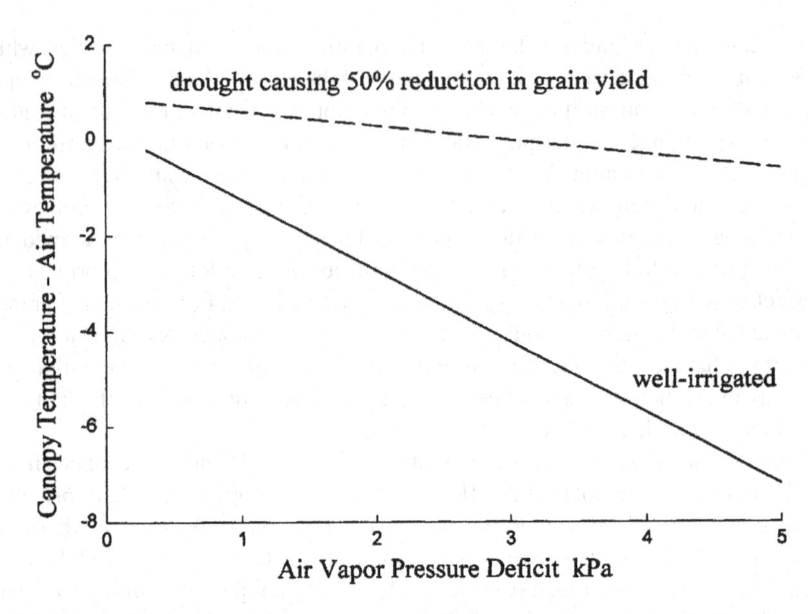

FIGURE 10.1 Differences between canopy and air temperatures just prior to irrigation, normalized by plotting as a function of the air vapor pressure deficit, for cowpea under well-watered conditions (solid line) and a longer interval between irrigations, causing periodic droughts that reduced grain yield by 50% (dashed line). (With kind permission from Springer Science+Business Media: *Irrig. Sci.*, Irrigation management methods for reducing water use of cowpea (*Vigna unguiculata* (L.) Walp.) and lima beans (*Phaseolus lunatus* L.) While maintaining seed yield at maximum levels, 6, 1985, 223–239, Ziska, L.H. et al.)

the root zone and amount of water available in the root zone. The advantages and disadvantages of drip systems are as follows:

1. The extent of soil evaporation will be less with a drip system than with an irrigation system that wets all of the soil surface. The differences will be greatest where the soil surface is not completely covered by the crop and where the soil surface is not wetted very much by the drip irrigation system. Note that there may be little difference in crop water use rate when using different irrigation systems where the ground cover exceeds about 80% and in most other cases when the soil surface is dry because the major component of crop water use is transpiration which often is not influenced by the type of irrigation system that is used.
2. Drip systems may have fewer weed problems than surface systems, since areas of the soil surface that are not wetted will be less supportive of weed growth.
3. Drip systems are often more effective on steep slopes, since they usually cause less soil erosion than most other irrigation systems.
4. Drip systems can complement the use of plastic mulches, such as with strawberry production.

5. Drip systems can be used for frequent applications of chemicals to the root zone if they are needed and if the irrigation system is very uniform.
6. In a few cases, very frequent applications of irrigation can enhance crop yield compared with conventional lower-frequency irrigation management. For cotton, either drip irrigation or surface irrigation with frequent applications during the peak fruiting period increased yield of cotton as compared with conventional surface irrigation involving longer intervals between irrigations (Radin et al., 1992). The increased yield was associated with a longer period of profuse flowering. The authors proposed that more frequent irrigation of cotton was desirable during the fruiting period, because roots were deteriorating during this stage due to diversion of carbohydrates to developing fruits rather than to the maintenance of the root system.
7. Drip systems can be expensive to install, so they often are restricted to perennial or high-value annual crops. An exception to this is where a drip tape system is installed below the surface of the ground and is used for several years for the drip irrigation of annual crops. For example, a system was developed for providing supplemental irrigation to maize or cotton grown on rows spaced 76 centimeters apart. Drip tape with an emitter (hole) every 61 centimeters was installed 41 centimeters deep in the soil. The drip tapes were placed below alternate inter-row spaces, and there was 152 centimeters between drip tape lines.
8. Drip systems typically result in incomplete wetting of the soil profile, the development of a restricted rooting system that does not fully exploit soil nutrients, and difficulties in controlling the level of salinity in the soil. If periodic rains do not leach salts, it may be necessary to rent an overhead sprinkler irrigation system on an occasional basis when growing perennial crops under drip irrigation to leach salts out of the rooting profile.

10.2 CLIMATIC ZONE DEFINITION BASED ON WATER

Climate, which is based on averages of weather variables over years, determines the types of crops, cropping systems, and management methods that are optimal. The aridity of a climate depends on the relative amounts of rainfall and the evaporative demand (ET_o). The evaporative demand is the potential crop water use and provides an indication of the level of water supply that annual crops need in the middle of their season of active growth if they are to achieve potential productivity (Chapter 7). Herein I provide definitions of climatic zones that are relevant to crop production, including consideration of rainfall and evaporative demand, and also consideration of temperature as discussed in Chapter 5. Note that my definitions may differ from the definitions of other people, as their definitions may have been designed to meet different objectives. Average rainfall and ET_o data for many locations in California may be found at www.ipm.ucdavis.edu. Monthly mean values of rainfall and ET_o for many other locations in the world may be found on the U.N.'s Food and Agriculture Organization website at www.fao.org/waicent/faoinfo/agricult/agl/aglw/climwat.

Arid zones (deserts) are where the average length of the crop growing season, as determined by the availability of water, is less than two to three months (e.g., there is no period during the year with three consecutive months having average rain/ET_o > about 0.6), and there is a very long dry season. Major deserts occur in a band between 15 degrees and 37 degrees latitude, including agricultural areas such as the Imperial Valley of California, which has an arid subtropical climate that is extremely hot in summer (Figure 10.2). Other agricultural areas in California, including Bakersfield in the southern San Joaquin Valley (Figure 10.3), and Riverside (Figure 10.4) also have arid subtropical climates, but they have significant rain in the late fall and winter.

On the equatorial side of the band of deserts, there are arid tropical agricultural zones such as the one around Louga, Senegal, in the Sahelian zone (Hall, 2017), which is just south of the Saharan desert and has a small amount of rain in summer (Figure 10.5). In the absence of water-harvesting systems, such as catchment areas to capture water or fallow farming techniques to store water in the soil, rain-fed cropping is either not possible or extremely unreliable in arid zones. However, these zones are excellent for irrigated cropping due to the lack of clouds and thus high *PFD*, which results in high potential productivity. Also, the dry conditions result in low incidence of many plant diseases and some insect pests during the reproduction and maturation of fruits and grains, which is particularly important for the yields and quality of some crops, such as cotton, cowpea, and most cereals.

Semiarid zones include regions where there are one or two distinct, short, rainy seasons with three to four consecutive months when plants can actively grow (e.g., average rain/ET_o > about 0.6, and average air temperature > 10°C) and one or two long dry seasons.

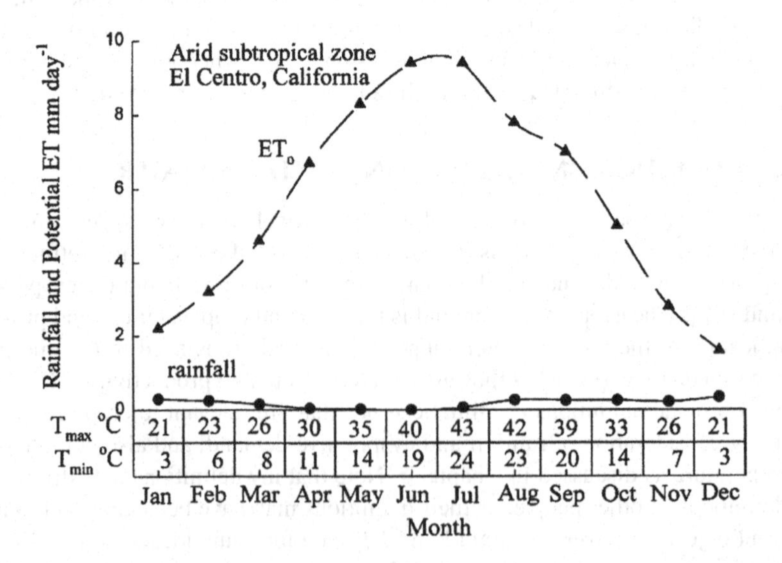

FIGURE 10.2 Average rainfall (solid line) and potential evapotranspiration (dashed line) for 1961–1990 and monthly means of daily maximum and minimum air temperatures for 1951–1980 at El Centro in the Imperial Valley, California, U.S.A. (location 32°46′N, 115°34′W, elevation –9 meters), an arid subtropical zone with an extremely hot summer. The average annual rainfall is 55 millimeters.

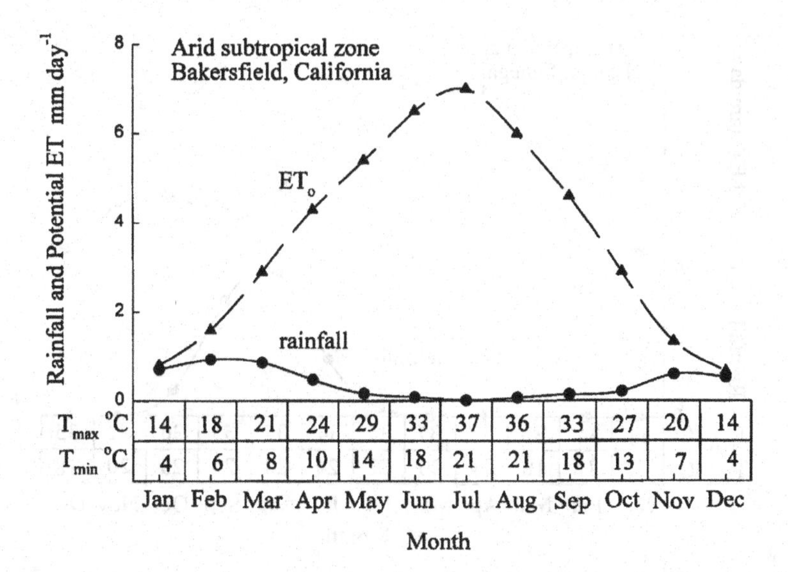

FIGURE 10.3 Average rainfall (solid line) and potential evapotranspiration (dashed line) for 1961–1990 and monthly means of daily maximum and minimum air temperatures for 1951–1980 at Bakersfield in the San Joaquin Valley, California, U.S.A. (Location 35°25′N, 119°3′W, elevation 151 meters), an arid subtropical zone with a hot summer. The average annual rainfall is 143 millimeters.

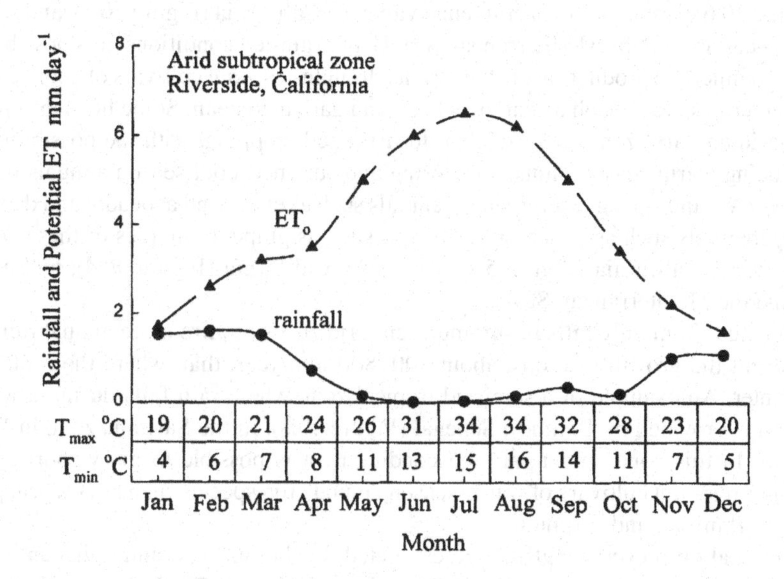

FIGURE 10.4 Average rainfall (solid line) and potential evapotranspiration (dashed line) for 1961–1990 and monthly means of daily maximum and minimum air temperatures for 1951–1980 at Riverside, California, U.S.A. (Location 33°58′N, 117°21′W, elevation 301 meters), an arid subtropical zone. The average annual rainfall is 250 millimeters.

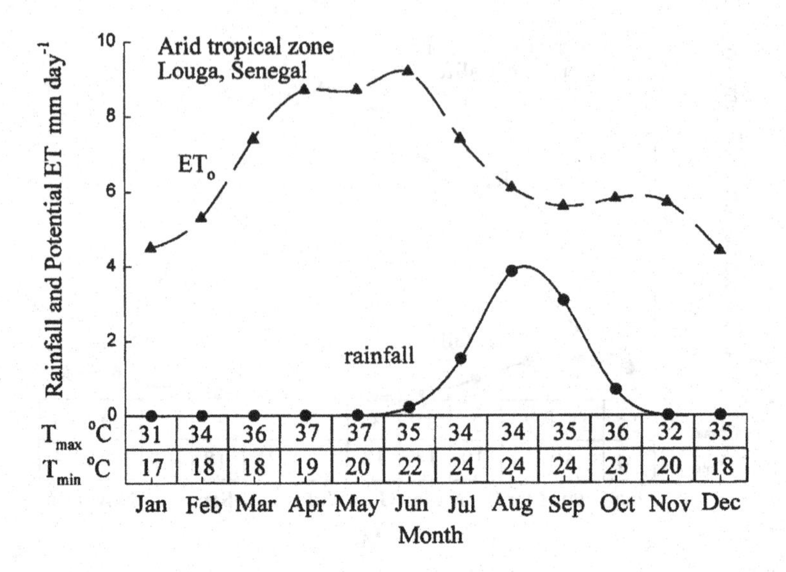

FIGURE 10.5 Average rainfall (solid line) for 1968–1999 and potential evapotranspiration (dashed line) and monthly means of daily maximum and minimum air temperatures for 1980–1996 at Louga, Senegal (location 15°37′N, 16°15′W, elevation 38 meters). This arid tropical zone had an average annual rainfall of 284 millimeters during this 32-year drought.

Where rain mainly falls in winter, the annual rainfall of semiarid zones is 300–450 millimeters, for example, Modesto in the San Joaquin Valley of California (Figure 10.6), Davis in the Sacramento Valley of California (Figure 10.7), and semiarid zones around the Mediterranean Sea. Under rain-fed conditions, it is possible to have commercial production of short-cycle, drought-resistant cultivars of cool-season annual crop species such as barley, wheat, and garbanzo bean. Some hot subtropical semiarid and arid zones are excellent for irrigated cropping, with the possibility of producing warm-season annuals in spring and summer; cool-season annuals in the fall, winter, and spring; evergreen perennials such as citrus and avocado; and deciduous perennials such as peach or vine crops such as grape. Examples of these zones are Fresno, California (Figure 5.2), Riverside, California (Figure 10.4), and areas around the Mediterranean Sea.

Where rain mainly falls in summer, semiarid zones require more annual rainfall to permit the growth of crops, about 400–800 mm/year, than where the rain falls in winter. An example of a semiarid tropical zone where rain falls during a warm season is provided by Bambey, Senegal (Figure 10.8), in the Savanna zone in West Africa. In this zone, under rain-fed conditions, it is possible to grow short-cycle, drought-resistant cultivars of warm-season annual crop species such as cowpea, peanut, pearl millet, and sorghum.

Optimal crop cycle lengths were estimated for the Savanna and Sahelian zones using a hydrologic budget method (Dancette and Hall, 1979). In this method, the cycle length was estimated that would enable annual grain crops to receive at least 80% of the ET_m level of crop water use in eight years out of 10. Hydrologic budget

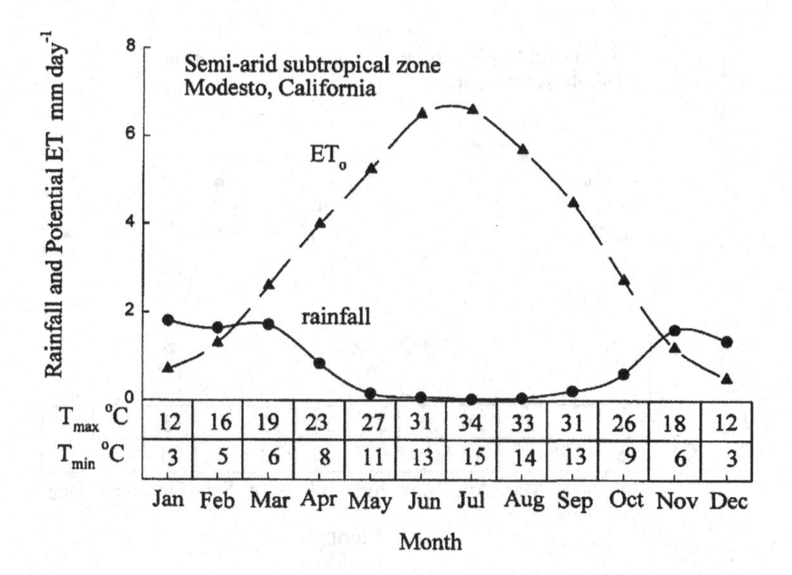

FIGURE 10.6 Average rainfall (solid line) and potential evapotranspiration (dashed line) for 1961–1990 and monthly means of daily maximum and minimum air temperatures for 1951–1980 at Modesto in the San Joaquin Valley, California, U.S.A. (location 37°39′N, 121°W, elevation 28 meters). This semiarid subtropical zone has an average annual rainfall of 308 millimeters.

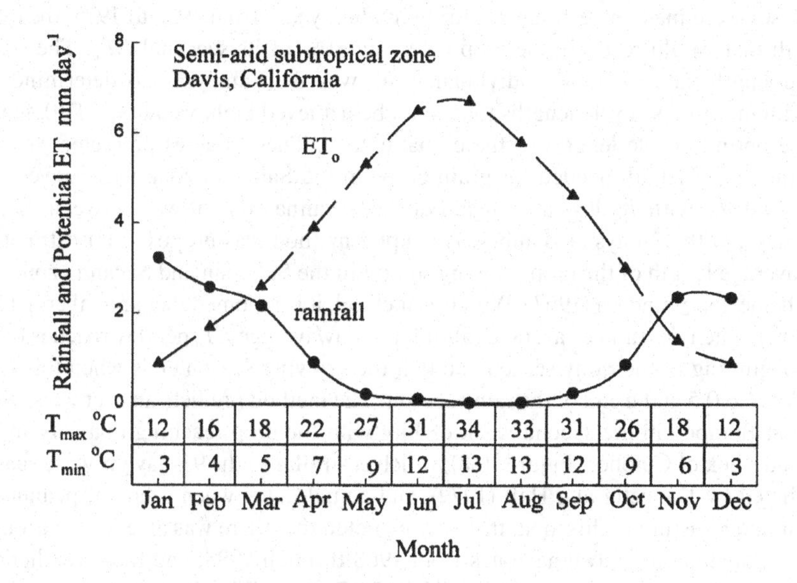

FIGURE 10.7 Average rainfall (solid line) and potential evapotranspiration (dashed line) for 1961–1990 and monthly means of daily maximum and minimum air temperatures for 1951–1980 at Davis in the Sacramento Valley, California, U.S.A. (location 38°32′N, 121°46′W, elevation 28 meters). This semiarid subtropical zone has an average annual rainfall of 457 millimeters.

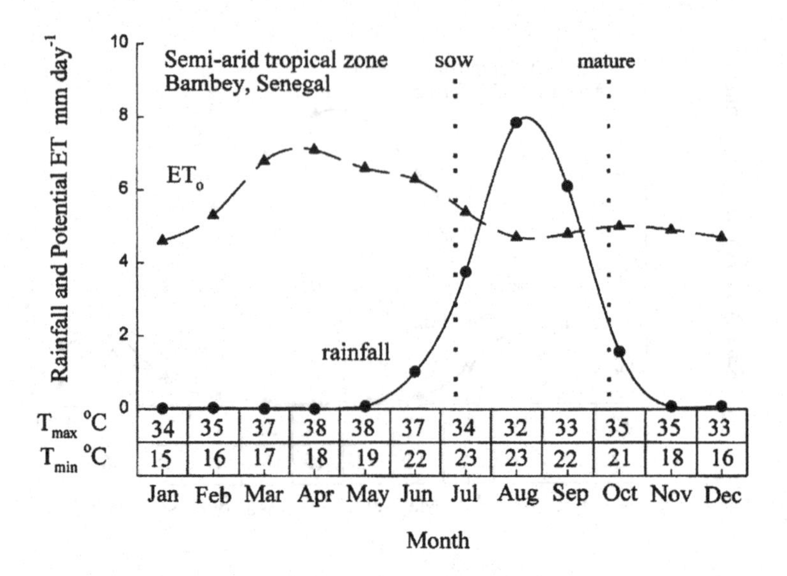

FIGURE 10.8 Average rainfall (solid line) for 1921–1980 and potential evapotranspiration (dashed line), estimated dates of sowing and crop maturity, and monthly means of daily maximum and minimum air temperatures for 1966–1975 at Bambey, Senegal (location 14°42′N, 16°28′W, elevation 17 meters). This semiarid tropical zone has an average annual rainfall of 631 millimeters.

analyses were made by estimating, for individual years from 1931 to 1975, the cycle length that would result in the crop receiving 80% of its seasonal ET_m. The set of cycle-length data for these individual years was then analyzed to determine the maximum of these cycle lengths that could be achieved eight years out of 10, which is the optimal cycle length for the climatic zone. These studies predicted that the optimal cycle length needed for grain crops in the Sahelian zone was between 60 and 80 days, with 78 days at Louga. For the Savanna zone, it was between 80 and 130 days, with 93 days at Bambey. A simpler method was proposed for estimating the average length of the crop growing season in the Sahelian and Savanna zones by Cochemé and Franquin (1967). With this method, it is assumed that annual crops can be sown when the ratio of average rainfall per day/average ET_o per day reaches 0.5 at the beginning of the rainy season and that the growing season ends when this ratio declines to 0.5 at the end of the rainy season. The method predicts an average sowing date at Bambey in the second week of July with the crop reaching maturity in the second week of October (Figure 10.8), which is similar to the 93-day growing season predicted by Dancette and Hall (1979), and actual dates when seeds of peanut are sown into moisture in this area. It should be noted that there was an extreme drought in the Sahelian and Savanna zones from 1968 through 1998, and these predictions were not effective during this period (Hall, 2017), as will be discussed later.

Hydrologic budget methods for estimating cycle length can be more generally reliable than methods based on only rainfall and ET_o data, such as the one proposed by Cochemé and Franquin (1967). They can be used to estimate the extent that water storage in the root zone prolongs the season of crop growth after the rainy season has ended. In some cases, this effect can be substantial. For example, the Yolo clay loam found around Davis, California (Figure 10.7), favors deep growth of roots and can hold large quantities of plant-available water. It has a $P_{v(FC)}$ of about 35% and a $P_{v(LL)}$ of about 10%. Crops of sorghum sown into this soil in May, after rainfall that brought the soil profile to field capacity, have produced grain yields of 8 ton/ha, even though they had no more rain or irrigation. In this instance, the sorghum developed a rooting system that was more than 2 m deep and extracted about 300 millimeters of water from the soil profile. Consequently, the area around Davis with the Yolo clay loam can be more favorable for rain-fed production of crops than would be predicted based solely on the rainfall and ET_o data.

In contrast, sandy soil areas near to Davis would not be favorable for rain-fed production of crops that grow into the summer. In some dry environments, however, sandy soils can be beneficial. The sandy loess soil around Louga, Senegal (Figure 10.5), has a $P_{v(FC)}$ of only 8% and a $P_{v(LL)}$ of only 4% and does not hold much water. The advantage of the sandy soil for this zone is that it takes only small amounts of rain to bring the soil to field capacity, and the small rains that fall during the cropping season can be used efficiently. Consequently, even though the Louga region is classified as arid by systems that are based on only rainfall and ET_o, it has been possible to obtain useful crops of cowpea under rain-fed conditions in some low-rainfall years since the late 1990s as a consequence of a collaborative research program that I directed (Hall, 2017).

Average weather data have been useful for predicting the crop species, cultivars, and management methods that can be effective in specific climatic zones, if the influences of soil types also are considered. There have been a small number of cases, however, where predictions based on the climate have not been effective, because the climate either was poorly described or changed. I will provide one extreme example of a change that occurred in a climate. The Sahelian and Savanna zones of Africa, which stretch from Senegal in the west to the Sudan in the east, have been subjected to a long drought that began in 1968. This drought was still present 30 years later in Senegal. For the 50 years prior to 1968, average annual rainfall in Louga had been 442 millimeters (Figure 10.9). The rain mainly came in July, August, and September and provided a growing season of about 90 days duration. This provided sufficient moisture that, in most years, local landraces of pearl millet and cowpea and local peanut cultivars which had cycles from sowing to harvest of 90–100 days produced some grain. During the 30-year period from 1968 to 1998, the average annual rainfall at Louga was only 276 millimeters (Figure 10.9), with rain mainly falling in only August and September (Figure 10.5). Hydrologic budget and agronomic analyses made in the 1970s predicted the rainfall at Louga since 1968 would provide only an average growing season of about 60 days but that a specific type of cowpea might have sufficient drought resistance to be effective in these harsh conditions.

Unfortunately, in the 1970s, hardy warm-season cultivars with a short cycle length of 60–70 days were not available for cowpea or any other crop species in the world. I directed a research program to breed cultivars of cowpea with the ability to produce substantial grain within 60 days from sowing. These cultivars would need to combine drought resistance during the vegetative stage we had discovered with early flowering on an erect habit. They would need to be planted at a high density of

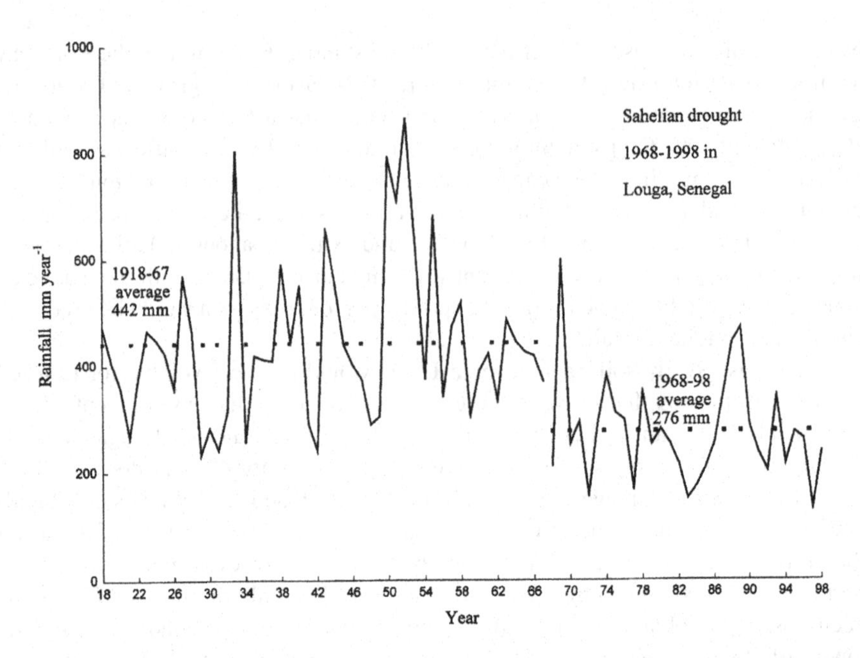

FIGURE 10.9 Average rainfall for 1918–1998 (solid lines) and averages for 1918–1967 and 1968–1998 (dotted lines) for Louga in the Sahelian zone of Senegal (location 15°37′N, 16°15′W, elevation 38 meters).

50 × 20 centimeters. This program was successful in developing breeding lines of this type. When planted at close spacing on a sandy soil they produced 1,000 kg/ha of dry grain by 60 days after sowing in a year at Louga with only 215 millimeters of rain (Hall and Patel, 1987). These breeding lines did not have sufficient resistance to the biotic stresses that occur in Senegal but, several years later, one of these breeding lines was released as the cultivar Ein El Gazal for use by farmers in the Sahelian zone of the Sudan around El Obeid (Elawad and Hall, 2002).

In most years from 1968 to the late 1980s, crops in the area around Louga failed to produce grain. A key feature of arid and semiarid zones is that the relative variability of the rainfall is very high (Figure 10.9), which makes agriculture very unreliable in these zones. The drought also affected the Savanna zone in Senegal in that the average rainfall at Bambey for the 47 years prior to 1968 was 670 millimeters, whereas from 1968 through 1998 it only was 466 millimeters. The drought in the Savanna zone had less severe effects on agriculture than the drought in the Sahel because the rainfall still provided sufficient water, in most years, to permit reasonable productivity by the available cultivars of pearl millet, peanut, and cowpea.

During the 1980s and early 1990s, two cowpea cultivars—Mouride and Melakh—were bred for the Sahelian and Savanna zones of Senegal that have cycle lengths of 60–70 days and resistance to drought, heat, and several pests and diseases (Cisse et al., 1995, 1997). These cultivars have provided greater and more stable grain yields in the Sahelian and Savanna zones than the landraces of cowpea that were being grown by farmers (Cisse et al., 1995, 1997). Mouride and Melakh are remarkable in that they are productive under water-limited conditions where all other cultivars of cowpea and all other crop species either die in the vegetative stage or at most produce only very small quantities of grain.

Predicting the cropping systems needed for the Sahelian zone in the twenty-first century is difficult because it is not known what the average rainfall will be. Will the annual rainfall be close to 442 millimeters, as it was from 1918 to 1967, or will it be closer to 276 millimeters, as it was from 1968 to 1998? Climate changes of the severity experienced in the Sahelian zone rarely have occurred in human history, but some climate changes may occur in the twenty-first century. Various models have predicted that the continuing increases in atmospheric [CO_2] will cause climate changes on a global level, including increases in temperature, increases in evaporative demand, increases in rainfall in some zones, and decreases in rainfall in some other zones (Yadav et al., 2011).

Tropical zones that are influenced by monsoons can have bimodal rainfall with two dry seasons. An example of a semiarid tropical zone with bimodal rainfall is provided by Katumani, Kenya (Figure 10.10). This location is near the equator but is cool for a tropical zone because of its intermediate elevation of 1,575 meters. The total annual rainfall is fairly high at 712 millimeters, but crops are subjected to frequent droughts. The monsoon rains fall in two distinct rainy seasons, with only about 300 millimeters per season, and farmers use each rainy season as a distinct cropping season, sowing annual crops in October and also in March. The main crops cultivated on small farms in this area are maize and common bean, but some sorghum and cowpea also are grown. Due to their greater drought resistance, cowpea and sorghum would be the better adapted crops for this zone. Presumably, farmers plant more of the drought-susceptible crops, maize and common bean, because they prefer their taste, and maize has greater resistance to seed-eating weaver birds than does sorghum. The tropical semi-arid zones are excellent for

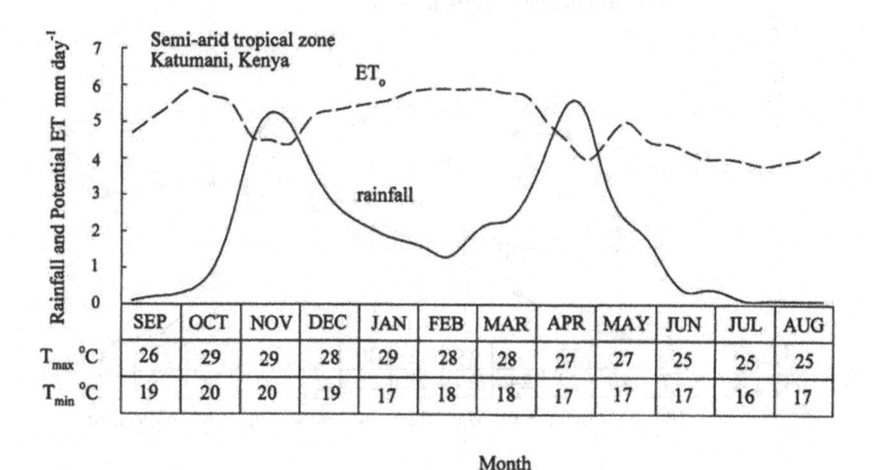

	SEP	OCT	NOV	DEC	JAN	FEB	MAR	APR	MAY	JUN	JUL	AUG
T_{max} °C	26	29	29	28	29	28	28	27	27	25	25	25
T_{min} °C	19	20	20	19	17	18	18	17	17	17	16	17

FIGURE 10.10 Average daily rainfall (solid line) and potential evapotranspiration (dashed line), and monthly means of daily maximum and minimum air temperatures for 1962–1980 at Katumani, Kenya (location 1°31′S, 37°43′E, elevation 1,575 meters). This semiarid tropical zone has an average rainfall of 712 millimeters.

irrigated production of crops such as sugar cane and rice, providing sufficient water is available from rivers, lakes, or wells.

Another type of semiarid zone is where a small amount of water is available each month (i.e., average rain/ET_o of 0.3–0.6) and occurs in Australia. This zone is suitable for pasture production under rain-fed conditions, and growing CAM plants such as the prickly pear cactus and various *Agave* species for producing sisal or tequila.

Subhumid zones usually have long rainy seasons with a rain-fed crop growing season of 5–10 months and a distinct but short dry season (e.g., where average rain/ET_o > about 0.6 and average air temperature > 10°C for 5–10 months). An example of a subhumid temperate zone is presented in Figure 5.5, which is a location where there is substantial rain-fed production of maize and soybean. In tropical subhumid zones (e.g., Figure 10.11) many types of crops can be commercially produced, including drought-susceptible annual crops such as maize and long-cycle annual crops such as cotton and pigeon pea. In temperate subhumid or humid zones, deciduous tree crops may not require supplemental irrigation if sufficient rain occurs during their season of active growth (Figures 5.5 and 10.14). In subtropical subhumid zones, drought-susceptible evergreen perennials, such as avocado and citrus, may require some supplemental irrigation during the dry season, as occurs in parts of Florida, which is on the boundary between subhumid and humid conditions (Figure 10.12).

Humid zones have significant rain nearly every month (e.g., average rain/ET_o > about 0.6 for 10–12 months of the year) and no distinct dry season (Figure 10.13). In tropical humid zones (e.g., areas with tropical rain forests), possible rain-fed cropping includes evergreen chilling-sensitive trees such as cocoa, rubber, mango, oil

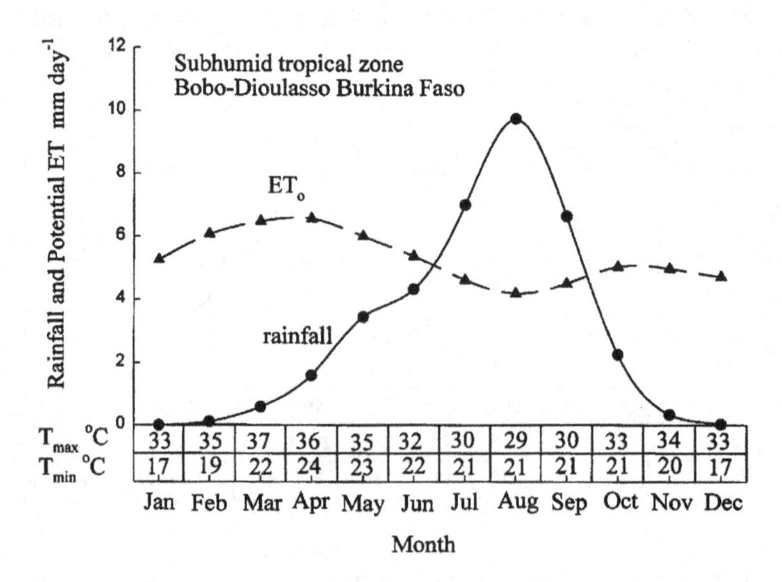

FIGURE 10.11 Average rainfall (solid line, potential evapotranspiration (dashed line), and monthly means of daily maximum and minimum air temperatures for 1920–1984 at Bobo-Dioulasso, Burkina (location 11°10′N, 4°19′W, elevation 432 meters). A subhumid tropical zone with average annual rainfall of 1,111 millimeters.

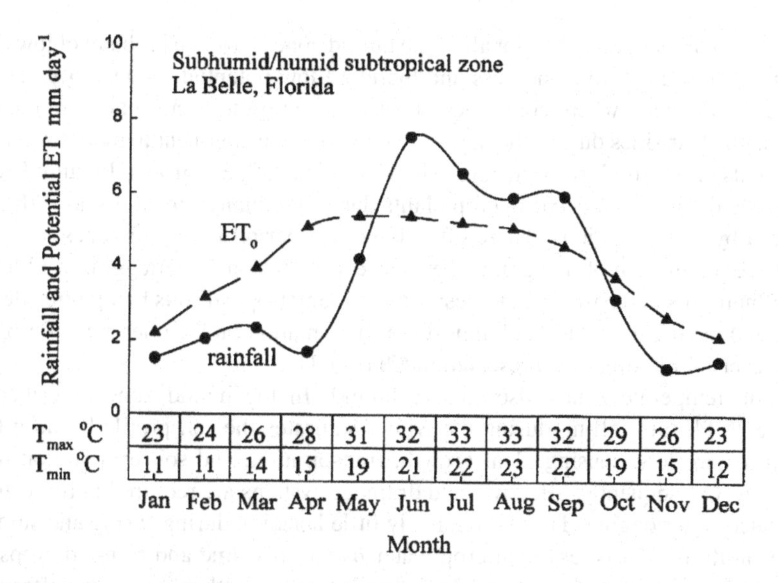

FIGURE 10.12 Average rainfall (solid line), potential evapotranspiration (dashed line), and monthly means of daily maximum and minimum air temperatures for 1966–1996 at La Belle, Florida, U.S.A. (location 26°45′N, 81°26′W, elevation 5 meters). This subhumid/humid subtropical zone has an average annual rainfall of 1,331 millimeters.

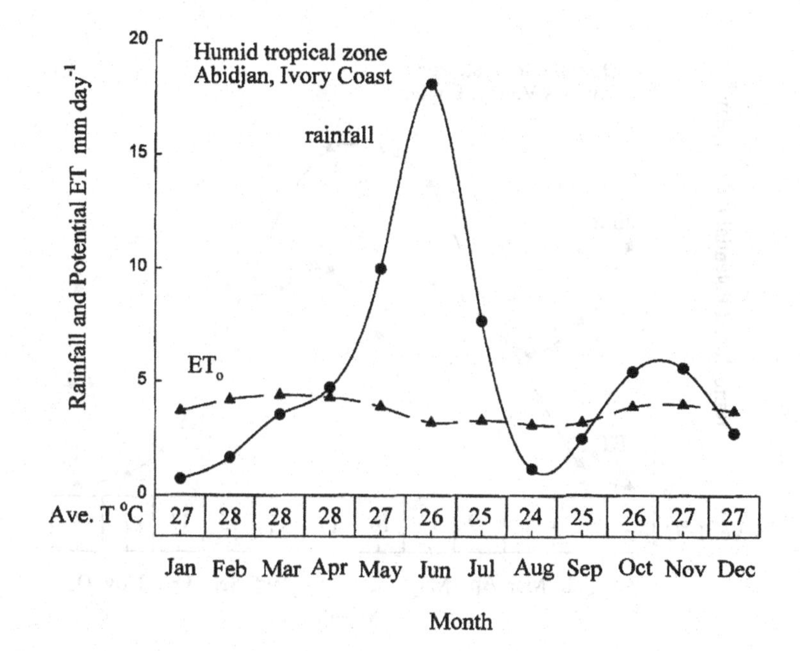

FIGURE 10.13 Average rainfall (solid line), potential evapotranspiration (dashed line), and monthly means of daily maximum and minimum air temperatures for 1923–1990 at Abidjan, Ivory Coast (location 5°15′N, 3°54′W, elevation 7 meters). This humid tropical zone has an average annual rainfall of 1,942 millimeters.

palm, and banana. Many but not all of the humid zones have the problem of low daily solar irradiance due to cloudiness and therefore low potential productivity per day. The exceptions are where convection storms cause rain to occur mainly in the evening with clear skies during the day, as occurs in some continental areas such as the highlands of North Mara, Tanzania, where Arabica coffee is grown. In humid zones extensive damage can occur to crop plants due to the many fungal diseases that are favored by wet conditions. There often is no dry period in humid zones to permit effective maturation of dry grains. Extensive soil erosion, waterlogging, and leaching of nutrients and bases occurs, resulting in many tropical soils being infertile, too acid, and having toxic levels of aluminum. Information on the tolerance of crops to high levels of aluminum is presented in Chapter 11.

Cool, temperate zones also can be humid. In the humid zone described by Figure 10.14, rain falling during the winter saturates the soil profile because temperatures are low, causing plant growth, transpiration, and soil evaporation to be very low. Consequently, effective field drainage systems are required in these zones to reduce waterlogging. There is relatively little leaching during spring and summer since rainfall often is less than crop water use at this time and rain-fed crops use the moisture stored in the soil profile. Cereal grains and other crops can suffer from fungal damage at harvest because there is no distinct and reliable dry season. For all humid zones, working the soil can be difficult due to the frequent rain. Soil structure

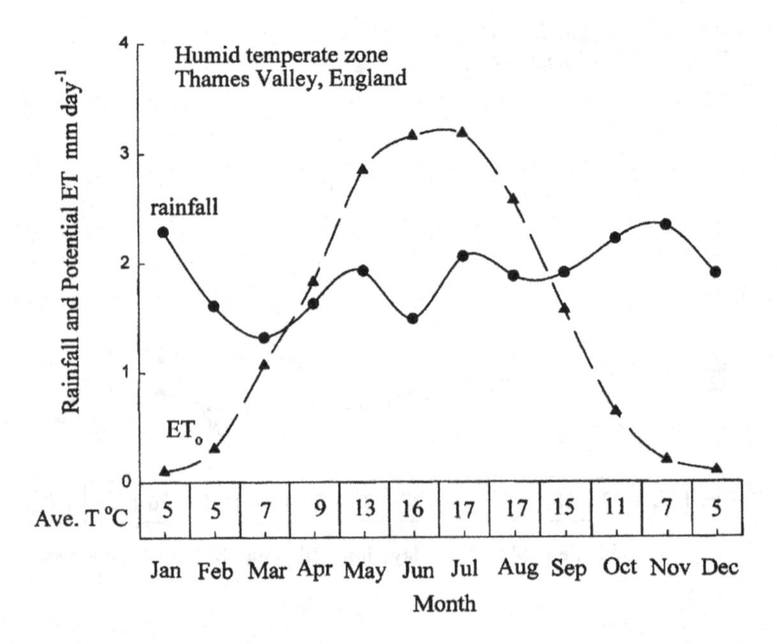

FIGURE 10.14 Average rainfall (solid line, potential evapotranspiration (dashed line), and monthly means of daily maximum and minimum air temperatures for the Thames Valley, England (location 52°1N, 0°W). This humid temperate zone has an average annual rainfall of 688 millimeters. (Data from Bunting, A. H., *Weather*, 30, 312–325, 1975.)

can be badly impaired if soil is plowed, cultivated, or simply driven on when it is too wet. Influences of hard soil on plant function are described in Chapter 12.

ADDITIONAL READING

Boyer, J. S. 1995. *Measuring the Water Status of Plants and Soils.* Academic Press, San Diego, CA, p. 178.

Elawad, H. O. A. and A. E. Hall. 2002. Registration of "Ein El Gazal" cowpea. *Crop Sci.* 42: 1745–1746.

Hall, A. E. 2017. *Sahelian Droughts: A Partial Agronomic Solution.* Nova Science Publishers, New York, p. 216.

McCutchan, H. and K. A. Shackel. 1992. Stem-water potential as a sensitive indicator of water stress in prune trees (*Prunus domestica* L. cv. French). *J. Amer. Soc. Hort. Sci.* 117: 607–611.

Radin, J. W., L. L. Reaves, J. R. Mauney, and O. F. French. 1992. Yield enhancement in cotton by frequent irrigations during fruiting. *Agron. J.* 84: 551–557.

Shackel, K. A., H. Ahmadi, W. Biasi, R. Buchner, D. Goldhamer, S. Gurusinghe, J. Hasey, et al. 1997. Plant water status as an index of irrigation need in deciduous fruit trees. *HortTechnology* 7: 23–29.

Ziska, L. H. and A. E. Hall. 1983a. Seed yields and water use of cowpeas *Vigna unguiculata* (L.) Walp., subjected to planned-water-deficit irrigation. *Irrig. Sci.* 3: 237–245.

Ziska, L. H. and A. E. Hall. 1983b. Soil and plant measurements for determining when to irrigate cowpeas *Vigna unguiculata* (L.) Walp., grown under planned-water-deficits. *Irrig. Sci.* 3: 247–257.

Ziska, L. H., A. E. Hall, and R. M. Hoover. 1985. Irrigation management methods for reducing water use of cowpea (*Vigna unguiculata* (L.) Walp.) and lima beans (*Phaseolus lunatus* L.) While maintaining seed yield at maximum levels. *Irrig. Sci.* 6: 223–239.

11 Crop Responses to Flooding, Salinity, and Other Limiting Soil Conditions

This chapter considers crop responses to flooding because its occurrence likely is being heightened by global climate change. Influences of flooding on rice are examined together with progress in breeding rice that is better able to withstand submergence by flooding. In Asia there are 47 million hectares of rain-fed lowland and other flood-prone areas that constitute 35% of the total rice-growing area (Mackill et al., 2012). Damage to rice due to submergence by flooding was estimated to cost farmers in South and Southeast Asia $1 billion US per year (Ismail, 2013).

Flood-prone areas experience different types of excess water: either direct heavy rains or floodwater from adjacent rivers or canals because of too much water in upper catchments or because of high tides in deltas (Mackill et al., 2012). Over the past 100 years, average global sea level has risen 10–15 centimeters and a further increase of about 50 centimeters is expected by 2100 (Mackill et al., 2012). Seawater temperature is increasing and can result in storms that produce more rainfall. Consequently, global climate change is predicted to increase the extent to which rice and other crops are damaged by submergence due to flooding and increased salinity.

This chapter also will consider crop responses to salinity. About 10% of the world's arable land area is affected by salinity, including delta areas where there is much rice production. Increased flooding of rice by seawater is predicted to occur due to global climate change; consequently, consideration will be given to breeding to improve the resistance of rice to salinity. Irrigation systems are prone to salinization. About half of the irrigation schemes in the world are now under the influence of salinization either due to low-quality irrigation water or to excessive leaching and subsequent rising water tables bringing saline groundwater close to root zones. Salt tolerance of many crops grown under irrigation will be considered, including the breeding of wheat for resistance to salinity.

Some limiting soil conditions can be solved by applying fertilizer. These are not considered in this chapter because there have been many books on mineral nutrition, such as Epstein (1972). Various whole-plant aspects of nitrogen and phosphorus nutrition are discussed in other chapters of this book.

This chapter examines problems caused by limiting soil characteristics that are difficult to solve through management: extremes in soil texture, high soil bulk density, and high boron and aluminum levels. It will be shown that these limiting conditions may be partially solved by choosing crop species and cultivars that are better adapted than others to the specific soil limitation.

11.1 FLOODING

Plants subjected to flooding suffer due to inadequate exchange of gases such as oxygen, carbon dioxide, and ethylene. Diffusion of gases in liquid water is 10^{-4} slower than in air (Chapter 4). Roots in particular can suffer from a shortage of oxygen, which is required for aerobic respiration. Ethylene is produced by plants during flooding and increases in its concentration have hormonal effects, including the promotion of substantial shoot elongation.

The physiological and morphological responses of plants to poor aeration in the root zone are complex and involve adaptive and damaging effects of the hormones ethylene and abscisic acid. Some examples are provided from the review of Jackson et al. (1993). Aerenchyma tissue consisting of extensive air spaces that provide pathways for the diffusion of oxygen from the shoot to the root is well developed in rice, and ethylene stimulates aerenchyma formation in roots and leaf bases of maize. Coleoptiles of a few wetland plants, such as rice, can grow when deprived of oxygen, whereas the tissues of most plants only survive a few hours when deprived of oxygen. Shoot tissues in well-aerated conditions respond to anaerobic root conditions due to changes in the hormonal signals coming from the roots. In poorly adapted plants, downward curving of petioles (epinasty) occurs with waterlogging and has been attributed to high levels of ethylene in the shoot. Waterlogged roots produce high levels of a precursor of ethylene that moves to the shoots in the xylem stream and is then converted to ethylene. Waterlogged roots also produce high levels of abscisic acid, which is transported in the xylem stream to the shoots and probably is responsible for the stomatal closure that can occur with waterlogging of roots. The stomatal closure and epinasty that result from waterlogging of the roots can be viewed as either damaging or adaptive in that the downward bending of leaves and stomatal closure reduce both photosynthesis and the tendency for the plant to lose water and become dehydrated. Flooding can reduce the supply of water to the shoot in that, even though the soil water potential is very high, root uptake of water can be inhibited by anaerobic conditions. For example, when alfalfa fields are surface irrigated during hot summer weather in soils that have low permeability, the roots can become anaerobic for a while, and the leaves wilt even though the soil is still very wet. A little later, once the water has drained down into the lower soil profile and the roots have become aerated, the leaves become turgid again.

Rice is better-adapted to growing under flooded conditions than many other plants because of the extensive aerenchyma tissue that enhances gaseous exchange between the plant shoot and roots. However, most varieties are damaged if they are subjected to excess flooding that results in complete plant submergence for more than four to seven days or if the plants are subjected to deep water for an extended time.

I will consider rain-fed rice subjected to three different types of excessive flooding: flash-flooding for a relatively short duration, ranging from a few days to two weeks; longer-term stagnant flooding of 20–50 centimeters depth through most of the season even without complete submergence; and deep-water flooding areas where water at a depth of over 50 centimeters to a few meters stagnates in the rice field for several months.

In the latter case there are areas of West Africa where water depth can exceed four meters. Special floating varieties of African rice (*Oryza glaberrima*) are cultivated that have shoots with the ability to elongate progressively as the water becomes deeper. These varieties, however, have very low grain yields. In parts of Asia where deep-water rice production has been practiced with special varieties of *Oryza sativa* during the rainy season, there has been a tendency to replace it by using different varieties of rice grown in the dry-season with irrigation using shallow tube wells.

Transient flash floods can occur at any growth stage and reduce yields by 10%–100%. For example, where direct seeding is practiced, excess flooding prior to emergence can result in complete crop failure because of the high sensitivity of rice to anaerobic conditions during germination. Some progress has been made in identifying rice landraces with greater tolerance to submergence during germination (Ismail, 2013). Note that traditionally much rice has been grown using seedlings transplanted from nurseries, but there now is a tendency to use direct seeding with tractor-pulled equipment. In California, a different system is used. Fields are leveled and shallow furrows are made in the surface. Then the fields are flooded with about 12 centimeters depth of water. Rice seeds are soaked to make them heavy enough to sink in water and then are flown onto the fields using airplanes. While the seedlings begin growing, the water is kept at 12 centimeters depth to reduce the emergence of weed seedlings.

In attempting to describe rice germination and emergence in these different systems, I designed an experiment for a high school project using soil in beakers and the Calrose rice variety. The experiment demonstrated that rice germinated and emerged from either seed placed two centimeters down in moist and drained soil with no water on top of the soil or with seed placed on top of soil with 12 centimeters depth of water on top of the soil and the seed. In contrast, rice did not germinate when the seed was placed two centimeters down in soil with 12 centimeters depth of water on top of the soil.

The most common and seriously damaging type of flooding of rice crops is the short-term inundation that often occurs. Most rice varieties are either killed or severely damaged if the period of inundation extends beyond four to seven days. In recent years this problem has been partially solved by the development and adoption of submergence-tolerant rice varieties that can survive two weeks or more of flooding (Mackill et al., 2012). The International Rice Research Institute (IRRI) and its national program collaborators made major contributions to this project, which has had substantial beneficial impacts on poor rice farmers.

The research that provided the foundation for the progress in breeding rice with this trait began in the 1970s when submergence-tolerant materials were discovered through systematic screening of germplasm collections in controlled-submergence tanks at IRRI. An important submergence-tolerant line was discovered: Indian landrace FR13A, which earlier had been reported by Indian scientists to have resistance to flooding. FR13A has been widely used as a donor of submergence tolerance; unfortunately, it has many undesirable agronomic traits and only slow progress was made in breeding until about 2003. Studies with FR13A led to the discovery of the *SUB*1 gene, which confers significant submergence tolerance. In addition, DNA markers were found that facilitated marker-assisted backcrossing, which was used to transfer the submergence tolerance gene into varieties with desirable agronomic traits without transferring the negative traits from FR13A (Mackill et al., 2012). The *SUB*1 gene suppresses elongation of the shoot when the plants are submerged, thus limiting anaerobic catabolism and leading to the preservation of carbohydrate reserves such that the plants recover rapidly once the water level has dropped (Ismail, 2013). The gene is an ethylene response factor and only is induced at the transcript level during submergence. Agronomic studies have been conducted with several pairs of lines with and without the *SUB*1 gene (Mackill et al., 2012). The gene did not affect yield or other agronomic traits when plants were not subjected to excessive flooding. The gene provided yield advantages of 1–3 tons/ha with submergence durations of 4–20 days at all growth stages from a week after seeding up to two weeks before flowering. Using the marker-assisted backcrossing, *SUB*1 has been transferred into numerous mega varieties that are popular among Asian farmers. As of 2013, five of these submergence-tolerant varieties had been released in five countries.

The submergence tolerance conferred by the *SUB*1 gene is only partial and, in some cases, stronger tolerance is needed that is present in FR13A. In addition, tolerance to submergence during germination and tolerance of longer-term stagnant flooding are needed and are being sought (Mackill et al., 2012).

11.2 SALINITY

Shannon (1997a) posed the question, "If life evolved in the sea, and if ancient seas were saline, why then are crop plants sensitive to salt?" He suggested that many years of natural selection in non-saline terrestrial environments removed the salt tolerance from the progenitors of most crop plants. In principle, much of the soil salinity could be reduced by installing drainage systems and leaching the soil profile with water of adequate quality. In practice, however, these amelioration methods may be too expensive or impossible, and there are many cases where it always will be necessary to grow crops that have some salt tolerance, such as in delta areas that occasionally become inundated by seawater.

When evaluating the salt tolerance of plants, it is important to recognize that the results can be affected by the composition of the salts used in salinizing the plant growth medium. Numerous agronomic experiments with plants growing in soils have been salinized by adding NaCl and $CaCl_2$ at a 2:1 molar ratio. Many experiments of this type were conducted by the USDA in the Salinity Laboratory (www.ussl.ars.usda.gov). In contrast, in many physiological experiments, plants were salinized with only NaCl. The advantages of using the mixture of NaCl and CaCl are (1) avoiding slow water permeability and waterlogging problems in the soil associated with soil sodicity that can occur when only NaCl is used, (2) having a chemical composition that is more similar to the salinities that occur in many crop lands than only NaCl, and (3) being less damaging to plants than only NaCl because the presence of Ca^{++} acts to protect plant membranes in the roots that are in contact with the soil solution. Another experimental approach involves irrigating sandy soils with seawater or seawater that has been diluted with freshwater, but this approach is limited to areas where it is a viable agronomic option (Epstein and Norlyn, 1977).

A special artifact may occur when using hydroponic systems in salinity experiments. The phosphate concentration used in hydroponics usually is much higher than occurs in soils. This is done to maintain an adequate capacity for supplying the phosphate needed by the plants. For soybean and many other species, there can be an adverse salinity × phosphate interaction in that the combination of salinity and high phosphate can be very damaging. Studies by Grattan and Maas (1988) indicate that this interaction is caused by synergistic detrimental effects of the combination of high phosphate and high chloride levels in leaves.

An important first step when planning agriculture for saline lands is to choose crop species with sufficient salt tolerance that they can be profitably grown. The salt tolerance of many plant species has been reviewed, mainly using data from agronomic studies with a 2:1 molar mixture of NaCl and $CaCl_2$ (Maas, 1986). When comparing different crop species, it is useful to define salt tolerance in terms of relative yield under saline compared with non-saline soil conditions and express them as a function of the salinity level in the part of the root zone where most of the water uptake occurs. Salinity level usually is expressed in terms of the electrical conductivity (*EC*) of an extract from a saturated soil paste. *EC* has units of deciSiemens per meter (dS m^{-1}) or, in earlier years, in milli-mho per centimeter. These have equivalent values. The relationship between the concentration of NaCl in mol m^{-3} (M) and *EC* in dS m^{-1} is given by Equation 11.1 (Richards, 1954):

$$\log_{10} M = 0.91 + 1.07 \times \log_{10} EC \qquad (11.1)$$

An approximate relation between *EC* and total concentration of various salts is as follows:

$$1 \text{ dS m}^{-1} = 10\ M\ salts = 10 \text{ mmol l}^{-1}\ salts = 700 \text{ mg l}^{-1}\ salts$$

where seawater has an *EC* of about 50 dS m^{-1}. When using *EC* as an index of salinity, it is assumed that plants respond primarily to the total concentration of salts rather than the concentrations or proportions of individual salt constituents, which is a good first approximation (Rhoades et al., 1992). The *EC* of an extract of a saturated soil paste is often used to attempt to standardize measurements, but it should be recognized that as soils dry, the soil solution becomes concentrated and *EC* can increase substantially.

Relative yield (Y_r) often exhibits a linear decrease after a threshold salinity has been reached (Figure 11.1), and salt tolerance has been defined in terms of two parameters: the threshold electrical conductivity (EC_t) and the percent decrease in relative yield per unit of electrical conductivity in dS m^{-1} above the threshold (the slope s) as shown in Equation 11.2:

$$Y_r = 100 - s \times \left(EC - EC_t\right) \text{ for } EC \geq EC_t \qquad (11.2)$$

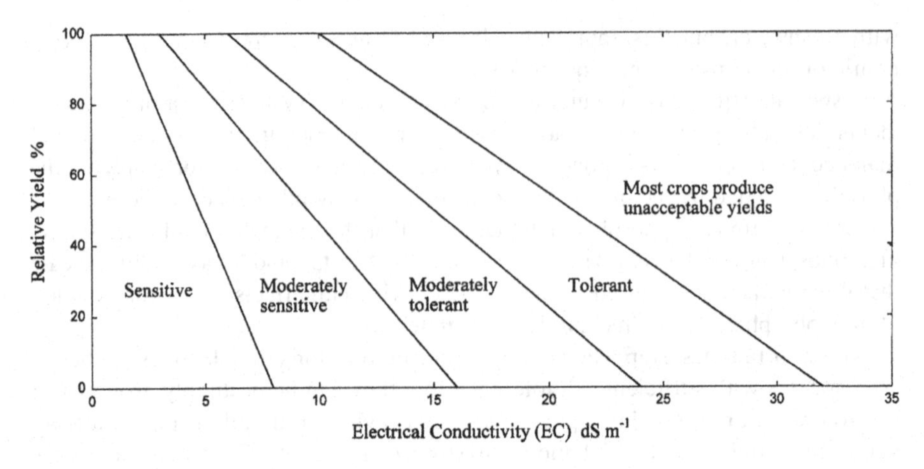

FIGURE 11.1 Relative yields in response to the electrical conductivity (*EC*) in deciSiemens per meter (dS m^{-1}) of an extract from a saturated soil paste with crop salt tolerances. (From Maas, E.V., *Appl. Agr. Res.*, 1, 12–26, 1986.)

Crop species have exhibited substantial differences in salt tolerance, based on their relative yields (Table 11.1). It should be noted, however, that these data were obtained with only one or a few varieties, so the extent to which they represent the average salt tolerance of the species is not known. The slope and threshold data can be used to calculate the electrical conductivity at which a 50% reduction in yield occurs. From these data, it is clear that the following crops are salt tolerant: barley, tall wheatgrass, cotton, safflower, sugar beet, Bermuda grass, sunflower, sudan grass, canola, asparagus, and date palm. The data for the electrical conductivity at which there is 50% emergence indicate that many crops, but not sugar beet, are more salt tolerant at this stage. In practice, this is important because evaporation from the soil surface causes upward movement of water and salt accumulation in the upper soil layer where the seeds are placed. Consequently, salinity levels can be higher during germination and emergence than later in the season, after irrigation has been used to leach salts downward. Most crop species are stunted by salinity during early vegetative growth. This has led to the hypothesis that salinity effects may be partially offset by sowing crops at closer plant spacing. Several crop species, including wheat, sorghum, and cowpea, are more tolerant to salinity after flowering than during the vegetative stage, making possible the use of more salty irrigation water during the last part of the growing season.

Salinity damages crops through several mechanisms, including (1) osmotic effects that make water less available due to decreases in water potential, (2) toxic effects of either Cl^- or Na^+, and (3) interference with plant nutrition due to competition between Na^+ and K^+ or Ca^{++}, or Cl^- and NO_3^{--}. Effects on water availability constitute a major mechanism. It can be estimated by calculating the solute potential using Equation 8.22 if the osmolarity is known. For example, seawater has about a 1 osmolal solution, which has a solute potential of about –24 bar. If the electrical

TABLE 11.1
Crop Species' Differences in Salt Tolerance

Crop Species	Slope, % per dS m^{-1}	Electrical Conductivity of Soil Extract, dS m^{-1}			Salt Tolerance Rating[a]
		Threshold	50% Yield	50% Emergence	
Cereals					
Barley	5.0	8.0	18	16–24	Tolerant
Wheat	7.1	6.0	13	14–16	Moderately tolerant
Sorghum	16.0	6.8	9.9	13	Moderately tolerant
Rice	12.0	3.0	7.2	18	Moderately sensitive
Maize	12.0	1.7	5.9	21–24	Moderately sensitive
Grain Legumes					
Cowpea	12.0	4.9	9.1	16	Moderately tolerant
Soybean	20.0	5.0	7.5	NA[b]	Moderately sensitive
Fava bean	9.6	1.6	6.8	NA	Moderately sensitive
Peanut	29.0	3.2	4.9	NA	Sensitive
Common bean	19.0	1.0	3.6	8.0	Sensitive
Industrial and Forage Crops					
Wheatgrass, tall	4.2	7.5	19	NA	Tolerant
Cotton	5.2	7.7	17	15	Tolerant
Safflower	6.0	7.5	16	12	Tolerant
Sugar beet	5.9	7.0	15	6–12	Tolerant
Bermuda grass	6.4	6.9	15	NA	Tolerant
Sunflower	5.0	4.8	15	NA	Tolerant
Sudan grass	4.3	2.8	14	NA	Tolerant
Canola	11.2	10.0	14	NA	Tolerant
Sugar cane	5.9	1.7	10	NA	Moderately tolerant
Alfalfa	7.3	2.0	8.8	8–13	Moderately sensitive
Flax	12.0	1.7	5.9	NA	Moderately sensitive
Clover	12.0	1.5	5.7	NA	Moderately sensitive
Vegetable Crops					
Asparagus	2.0	4.1	29	NA	Tolerant
Red beet	9.0	4.0	10	14	Moderately tolerant
Zucchini	9.4	4.7	10	NA	Moderately tolerant
Artichoke	11.5	6.1	10	NA	Moderately tolerant
Celery	6.2	1.8	9.9	NA	Moderately tolerant
Spinach	7.6	2.0	8.6	NA	Moderately sensitive
Broccoli	9.2	2.8	8.2	NA	Moderately sensitive
Tomato	9.9	2.5	7.6	7.6	Moderately sensitive
Cabbage	9.7	1.8	7.0	13	Moderately sensitive

(Continued)

TABLE 11.1 (*Continued*)
Crop Species' Differences in Salt Tolerance

		Electrical Conductivity of Soil Extract, dS m^{-1}			
Crop Species	Slope, % per dS m^{-1}	Threshold	50% Yield	50% Emergence	Salt Tolerance Rating[a]
Turnip	9.0	0.9	6.5	NA	Moderately sensitive
Squash, scallop	16.0	3.2	6.3	NA	Moderately sensitive
Cucumber	13.0	2.5	6.3	NA	Moderately sensitive
Sweet potato	11.0	1.5	6.0	NA	Moderately sensitive
Irish potato	12.0	1.7	5.9	NA	Moderately sensitive
Melon	14.3	2.0	5.5	NA	Moderately sensitive
Lettuce	13.0	1.3	5.1	NA	Moderately sensitive
Pepper	14.0	1.5	5.1	NA	Moderately sensitive
Radish	13.0	1.2	5.0	NA	Moderately sensitive
Carrot	14.0	1.0	4.6	NA	Sensitive
Onion	16.0	1.2	4.3	5.6–7.5	Sensitive
Strawberry	33.0	1.0	2.5	NA	Sensitive
		Tree, Vine, and Cane Crops			
Date palm	3.6	4.0	18	NA	Tolerant
Olive	NA	NA	NA	NA	Moderately tolerant
Grape	9.6	1.5	6.7	NA	Moderately sensitive
Grapefruit	16	1.8	4.9	NA	Sensitive
Orange	16	1.7	4.8	NA	Sensitive
Prune	18	1.5	4.1	NA	Sensitive
Almond	19	1.5	4.1	NA	Sensitive
Peach	21	1.7	4.1	NA	Sensitive
Blackberry	22	1.5[c]	3.8	NA	Sensitive
Boysenberry	22	1.5	3.8	NA	Sensitive
Apricot	24	1.6	3.7	NA	Sensitive

Source: Maas, E. V., *Appl. Agr. Res.*, 1, 12–26, 1986; Shannon, M. C., *Adv. Agron.*, 60, 75–120, 1997a.

[a] EC for 50% yield ≥ 14 for tolerant, < 14 and ≥ 9 for moderately tolerant, < 9 and ≥ 5 for moderately sensitive, and < 5 for sensitive.

[b] NA = not available.

conductivity of the soil solution (EC_s) is known, the solute potential (ψ_s) can be estimated using Equation 11.3 (Richards, 1954):

$$\log_{10}(-\psi_s) = -0.5115 + 1.0871 \times \log_{10} EC_s \tag{11.3}$$

with ψ_s in bar and EC_s in dS m^{-1}. Since root medium salinity (soil water potential) and atmospheric humidity both influence plant water status (Equation 8.17), it is possible that salt tolerance depends on atmospheric humidity. The salt tolerance of sensitive crops, such as common bean and onion, was greater in more humid conditions, whereas the salt tolerance of more tolerant crops, such as cotton and red beet, was not influenced by humidity (Hall, 1982b).

Accumulations of Cl^- and Na^+ cause injury to leaves of many woody species. Rootstocks have been selected that provide some salt tolerance in that they restrict the uptake of Cl^- and Na^+. Maas (1986) provides a table of the maximum permissible Cl^- levels in soil water for some different rootstocks and cultivars. Herbaceous species can be damaged by the uptake of Na^+. The ability of roots to favor uptake of K^+ over Na^+ is important, and genes for this trait have been discovered in wheat and its relatives (Shannon, 1997b).

Crops irrigated by sprinkler systems often are subjected to additional salt damage when the foliage is directly wetted by saline water (Rhoades et al., 1992). Yields of bell peppers were reduced by 59% more when irrigation was applied with sprinklers rather than a drip system. The irrigation water had an EC of 4.4 dS m^{-1}. Similar results were obtained with Irish potato. Susceptibility to foliar injury varies considerably among crop species (Table 11.2).

Leaves of deciduous fruit trees absorb Na^+ and Cl^- readily and are severely damaged. Citrus leaves absorb these ions at a slower rate, but still fast enough to cause damage. Avocado and strawberry, which are very sensitive to soil salinity, absorb salt so slowly into the leaves that foliar absorption is negligible. Some herbaceous species that are not particularly sensitive to N^+ or Cl^- in the root zone (e.g., barley, cotton, and sugar beet) can be injured by sprinkling with irrigation water containing more than 10–20 mol m^{-3} Na^+ or Cl^- (Table 11.2).

Substantial salt tolerance is present in some crop species (Table 11.1), and these species can be productive under saline conditions. Consequently, it may be possible to breed cultivars with greater salt tolerance for those crop species currently classified as being sensitive or moderately sensitive to salinity. Unfortunately, as Blum (1988) asserted, "It is a matter of record now that, in spite of some 30 years of research and publication on plant resistance to salinity, there are hardly any cases

TABLE 11.2
Susceptibility of Crops to Foliar Injury from Saline Sprinkling Waters

Na^+ or Cl^- Concentrations (mol m^{-3}) in Sprinkling Water Causing Foliar Injury			
<5	5–10	10–20	>20
Almond	Grape	Alfalfa	Cauliflower
Apricot	Pepper	Barley	Cotton
Citrus	Irish potato	Maize	Sugar beet
Plum	Tomato	Cucumber	Sunflower
		Safflower	
		Sesame	
		Sorghum	

Source: Maas, E. V., *Appl. Agr. Res.*, 1, 12–26, 1986, © Springer Verlag.

of crop cultivars bred for salinity resistance and used as an economic solution in saline agricultural systems."

With the passage of an additional 30 years the situation has improved. Substantial progress has now been made in breeding rice and wheat cultivars with resistance to salinity (Ismail and Horie, 2017), which will be discussed later in this chapter.

Some of the difficulties, opportunities, and progress made in breeding for salt tolerance have been reviewed by Shannon (1997a, b). The absolute yield of crops under saline conditions is important in that it determines their profitability. Many salt-tolerant cultivars can have less absolute yield under saline conditions than other cultivars of the same species that have higher yield potential but less salt tolerance in terms of their relative yields under saline compared with non-saline conditions. What are needed for crop production are cultivars with resistance to salinity, that is, cultivars that have greater absolute yield under the salinity stress occurring in the target production environment than current cultivars.

In general, efficient empirical methods for screening for either salt resistance or salt tolerance are not available. Germination and emergence under salt stress can be screened efficiently but, in some cases, genotypic differences in ability to germinate or emerge have not been associated with ability to grow and yield under salt stress. Sugar beet is an exception where screening for salt tolerance during emergence might be useful. Sugar beet is more sensitive to salt at emergence than during subsequent stages, and the sensitivity at emergence appears to represent a weak link in its adaptation to saline soils. Saline field conditions usually have highly variable salinity and are not effective for screening large numbers of genotypes to detect differences in salt resistance. Special greenhouses with sand culture or hydroponic systems that provide uniform salinity in the root zones can be effective in detecting genotypic differences in salt tolerance. However, greenhouse environments differ substantially from most target production environments and therefore may not be useful for detecting genotypic differences in salt resistance. Another problem confronting empirical breeding for salt resistance is that it appears to be a multigenic, quantitative trait with low heritability (Shannon, 1997a, 1997b).

At the whole-plant level, salt tolerance depends on the ability of plants to control the transport of salt at five sites (Munns et al., 2002):

1. Selectivity of uptake of ions by root cells. It is still unclear which cell types control the selectivity of ion uptake from the soil solution.
2. Selectivity in loading of ions into the xylem. There is evidence for a preferential loading of K^+ rather than Na^+ by the cells of the stele.
3. Removal of salt from the xylem in the upper part of the roots, the stem, petiole, or leaf sheaths. In many species, Na^+ is retained in the upper part of the root system and in the lower part of the shoot.
4. Loading of the phloem. There is little re-translocation of Na^+ or Cl^- in the phloem, particularly in the more tolerant species. This ensures that salt is not exported to growing tissues of the shoot.
5. Excretion through salt glands or bladders. Only halophytes have these specialized cell types.

Selection for physiological traits that contribute to salt resistance has been recommended (Noble and Rogers, 1992). Since salts in the root zone impose an osmotic stress, breeding for enhanced osmotic adjustment by the plant may confer some salt resistance, and some progress has been made in breeding for osmotic adjustment (Shannon, 1997b; also refer to Chapter 9). However, excessive uptake of Na^+ or Cl^- could subject the plant to specific ion toxicities. In this regard, some progress has been made in selecting plants with root systems that have greater selectivity for K^+ over Na^+ compared with other plants (Shannon, 1997b). Also, progress has been made in breeding rootstocks (Mass, 1986) and legumes with restricted Cl^- accumulation in shoots (Noble and Rogers, 1992). Osmotic adjustment that results in accumulation of inorganic ions in the protoplasm is likely to detrimentally influence the structure and functioning of macromolecules in the protoplasm. Consequently, it has been hypothesized that osmotic adjustment in the protoplasm should involve accumulation of small organic compounds that are compatible with plant function. Approaches for the metabolic engineering of salt resistance using betaines and related osmo-protectants is discussed by McNeil et al. (1999), who point out that different osmo-protectants may be effective in different environments. For example, they suggest that choline-O-sulfate may be more effective than glycine betaine in high sulfate saline conditions because its synthesis also could detoxify the sulfate anion.

Cell culture has been used to attempt to select for salt resistance. This approach was justified based on the argument that salt resistance may depend on membrane and cellular properties and be little affected by properties associated with higher levels of plant organization. Several potential problems should be recognized when taking this approach, and these are discussed by Blum (1988). Some progress has been made. Cell lines have been selected that can withstand and grow when subjected to high concentrations of salt. Plants have been regenerated from these cell lines, but they had poor agronomic performance (Stavarek and Rains, 1984). Blum commented in 1988, "The development of a practical contribution by this method in the form of a viable commercial, resistant cultivar, or at least a valuable resistant parental line, is still awaited, but may not be far away." Some 30 years later, I am not aware of any salt-resistant cultivars that have been developed using this approach. Selection in cell culture is powerful because large numbers of cells can be rapidly screened in a uniform environment. But what is the genetic makeup of these cells? Also, the cell culture approach may not be successful for any resistance trait that is strongly dependent on the emergent properties of organs or whole plants, and salt resistance may depend on emergent as well as membrane and cellular properties.

There is much need for breeding rice cultivars that have a greater resistance to salinity (Ismail and Horie, 2017). Current cultivars with good yield potential are moderately sensitive to salinity (Table 11.1) with low thresholds of 1.9–3.0 dS m^{-1}, above which growth and yield are affected. In spite of this sensitivity to salinity much rice is grown on salt-affected soils, including in delta areas. Although salt tolerance in rice is complex and developmentally regulated, progress has been made in recent years in developing cultivars with resistance to salinity using conventional breeding methods. This success resulted from several factors, including the availability of several tolerant landraces for use as donors; phenotyping protocols that made possible the evaluation of large populations in a relatively short time; and progress in understanding the genetics and physiology of tolerance. As a consequence of breeding and extension, several salt-resistant rice cultivars were recently released and commercialized in several Asian countries (Ismail and Horie, 2017). The development of these salt-resistant cultivars took 10–15 years of rigorous evaluation of many breeding lines and incurred high costs, which has made marker-assisted selection attractive. Studies now have identified QTLs associated with salt tolerance in rice that have major effects, including a locus on chromosome 1 that contains *Saltol* which has made marker-assisted selection an attractive new method because it reduces breeding time and costs (Ismail and Horie, 2017). Marker-assisted selection is being used to transfer the *Saltol* locus into several popular varieties and to combine it with the *SUB*1 gene that confers some tolerance to submergence.

There is much need to breed wheat cultivars with greater resistance to salinity even though current cultivars are moderately tolerant to salinity (Table 11.1). Wheat is the major staple food crop of the world and some irrigation schemes where it is being cultivated are becoming salinized. Conventional breeding has resulted in moderate progress in breeding wheat cultivars with greater resistance to salinity in several countries, including India, Pakistan, Australia, and Egypt (Ismail and Horie, 2017). Salt-tolerant wheat is efficient in excluding Na^+ from shoots, especially from young tissue and leaves. This exclusion mechanism reduces the Na^+ load and maintains a greater and more favorable K^+/Na^+ ratio in plant tissue and results in reductions in leaf death and greater growth and grain yield under salt stress. Bread wheat generally has more effective exclusion mechanisms and salt tolerance than durum wheat. However, progress now is being made in breeding to improve the salt resistance of durum wheat (Munns et al., 2002). Alleles have been identified that are associated with salt exclusion and other mechanisms of salt tolerance in bread and durum wheat. Marker-assisted selection to combine *Nax*1, *Nax*2, *and Kna*1 into cultivars with high yield potential might increase the salinity-resistance of both bread wheat and durum wheat and make the breeding programs faster and more efficient (Ismail and Horie, 2017). Another approach is to grow triticale or barley that has greater salinity-resistance than wheat.

The agronomic value of breeding cultivars with greater salt resistance has been questioned. Richards (1995) argued that breeding to enhance grain yield under non-saline conditions may be more effective for both non-saline and saline conditions than trying to enhance salt resistance through breeding. Also, he has compared dry matter production of various crops in soils having different salt concentrations and a limited water supply. He found that current salt-resistant barley and wheat cultivars produced more dry matter under saline conditions than wild relatives that have greater salt tolerance. A sunflower cultivar was similarly more productive than a salt-tolerant relative. Consequently, there appear to be cases where trying to breed cultivars with additional salt resistance may not be very successful, and the greatest opportunities exist for improving crop production are by reducing salinity through management methods and by breeding to increase yield potential.

An alternative approach would be to enhance the domestication of minor crop species that already have considerable salt tolerance. One possibility is quinoa that can be grown in water having a range of salinities up to that of seawater. Quinoa produces nutritious seeds that suffice as a staple food, are gluten free, and have a high content of high-quality protein (Adolf et al., 2013). Quinoa was cultivated for thousands of years as a staple food in the Andes of South America. In recent years, a world market has developed for quinoa seeds as a specialty food. Peru and Bolivia are the greatest producers of quinoa but on-farm yields typically are low at < 1,000 kg ha^{-1} (Adolf et al., 2013). Due to the low productivity and incompletely developed market, the cost of quinoa grain is 10 times higher than the cost of wheat grain. Consequently, quinoa does not have much potential as a food for poor people at this time, though it has some potential for poor farmers who can take advantage of the high prices to sell the quinoa they produce and then buy either wheat or rice grain for home consumption. If breeding and agronomic research were devoted to quinoa, it is likely that the grain yield could be increased substantially because relatively little research has been conducted on this crop.

The halophyte salicornia has substantial salt tolerance and potential as a crop for saline environments. When irrigated with seawater, it can germinate, actively grow, and produce abundant seed with a high oil and protein content that, with modifications, might be useful as a livestock feed (Glenn et al., 1991). Whether salicornia will be adopted by many farmers depends on its yields and profitability. It was reported to have seed and biomass yields, when grown with seawater that equaled or exceeded those of soybean and sunflower grown on fresh water (Glenn et al., 1991). This report is misleading, however, in that, even though salicornia did produce high seed yields of 139–246 g m^{-2}, the growing season was 6–10 months, which is much longer than the growing seasons of crops of soybean and sunflower, and thus the costs per unit of production would have been high. Other potential crop plants for saline environments include halophytes that are able to provide useful browse for livestock.

11.3 BORON TOLERANCE

Boron is an essential plant nutrient, but it can become toxic to some plants when soil-water concentrations only slightly exceed those required for optimum plant growth. Generally, toxic concentrations in the soil are found only in arid zones and may occur together with salinity. The primary source of boron is the use of irrigation water from wells that contain high levels of boron, but also some soils derived from boron-containing sedimentary or igneous rocks already have high levels of boron. In general, boron is more difficult to leach from soils than ions such as Na^+ and Cl^-. Where well water used for irrigation has a significantly high boron concentration, crop species must be chosen that have sufficient tolerance. Foliar symptoms of boron toxicity can be distinctive for many species involving chlorosis and necrosis of the margins of leaves and necrotic spots on central parts of leaves. Leaf symptoms of boron toxicity may not be completely effective in diagnosing tolerance, however, in that some species can exhibit leaf symptoms but no decline in yield, and others can exhibit declines in yield but no leaf symptoms (Maas, 1986). Threshold values for boron concentrations in soil water above which decreases in yield have occurred are presented in Table 11.3. In addition, Maas (1986) provides a table where some citrus and stone-fruit rootstocks are ranked in order of increasing boron accumulation and transport to scions.

TABLE 11.3
Boron Tolerance Limits for Crop Plants

Crop Species	Boron Yield Threshold, g m^{-3}	Crop Species	Boron Yield Threshold, g m^{-3}
Very tolerant		*Sensitive*	
Asparagus	10–15	Peanut	0.75–1
Celery	6–10	Lima bean	0.75–1
Onion	6–10	Common bean	0.75–1
Cotton	6–10	Strawberry	0.75–1
Sorghum	6–10	Lupine	0.75–1
Tolerant		Sesame	0.75–1
Sugar beet	4–6	Mung bean	0.75–1
Red beet	4–6	Sunflower	0.75–1
Parsley	4–6	Wheat	0.75–1
Vetch	4–6	Sweet potato	0.75–1
Alfalfa	4–6	Pecan	0.5–0.75
Tomato	4–6	Walnut	0.5–0.75
Garlic	4–6	Grape	0.5–0.75
Moderately tolerant		Fig, Kadota	0.5–0.75
Cauliflower	2–4	Persimmon	0.5–0.75
Melon, musk	2–4	Plum	0.5–0.75
Squash	2–4	Cherry	0.5–0.75
Clover, sweet	2–4	Peach	0.5–0.75
Mustard	2–4	Apricot	0.5–0.75
Tobacco	2–4	Orange	0.5–0.75
Artichoke	2–4	Grapefruit	0.5–0.75
Maize	2–4	Avocado	0.5–0.75
Oats	2–4	*Very sensitive*	
Barley	2–4	Blackberry	<0.5
Bluegrass, Kentucky	2–4	Lemon	<0.5
Turnip	2–4		
Cabbage	2–4		
Cowpea	2–4		
Moderately sensitive			
Cucumber	1–2		
Irish potato	1–2		
Radish	1–2		
Carrot	1–2		
Pea	1–2		
Pepper	1–2		
Broccoli	1–2		
Lettuce	1–2		

Source: Maas, E. V., *Appl. Agr. Res.,* 1, 12–26, 1986 © Springer Verlag; Maas, E. V. and Grattan, S. R., Crop yields affected by salinity, *Agricultural Drainage,* Agronomy Monograph no. 38, American Society of Agronomy, Madison, WI, pp. 55–108.

11.4 ALUMINUM TOLERANCE

Aluminum in soil solution can reach toxic levels when the soil is acid with pH < 5.5. There are greater detrimental effects when soil calcium and organic matter also are low. Aluminum toxicity is the major factor limiting crop productivity on acid soils, which comprise about 40% of the world's crop land (Kochian, 1995). Adding amendments, such as calcium compounds, to the upper part of the soil profile to raise the soil pH may not be economical in some cases. Also it may not solve the problem, especially where the subsoil has a low pH and is not affected by the amendments.

Crop resistance to high aluminum, defined as where a cultivar has greater yields than another cultivar in soils with high aluminum, has provided a solution. Crop species vary in their tolerance to high aluminum levels in the soil solution, where tolerance is defined on the basis of relative shoot dry matter in the presence of aluminum compared with optimal root zone conditions (Table 11.4). These data must be interpreted carefully, since there is substantial variation in tolerance to aluminum within many crops species (Foy, 1988). Additional aluminum tolerance ratings for several pastoral grasses and legumes are provided in Wheeler et al. (1992) and Wheeler (1995).

The initial and most dramatic symptom of aluminum toxicity is inhibition of root elongation, and the root apex is the site of aluminum toxicity (Kochian, 1995).

TABLE 11.4
Crop Species' Differences in Aluminum Tolerance

Crop Species	Cultivars	Aluminum Tolerance Rating[a]
Rice	CT7244-9-2-1-152-1, AC 165, CT6156-33-11-1-43, P5598-1-1-3P-4MP	Tolerant
Maize	P3362, P3475, P3553, P3585, P3902	Tolerant
Lima bean		Tolerant
Common bean	The Prince, Tendergreen, Yatescrop 37067	Moderately tolerant
Squash	Delica	Moderately tolerant
Wheat, bread	BH1146, Carazinho	Moderately sensitive
Common bean	Red Kidney, Black Turtle, Haricot	Moderately sensitive
Oat, white	Blackbutt, Swan	Moderately sensitive
Wheat, bread	Cardinal, Waalt, Atlas–66	Sensitive
Oat, white	Coolabah, Carbeen, Camellia, West	Sensitive
Asparagus	Lucullus	Sensitive
Onion	Pukekohe, Longkeeper	Sensitive
Wheat, bread	Warigal, Sonora–63, Scout	Very sensitive
Barley	Kearney, Dayton	Very sensitive
Oats, red	Acacia	Very sensitive

Source: Wheeler, D. M., *J. Plant Nutr.*, 18, 2305–2312, 1995; Wheeler, D. M. et al., *Plant Soil,* 146, 61–66, 1992.

[a] Aluminum (Al^{+++}) concentrations in μM in the soil that would reduce shoot yields by 50% are > 10 for tolerant, 5–10 for moderately tolerant, 2–5 for moderately sensitive, 1–2 for sensitive, and < 1 for very sensitive.

Aluminum resistance and tolerance depend on two types of mechanisms: exclusion of aluminum from the symplasm in the root apex, and detoxification of aluminum once it has entered the symplasm of the roots (Blum, 1988). Some major genes and effective screening methods are available for aluminum tolerance, and they also appear to confer aluminum resistance (Blum, 1988; Foy, 1988). The techniques include using hydroponics or sand or soil systems. Hydroponic systems may mainly select for mechanisms that operate within the root apex.

Substantial progress has been made in breeding for resistance to high aluminum levels in several crop species. Some of the aluminum resistance in wheat involves exclusion of aluminum from the root apex (reviewed by Kochian, 1995). The mechanism appears to involve release of dicarboxylic acids by the roots that is induced by high aluminum and protects the root apex by chelating Al^{+++}. Resistant lines release much more organic acid than susceptible lines. Sorghum usually has been classified as being sensitive to soil acidity and aluminum, but substantial progress has been made in breeding resistant cultivars of sorghum (Clark and Duncan, 1993).

Comparing the most effective strategies for solving problems due to salinity, high boron, and high aluminum levels illustrates some contrasts. The most effective first step in solving salinity problems is to use management methods to reduce the level of salinity, and then to choose cropping systems and crop species that have sufficient salt resistance to be suited to the sustainable salinity levels that are achieved. Where boron levels in well water are high, the first step is to choose a crop species that can tolerate the level of boron in the soil that is likely to result when this well water is used to irrigate the crop. Then, management methods are used to prevent excessive accumulation of the boron in the root zone. With acid soils, where aluminum levels in soil water are high, the first step is to choose the crop species best adapted to the overall socioeconomic and climatic conditions, and then breed to incorporate the level of aluminum resistance that is needed. Use of amendments to raise the soil pH also should be considered where they can be applied economically and complement the use of aluminum resistant cultivars.

11.5 EXTREMES OF SOIL TEXTURE AND HIGH SOIL BULK DENSITY

A loamy soil with good structure can provide good growing conditions for all crops, and its only limitation may be that the land is expensive. In contrast, soil with a high proportion of a swelling clay can have poor internal drainage and aeration. Such land may be ideal for growing paddy rice, since water does not drain through it. Alternative crops to rice for this type of land are few and include certain pasture grasses, sorghum, and cotton.

Many crops can be very sensitive to anaerobic conditions in the root zone. Most fruit trees require a free draining soil. Avocado is particularly sensitive to anaerobic conditions due to enhanced attacks by the root rot disease *Phytophthora cinnamomi* Rands, which is a major problem for avocado. Several crops grow more slowly and produce less grain when subjected to only occasional waterlogging, including grain legumes (Hodgson et al., 1989), pearl millet, and maize. With more extreme waterlogging and high-pH calcareous soils, several crops (including grain legumes and maize) can exhibit leaf chlorosis. The problem appears to involve Fe deficiency, which is induced by the combination of high pH and waterlogging (Hodgson et al., 1992). The problem has been solved by either supplying chelated Fe to the crop (Hodgson et al., 1992) or by irrigating only alternate furrows and cultivating to enhance the oxygen supply to at least part of the root system. In soybean, genotypic variability exists for using Fe from high-pH soils. Brown et al. (1967) evaluated two parents and iso-lines of soybean that differed in ability to use Fe from high-pH media. Genotypic ability to absorb and translocate Fe was associated with the release of compounds into the rhizosphere, which reduced the pH of the rooting medium and reduced iron from the ferric to the ferrous state. The release of compounds by the roots was induced by Fe deficiency. For soybean, Fe deficiency in high-pH soils has been partially solved by selecting cultivars that are more efficient in taking up and translocating Fe when subjected to this soil condition (Fehr, 1982; Clark and Duncan, 1993). Substantial progress also has been made in breeding common bean cultivars with resistance to Fe deficiency, and some progress has been made with sorghum (Clark and Duncan, 1993).

Very sandy soils have good aeration, but they have low water-holding capacity and can be very infertile. Low levels of soil nitrogen may be advantageous when growing specific crops, such as malting barley for brewing, because grain quality can be enhanced by low nitrogen. In other cases, it may be necessary to grow crops that are better adapted to infertile soils, such as finger millet and cassava, rather than crops such as maize and Irish potato that require fertile soils if they are to be adequately productive. Tree and vine crops may grow well on very sandy soils in arid and semiarid zones if they also are irrigated with a drip system that can frequently supply the water and inorganic nutrients needed by the plants.

Plant roots encounter mechanical resistance to their penetration through soil. This resistance is higher with natural high levels of soil bulk density associated with differences in soil structure and with increases associated with soil drying and due to soil compaction by tillage equipment. Crop yields have been decreased as much as 19% to 55% by soil compaction (Smucker and Allmaras, 1993). Soil compaction occurs when soils are cultivated that are too wet or when soils are cultivated with equipment that applies too much force in an inappropriate manner. Subsoil compaction is particularly detrimental because it may not be ameliorated by natural events such as freezing and thawing, wetting and drying, and bioactivity such that it may become more pronounced over time. The resistance can be so great that roots do not penetrate the soil layer. At intermediate levels of soil resistance, leaf expansion rate and stomatal conductance are reduced (Masle and Passioura, 1987; Passioura and Stirzaker, 1993). The authors hypothesized that these effects were not mediated by effects of soil conditions on uptake of water or nutrients and that they may have been caused by feedforward responses involving changes in hormonal signals from roots to shoot, but that feedback responses also probably occur at a later time (refer to Chapter 9 for a discussion of feedforward and feedback control systems in plants). Relatively little is known about species' differences in rooting response to high-strength soils. Genotypic differences in rooting responses have been reported among wheat and barley cultivars and landraces (Masle, 1992). Genotypes with less sensitivity to high-strength soils, in terms of relative growth rate, had slower rates of net assimilation and root growth in low-strength soils.

Plows are important tools for managing the environments of roots. For example, a substantial part of the Peanut Basin of Senegal has soil with a high bulk density such that root growth is impeded. Research at the National Center for Agronomic Research near Bambey has demonstrated that plowing with a mold-board plow just after the first major rainfall of the cropping season partially ameliorates this problem. Crops of peanut, pearl millet, and cowpea sown after the plowing exhibited enhanced root growth and greater grain yields than crops sown on land that had not been plowed. Also, there are other circumstances where mold-board plowing is useful, such as with mixed farming enterprises that combine the rearing of livestock with crop cultivation. Mold-board plowing is useful for incorporating farmyard manure or incorporating pastures prior to planting them with field crops. Irish potatoes are particularly responsive to soil with the good structure that can result from growing a pasture crop for several years or to the enhanced fertility and improved soil structure that can result from the incorporation of manure.

There are some circumstances, however, where minimal tillage in which the soil is not inverted by plowing is desirable. In these cases a robust planter is used to sow seed into the stubble of the previous crop and to make a small strip of tilled soil that promotes seed germination and the growth and establishment of seedlings. Varieties with herbicide tolerance are usually used with this minimum-tillage approach so that when weeds emerge they can be killed by spraying an herbicide over the weeds and the crop. This minimum-tillage approach is particularly useful when sufficient stubble and other crop residues are maintained on the surface of the soil and prevent or at least substantially reduce erosion of the soil by wind or rain. Note that topsoil is a valuable resource that is only produced slowly. It takes about 200 years to produce one centimeter depth of soil by weathering of the subsoil, and the subsequent action of earthworms and other organisms. A single wind storm can blow away one centimeter depth of soil if the soil surface is not protected. Another value of the minimum-tillage approach is that it results in less compaction of soil, which can occur when land is plowed or tilled. I recall a farm that I visited many years ago in the Central Valley of California. The farmer had grown crops on permanent soil beds with minimal tillage for many years. Access by tractors and other equipment only occurred on a few specific access ways between the beds. When I examined the permanent beds I discovered that the soil structure was excellent and that I could easily push my fingers into the soil.

The overall conclusion is that I cannot make general recommendations concerning the value of plowing. In some cases it can be useful, while in other circumstances minimal tillage that does not include plowing is desirable, especially if it retains substantial amounts of crop residues on the soil surface and does not compact the soil.

ADDITIONAL READING

Clark, R. B. and R. R. Duncan. 1993. Selection of plants to tolerate soil salinity, acidity and mineral deficiencies, in D. R. Buxton et al. (Ed.), *International Crop Science I*. Crop Science Society of America, Madison, WI, pp. 371–379.

Epstein, E. 1972. *Mineral Nutrition of Plants: Principles and Perspectives.* Wiley, New York, p. 412.

Epstein, E. and J. D. Norlyn. 1977. Seawater-based crop production: A feasibility study. *Science* 197: 249–251.

Hodgson. A. S., J. F. Holland, and P. Rayner. 1989. Effects of field slope and duration of furrow irrigation on growth and yield of six grain-legumes on a waterlogging-prone vertisol. *Field Crops Res.* 22: 165–180.

Hodgson, A. S., J. F. Holland, and E. F. Rogers. 1992. Iron deficiency depresses growth of furrow irrigated soybean and pigeon pea on vertisols of northern N.S.W. *Aust. J. Agric. Res.* 43: 635–644.

Ismail, A. M. and T. Horie. 2017. Genomics, physiology, and molecular breeding approaches for improving salt tolerance. *Annu. Rev. Plant Biol.* 68: 405–434.

Kochian, L. V. 1995. Cellular mechanisms of aluminum toxicity and resistance in plants. *Annu. Rev. Plant Physiol. Plant Mol. Biol.* 46: 237–260.

Maas, E. V. 1986. Salt tolerance of plants. *Appl. Agr. Res.* 1: 12–26.

Mackill, D. J., A. M. Ismail, U. S. Singh, R. V. Labios, and T. R. Paris. 2012. Development and rapid adoption of submergence-tolerant (*SUB*1) rice varieties. *Adv. Agron.* 115: 299–352.

Munns, R., S. Husain, A. R. Rivelli, R. A. James, A. G. Condon, M. P. Lindsay, E. S. Lagudah, D. P. Schachtman, and R. A. Hare. 2002. Avenues for increasing the salt tolerance of crops, and the role of physiologically based selection traits. *Plant Soil* 247: 93–105.

Rhoades, J. D., A. Kandiah, and A. M. Mashali. 1992. *The Use of Saline Waters for Crop Production.* FAO Irrigation and Drainage Paper No. 48. FAO, United Nations, Rome, Italy, p. 133.

Richards, R. A. 1995. Improving crop production on salt-affected soils: By breeding or management? *Expl. Agric.* 31: 395–408.

Shannon, M. C. 1997a. Adaptation of plants to salinity. *Adv. Agron.* 60: 75–120.

Shannon, M. C. 1997b. Genetics of salt tolerance in higher plants, in P. K. Jaiwal, R. P. Singh, and A. Gulati (Eds.), *Strategies for Improving Salt Tolerance in Higher Plants.* Oxford & IBH Publishing, New Delhi, India, pp. 265–289.

Wheeler, D. M., D. C. Edmeades, R. A. Christie, and R. Gardner. 1992. Effect of aluminum on the growth of 34 plant species: A summary of results obtained in low ionic strength solution culture. *Plant Soil* 146: 61–66.

12 Interaction of Crop Responses to Pests and Abiotic Factors

Some consideration of crop responses to pests and diseases is warranted because they can interact with crop responses to abiotic factors and involve the evolution of volatiles by a plant and the detection of volatiles emitted by other plants (Dicke et al., 2003). I will provide examples of these interactions and some principles concerning plant responses to abiotic and biotic factors. Earlier research has tended to emphasize specific mechanisms of plant resistance to specific pests and diseases in a view that is dominated by consideration of only plant defensive attributes. A broader integrated systems approach can be useful—one that puts plant defense within the context of whole-plant function and the life history traits and specializations of the different pests and diseases (Coleman et al., 1992).

12.1 CROP PHENOLOGY AND THE ESCAPE OR AGGRAVATION OF PEST PROBLEMS

Breeding to enhance adaptation to drought in the Sahelian and other semiarid zones has involved selection for early flowering and maturation as an adaptation to the short rainy seasons (Hall, 2017). For cowpea, the early cultivars that were developed not only escape drought, they also partially escape attacks from flower thrips, which can be a major insect pest (Hall et al., 1997). Except for the extreme droughts that often occur, the Sahelian zone is more suitable for growing cowpea than the wetter Savanna zones because of the reduced populations of insect pests, such as flower thrips, pod sucking bugs, and pod borer in the Sahelian zone.

Early pearl millet cultivars also were bred for the Sahelian zone, but a major problem occurred in extending them to farmers. When only one or a few farmers planted a new early pearl millet cultivar, the small area became a focal point for attacks by seed-eating weaver birds. Other sources of food were not available for the birds at that time, since the current cultivar of pearl millet that was grown on a large area has a 10-day-longer cycle length and became mature later than the new cultivar. Most pearl millet plants emerge on the same date in adjacent areas in this zone, since they are sown into dry soil prior to the start of the rainy season and emerge on the same date after the first rain. The solution to this problem is to have an extension program where there is a mass distribution of the seeds of the new cultivar of pearl millet to many farmers. With a large area planted, the attacks by weaver birds will be spread over many fields and cause less damage to grain yield per area or farm.

In contrast to the practice with pearl millet, cowpea is sown into moisture in the Sahelian and Savanna zones. The optimal sowing date is just after the first major rainfall event of the rainy season. Unfortunately, this rainfall event also causes the adult moth of *Amsacta moloneyi* Druce to begin emerging from diapause. The hairy caterpillars produced from the eggs of this moth become major pests, attacking young seedlings of cowpea. Defoliation makes it more difficult for the cowpea seedlings to survive the other stresses that often occur. Dry sowing cowpea may enable the crop to better withstand the attacks of hairy caterpillar because the crop emerges a few days sooner than a crop sown after the rain, and the seedlings are bigger and can outgrow the attacks of the insect. The disadvantage of dry sowing is that a small rainfall event may cause seeds to germinate while there is little moisture in the soil profile, and the plants will not become established. An important advantage of early sowing in both the Sahelian and the Savanna zones is that the plants take up the nitrogen that has mineralized during the dry season and thereby prevent it from being leached and lost (Bunting, 1975). Low soil nitrogen is a major factor that limits crop productivity in these zones (Breman and de Wit, 1983).

The development of garbanzo bean cultivars that are suitable for late fall or winter rather than spring sowing in Mediterranean climates has the potential to substantially increase grain yield because the crop is subjected to a more suitable thermal environment for a longer duration, thereby making more complete use of the available solar radiation (Saxena, 1987). Major constraints for fall and winter sowing of garbanzo bean in the San Joaquin Valley of California (Figure 10.6) are that the crop is subjected to high levels of virus diseases and ascochyta blight. The latter disease is favored by the cool weather and canopy wetness that occur in the winter months.

When changing the phenology of crop plants by either breeding or management to enhance crop adaptation to the abiotic environment, it is important to recognize that pests and beneficial organisms also have life cycles that are synchronized with the abiotic and biotic environments. In some cases, modifications to crop phenology, such as by changing sowing and harvesting dates, have been used to decrease the extent of pest attacks (Letourneau, 1997). In such cases, it is important to ask whether the phenology is now optimal in relation to the abiotic environment, such as by providing high levels of *PFD* and optimal temperatures during critical reproductive stages and a cropping season of sufficient duration. For example, early harvest of cotton can decrease the populations of insect pests that survive the following winter and thereby reduce insect pest problems the following year (Letourneau, 1997). But this method has both costs and benefits; some potential yield is lost, but insect pest management is enhanced, and input costs can be reduced both in that year and the following years.

12.2 CROP RESISTANCE TO PESTS

Crop resistance to pests is defined as the ability of a cultivar to be more productive or survive better than another cultivar when a specific pest is present in the environment, analogous to the definitions of drought resistance in Chapter 9, and flooding and salt resistance in Chapter 11. A major feature of most natural plant communities and many unsprayed crop plants is that they exhibit relatively little damage from insect pests, even though there are many potential insect pest species in their environment. In natural systems, physical and chemical defenses as well as nutrient deficiencies present major hurdles for phytophagous arthropods (Letourneau, 1997) and other herbivores. But, in some instances, a single major pest species can devastate crops. Most crops do have, however, considerable resistance to many potential insect and mite pests.

Mechanisms of resistance to insect pests have been defined as involving antixenosis (non-preference), antibiosis, and tolerance (Smith, 1989). In some cases, it may not be possible to separate antixenotic effects from antibiotic effects. For example, if an insect confined to a resistant plant gains weight at a slower rate than an insect on a susceptible plant, one might assume that it is due to the presence of antibiotic factors in the resistant plant. The slower weight gain may be caused, however, by less feeding due to the presence of physical or chemical feeding deterrents that cause the insect to have aberrant behavior (Smith, 1989). Plant resistance also can be complex and involve tritrophic systems. For example, when cabbage and maize are attacked by certain caterpillars, their leaves can produce volatile chemicals that attract parasitic wasps that attack the caterpillars (Lambers et al., 1998). In addition, plants may benefit from damage-related signaling of volatiles by neighboring plants, which causes them to either produce chemical defenses or recruit carnivorous arthropods to act as *bodyguards* (Dicke et al., 2003).

Antixenosis as a resistance mechanism is where the pest prefers other cultivars or species as hosts. This mechanism has particular relevance to mobile pests and environments where the pest has many options when choosing a host plant. The presence of thorns, hairs, or a waxy, thick cuticle may deter specific herbivorous pests. Plants in arid environments often have these traits, and they have been called *xeromorphic structures* (i.e., drought-adapted structures), but it is more likely that their adaptive advantage comes from preventing herbivory rather than enhancing adaptation to drought. There is relatively little plant biomass in arid environments, resulting in strong selection pressure for traits that reduce herbivory.

The environmental cues used by pests in choosing host plants can include the quality and intensity of reflected radiation, volatile chemicals, tactile sensations, or taste. When studying effects of mulches with different reflective properties on plant performance, as was discussed in Chapter 6, it is important to recognize that reflected light of specific wavelengths can deter aphids and may therefore also reduce the extent of their transmission of plant pathogenic viruses (Smith, 1989). An example of antixenosis is that lygus bugs prefer alfalfa over cowpea, and cowpea over cotton. The lygus bugs do little damage to alfalfa grown for hay; consequently, it is useful to manage alfalfa so that the lygus bugs remain in the alfalfa fields. In California, this has been achieved by cutting strips of the alfalfa in a rotation so that uncut plants, which are preferred hosts of lygus, always are present in every field. Unfortunately, some growers cut all of their alfalfa fields, and the lygus bugs are driven out and may enter fields of cowpea where they destroy floral buds and severely reduce grain yield. A possible resistance strategy for cowpea would be to breed cultivars with sufficient antixenosis that the lygus bug strongly prefers both alfalfa and cotton. This strategy could be effective in California, where a relatively small area of cowpea is grown in a mosaic of small fields surrounded by large fields of cotton and alfalfa.

Plant morphological traits can have deterrent effects and influence the suitability of plants as hosts for herbivores. Many crop plants have different types of hairs on their leaves that have either negative or positive effects on insect pests such as: having hairs with chemicals that have negative effects on the insects or hairs that impede mobility of the insects—or having hairs that aid oviposition by the insects or protect the insects from parasites or predators (Letourneau, 1997). Cotton cultivars with hairy leaves have exhibited greater resistance to lygus bugs and cabbage looper but greater susceptibility to cotton fleahopper and tobacco budworm (reviewed by Smith, 1989). Leaf hairs also increase the leaf boundary layer resistances to the transfer of water vapor, carbon dioxide, and heat (these resistances are discussed in Chapters 7 and 8). Increases in leaf boundary layer resistances due to the presence of leaf hairs can decrease rates of transpiration and photosynthesis, but the only significant effects may occur where the hairs are very long and dense and associated with changes in leaf temperature (Jones, 1992). Effects of specific types of hairs on leaf temperature could be determined by a combination of experimental studies and modeling. First, the differences in temperature between leaf and air would be determined under nighttime conditions when the stomata are closed for the hairy leaves and leaves where the hairs have been shaved off. These data could then be used to estimate boundary layer resistance properties (parameter c in Equation 8.4) of hairy and non-hairy leaves. Then a steady-state model such as Equation 7.10 could be used to predict the temperature differences of hairy and non-hairy leaves over a range of environmental conditions.

Another morphological trait that can affect insects is where pods are displayed above rather than in the canopy of leaves. Cowpea cultivars that have pods displayed above the canopy are not good hosts for pod borers that oviposit on pods at the point where they touch leaves, since very few pods touch leaves. Unfortunately, there is a trade-off in that pods which are above the canopy also shade leaves and reduce canopy photosynthesis and potential plant productivity (refer to Chapter 4 for additional discussion of this topic).

Antibiosis as a resistance mechanism is where the plant is not a good host for the pest such that the growth and reproductive rates of the pest are slow. Plants produce many compounds that are not involved in primary catabolic or biosynthetic pathways which are commonly called *secondary metabolites*. Some of these secondary metabolites have negative effects on herbivores as either toxins or compounds that reduce the digestibility or palatability of the plant as a food source and have deterrent effects.

The advantage of toxins as mechanisms of pest resistance is that they can be effective at very low concentrations and thus require little investment by the plant. A disadvantage of toxins is that they can reduce the food value of plants to people. For example, *bitter* cultivars of cassava have hydrocyanic acid (HCN) widely distributed in their tubers. The HCN provides protection from herbivores such as wild pigs and warthogs. Humans eating the tubers of *bitter* cultivars must destroy the HCN by boiling, roasting, or fermentation. In the *sweet* cultivars of cassava, which are widely grown, the HCN is mainly in the exterior of the tuber, and much of it can be removed by peeling. *Sweet* cultivars, however, are less effective than *bitter* cultivars in deterring feeding by wild pigs and warthogs.

Some toxins specifically affect certain insect pests, such as the toxic glycoproteins present in the bacterium *Bacillus thuringiensis* (Bt), and are not known to affect people. Cultivars of cotton and maize that produce these toxic Bt glycoproteins and thereby resist certain insect pests have been developed by genetic engineering. A disadvantage of toxins as mechanisms of pest resistance is that some pests can evolve so that they overcome the toxin and thereby gain a food source that is not available to other pests (Lambers et al., 1998). Where a single resistant crop cultivar is grown over a large area, the evolution of insect pest biotypes that overcome the resistance can become a major problem. Where the resistant crop cultivar is grown in a mosaic of fields surrounded by fields of susceptible cultivars and other crop species and weeds that also act as hosts to the insect pest, the evolution of insect pest biotypes that overcome the toxin-based resistance is less likely to occur. In some cases, farmers who are growing genetically engineered cultivars that have the toxic Bt glycoproteins also are required to grow a specified minimum area of a cultivar that does not have the toxic Bt glycoproteins to slow down the evolution of Bt-resistant insect pest biotypes (Barton and Dracup, 2000). Cowpea lines with a Bt gene now have been bred (Popelka et al., 2006) and have shown resistance to the legume pod borer, which is a major pest of cowpea in the Savanna zone of West Africa, but they have not yet been deployed as cultivars.

Some consumers strongly oppose the use of genetically engineered crops to produce food, especially in Europe and in particular England. The English are major consumers of baked beans, the bean type produced in Michigan. The bean growers in Michigan and California strongly oppose the use of genetic engineering in breeding bean varieties because it could have major negative effects on overseas markets for their beans. There is similar opposition to the use of genetically engineered wheat in foods. Plant breeders must be keenly aware of consumer needs and interests, especially their views on genetic engineering. I am not aware of significant opposition among consumers to varieties of crops bred using marker-assisted selection.

Some compounds that reduce the digestibility and palatability of plants as food sources for pests, such as phenolics, are present at relatively high concentrations in some plants and require a substantial investment in carbohydrate. This defense mechanism can be effective against a large number of pests and can be more durable than defense based on toxins. Mechanisms that involve a substantial investment of carbohydrate in protective compounds and structures tend to occur in slow-growing species, and especially those with evergreen leaves (Lambers et al., 1998). Fast-growing plant species mainly partition carbohydrate to compounds and structures that contribute to photosynthesis and growth (Evans, 1998, and the discussion in Chapter 4). Most crop species are fast growing and would be expected to have relatively small quantities of defensive compounds in their leaves. This pest resistance mechanism would not appear suitable for protecting plant parts that are eaten by people. Brown and red sorghum grains that contain phenolics, however, are used for making beer in Africa and for feeding livestock in the United States. The tannins add flavor to the beer and reduce molding during the various brewing processes. Substantial tannins in the grains of sorghum also partially deter feeding by weaver birds. The sorghum cultivars that are most popular for use as a staple food by people in Africa and Asia have white grains with low quantities of tannins.

The suitability of plants as hosts for herbivorous insects has been assessed using the C/N ratio of tissue. A lower ratio is considered to indicate a better host, providing it reflects a higher concentration of amino acids in the food ingested by the insects. In the elevated atmospheric [CO_2] levels expected to occur late in the twenty-first century, non-leguminous species could develop higher C/N ratios, especially in environments where soil nitrogen is limiting. A review indicates several cases where elevated [CO_2] raised C/N ratios and lowered the quality of plant species as hosts for several species of herbivorous insects (Coviella and Trumble, 1998). Applying nitrogen fertilizer often decreases the C/N ratio and could overcome the effect of elevated [CO_2]. Nitrogen fertilization does not necessarily enhance the food value of foliage for herbivores. For example, nitrogen fertilization can enhance levels of nitrogen-based defensive compounds, such as alkaloids, glucosinolates, and cyanogenic glycosides (Letourneau, 1997). Whether this occurs can depend on the physical environment. Alkaloid level in potato leaves was increased with nitrogen fertilizer when the weather was cool and cloudy but was lower in nitrogen-fertilized plants when the weather was warm and sunny (Letourneau, 1997). A possible explanation of the response in the warm sunny environment is that enhanced photosynthesis and growth by the nitrogen-fertilized plants resulted in a dilution of the levels of secondary metabolites in the leaves. It has been hypothesized that nitrogen often is a critical limiting nutrient for arthropod herbivores and that nitrogen fertilization usually enhances damage by leaf-chewing insects and mites. A review by Letourneau (1997) indicates this effect may be pronounced with potted plants but not apparent in the large-scale field environments typical of farms due to tritrophic, microclimate, and possibly other effects. The scale of experimental studies on plant resistance to insects can strongly influence the results and, as with crop water use (as was discussed in Chapter 2), large-scale field studies may be most effective in producing predictions that are relevant to farming.

Abiotic stresses can increase, decrease, or have no effect on plant susceptibility to pests and diseases. For example, cottonwood sapling cuttings were fumigated with ozone for five hours and then bioassayed with various pests and diseases (Coleman et al., 1992). The population of a rust disease and the damage it caused were less on fumigated plants, partially offsetting the direct detrimental effects of ozone on plant growth. In contrast, a leaf spot disease and an aphid had similar detrimental effects on fumigated plants and plants that were not fumigated. In this case, the disease, insect pest, and ozone had additive detrimental effects on the plants that could be very damaging. With a beetle, ozone fumigation resulted in increased feeding but reduced oviposition, so the overall effect is complex and requires analysis of the life history traits of the pest.

Drought during the vegetative stage can substantially enhance attacks on cowpea by lesser corn stalk borer and charcoal rot. In contrast, drought during the early flowering stage may reduce attacks on floral buds of cowpea by lygus bug. The senescence of many cowpea cultivars that occurs after producing the first flush of pods probably is caused by a combination of both carbohydrate starvation of roots and attacks by *Fusarium solani* f. sp. *phaseoli* (Ismail et al., 2000). The delayed-leaf-senescence trait that overcomes this senescence and confers adaptation to drought and other stresses during the flowering stage probably enhances partitioning of carbohydrate to roots, thereby conferring resistance to the pathogen but reducing potential productivity (Ismail et al., 2000).

Many crop plants are attacked by root knot nematodes that damage root systems and, in extreme cases, kill plants. Major resistance genes have been discovered, such as the M_{i-1} gene of tomato. The M_{i-1} gene was discovered in a related wild species, *Lycopersicon peruvianum*, and then transferred to tomato by conventional hybridization. Unfortunately, the M_{i-1} gene can become ineffective at high soil temperatures, as can genes that confer resistance to root knot nematode in alfalfa, sweet corn, and cotton. Recently, genes conferring heat-stable resistance were discovered in other races of *Lycopersicon peruvianum* (Veremis and Roberts, 2000). When studying plant resistance to pests, it would appear important to conduct evaluations in a range of target production environments.

Plant resistance can be induced in that it can become stronger after the plant has been subjected to the stress. This is similar to the acclimation that can occur to abiotic stresses such as drought and heat. The adaptive significance of the induction is that it allows the organism to obtain the benefits of defense when it is needed, while reducing the costs of the *strategy* when defense is not needed. The cost of induced defenses is that some damage must be tolerated by the plant until the defensive compounds are produced in adequate quantities and appropriate locations. Induction can involve systemic movement of chemical signals within plants or volatile chemical signals between plants (Lambers et al., 1998). Interactions may be present in the induction of resistance to some biotic and abiotic stresses that are either positive or negative. Interactions in resistance can occur because of relationships among various signaling systems involving salicylic acid, jasmonic acid, abscisic acid, and ethylene (Bostock, 1999). Plants treated with salicylic acid and related compounds produce several pathogenesis-related compounds, and this treatment also induces resistance to some pathogens and can cause partial stomatal closure that impedes the penetration of tissue by some pathogens and influences plant-water relations. Jasmonic acid and its volatile ester methyl jasmonate induce proteinase inhibitors and other factors such as polyphenol oxidase and lipoxygenase, and they can protect plants from insect pest herbivores. Exogenous applications of abscisic acid have very similar effects on plants as soil drought, including causing partial stomatal closure (Table 10.1 in Jones, 1992). Abscisic acid accumulates to high levels in plants when they are subjected to drought and induces gene expression, including the production of novel proteins that may protect plant tissues that undergo severe dehydration (Bray, 1993). Activity of the jasmonic acid signal pathway was shown to be lower in mutants of tomato and potato that are deficient in abscisic acid (Bostock, 1999). Ethylene is involved in some responses of plants to flooding (Chapter 11) and also can positively regulate jasmonic acid levels in plants, which can induce resistance to some insect pests and pathogens (Bostock, 1999). Ethylene may be synergistic with jasmonic acid in the induction of the protein osmotin (Bostock, 1999). The production of osmotin is induced by exposure of plants to salinity, and osmotin also has antifungal activity (Kononowicz et al., 1993). The issue is complex, however, in that exposure of plants to salinity or exogenous abscisic acid can enhance their susceptibility to some pathogens (Bostock, 1999). The induction of heat-shock proteins also can provide some tolerance to several different types of stresses (Sabehat et al., 1998).

Induced resistance to plant pathogens and insect pests is viewed as a desirable crop protection strategy. Compounds have been synthesized that strongly induce resistance, such as benzothiadiazole, which has greater effects in inducing plant resistance to disease than salicylic acid (Bostock, 1999). Crucial to the success of induced resistance in crop production is an understanding of the interactions, whole-plant responses, and trade-offs likely to be present when modifying these complex signaling systems (Bostock, 1999).

A recent study demonstrated the surprising effect that chemical and morphological defenses induced in an annual plant as a consequence of exposure to a caterpillar also were present in seedlings grown from seeds produced on the induced plants (Agrawal et al., 1999). The occurrence of maternally induced defense has implications for studies where genotypes are screened for differences in resistance. When screening plant genotypes for resistance to specific biotic or abiotic stresses, it would appear important to consider the following factors:

1. Seed of the different genotypes should have been produced in the same nursery to attempt to provide the same mother-plant environment.
2. Plants should be exposed to the stress in a manner that enables differentiation between constitutive and induced plant resistance.
3. Prior exposure to other types of stresses may influence resistance to a specific stress.
4. It may be necessary to conduct the screening in more than one abiotic environment.
5. When screening plant genotypes for resistance to a mobile pest, conditions should be established so that the possibility of expressing differences in preference are either included in the test or excluded. For example, if seedlings of different genotypes are growing close together and are then exposed to insect pests, and more insects reproduce on one genotype, it is not clear whether the other genotypes have antibiosis or antixenosis. If the tests are conducted for an extended period and the pests kill some seedlings and then move to other seedlings, measurements made at the end of the test may not provide reliable predictions concerning the resistance or susceptibility of uniform populations of these plant genotypes under the large-scale conditions of farmers' fields. Tests where insects are placed on caged leaves or plants exclude the possibility of testing for differences in preference. Insect pest performance in caged tests can, however, be influenced by mechanisms that influence both insect biology (antibiosis) and behavior (antixenosis) (Smith, 1989). For example, in no-choice conditions, phenolics can influence both feeding behavior and digestion of food. Also, certain types of leaf hairs can have both antixenotic and antibiotic effects. Through evolution of the pest, new plant traits that initially have antibiotic effects on the pest may eventually also have antixenotic effects on the pest.

Crop tolerance as a resistance mechanism is where the plant is attacked and provides an adequate host for the pest but suffers less reduction in yield than other cultivars. Landraces that produce a large area of leaves in relation to the size of their sink of fruits may suffer little reduction in fruit (or grain) yield from attacks by herbivores that reduce some of the leaf area (also refer to Chapter 4 for a discussion of this topic). Modern cultivars that have been bred to have a high productivity of fruits and grain also have a balance between the sizes of their photosynthetic sources and reproductive sinks that is optimal in pest-free environments. Consequently, the modern cultivars may have less tolerance to herbivores than the landraces. In contrast, there are cases where tolerance to insects may be associated with traits that enhance productivity in optimal conditions. A plant that is well adapted to the abiotic environment will grow vigorously during the vegetative stage and may be affected less by insect pest herbivores than a less vigorous plant.

Antibiosis, antixenosis, and crop tolerance can be combined and have additive effects in increasing crop resistance to pests. Some individual mechanisms of pest resistance, however, should be optimized by plant breeding in that they have both beneficial effects and costs to plant adaptation.

Global climate change probably will influence plant infestations by pests and diseases. In the Mediterranean climatic zone of the Central Valley of California, warmer temperatures may decrease the extent to which insect pest populations are killed during the winter and result in higher levels of insect pest populations and plant disease infestations that are transmitted by insects during the subsequent cropping season.

ADDITIONAL READING

Agrawal, A. A., C. Laforsch, and R. Tollrian. 1999. Transgenerational induction of defences in animals and plants. *Nature* 401: 60–63.

Bostock, R. M. 1999. Signal conflicts and synergies in induced resistance to multiple attackers. *Physiol. Mol. Plant Pathol.* 55: 99–109.

Coleman, J. S., C. G. Jones, and V. A. Krischik. 1992. Phytocentric and exploiter perspectives of phytopathology. *Adv. Plant Pathol.* 8: 149–195.

Coviella, C. E. and J. T. Trumble. 1998. Effects of elevated atmospheric carbon dioxide on insect-plant interactions. *Conserv. Biol.* 13: 700–712.

Dicke, M., A. A. Agrawal, and J. Bruin. 2003. Plants talk, but are they deaf? *Trends Plant Sci.* 8: 403–405.

Hall, A. E. 2017. *Sahelian Droughts: A Partial Agronomic Solution*. Nova Science Publishers, New York, p. 216.

Lambers, H., F. S. Chapin III, and T. L. Pons. 1998. *Plant Physiological Ecology*. Springer-Verlag, New York, p. 540.

Letourneau, D. K. 1997. Plant-arthropod interactions in agroecosystems, in L. E. Jackson (Ed.), *Ecology in Agriculture*. Academic Press, San Diego, CA, pp. 239–290.

Smith, C. M. 1989. *Plant Resistance to Insects: A Fundamental Approach*. Wiley, New York, p. 286.

13 Consideration of Crop Responses to Environment in Plant Breeding

In the previous chapters, different topics have been discussed that are relevant to plant breeding. I will now integrate these topics to illustrate how an understanding of crop responses to environment can guide an ideotype approach to plant breeding. I will use a broad definition of an *ideotype* as being a plan of the phenotype of a cultivar that will perform optimally in a specific set of climatic, soil, biotic, and sociocultural conditions. This definition of ideotype is similar to that of Sedgley (1991). The ideotype approach was first proposed by Donald (1968), who used a more restrictive definition that mainly emphasized crop performance in optimal environments.

In his landmark paper, Donald (1968) hypothesized that a successful crop ideotype would be a weak competitor in relation to its mass if it were to be well adapted to the high plant densities required to achieve maximum yields in optimal environments. He pointed out that efficient production of dry matter by a population of genetically identical crop plants requires that individual plants make maximum use of the resources in their environments while encroaching to a minimal degree on the environments of their neighbors. This hypothesis may not have been used very much in guiding plant breeding programs for optimal environments, but it appears to be valid.

Modern high-yielding cereal and grain legume cultivars are shorter and more erect with greater harvest index, and they require more careful weed management than the cultivars they replaced. These traits suggest they are weak competitors in relation to their mass. Even newer semi-dwarf cultivars of wheat may be less competitive than older semi-dwarf cultivars. Reynolds et al. (1994a) evaluated the responses to diminished aerial and soil competition of semi-dwarf spring wheat cultivars in field conditions. Grain yield response to diminished competition was negatively correlated with grain yield under intense competition. Newer semi-dwarf wheat cultivars with greater grain yield under the high plant densities typical of commercial production were less competitive than older semi-dwarf wheat cultivars. In seeking explanations for the differences among the cultivars, they observed that grain yield was negatively correlated with plant height and positively correlated with harvest index. There also was a positive correlation between light-saturated net photosynthesis per unit leaf area and grain yield under high plant densities. Note that a cultivar with smaller leaf size but higher photosynthetic capacity per unit leaf area could be less competitive but more productive in dense canopies. Root mass relative to total plant mass had a weak negative correlation with grain yield under high plant densities. A cultivar with a relatively small root system could be less competitive in the soil environment but more productive under high plant densities in optimal soil environments in that it would *invest* less carbohydrate in its root system. In general, much less is known about root ideotype traits than ideal canopy traits; however, the ability of ambrosia root systems to detect and avoid other ambrosia root systems would appear to be a communal ideotype trait (Mahall and Callaway, 1992).

The ideotype approach to plant breeding has been criticized. An analysis of the criticisms and usefulness of the ideotype approach was made by Hamblin (1993), who concluded, "The major benefits are now seen as being conceptual and analytical rather than in direct yield improvements as in many particularly stressful environments progress has remained painstakingly slow.... Donald's 1968 paper was a landmark in plant breeding, and breeders will continue to try to develop ideotypes as breeding objectives, even if these are not stated publicly." The improved understanding that is being gained of crop responses to environment should provide ideotype approaches to breeding with opportunities to contribute to increasing yield in the future, especially if a broad definition of ideotype is used, such as the one suggested in this chapter.

Crop physiology and ecology provide information on plant function and environment that, in principle, could be used to determine the suites of traits and their levels that should be adaptive in specific environments. Unfortunately, past contributions of crop physiology to plant breeding have been modest (Jackson et al., 1996). In earlier years, crop physiology mainly was successful in explaining how yield increases had been achieved through empirical plant breeding and did not provide much useful guidance concerning traits that should be selected in plant breeding programs (Evans, 1993). Now there are increasing demands for different types of crop products that mainly must be met by increased productivity; yet, empirical breeding may be having less success in enhancing yield potential (Chapter 1). Also, there are more opportunities for increasing productivity by using the improved understanding of crop responses to environment and plant physiology that have been obtained in recent years. Consequently, it is imperative that crop physiology now be used to define crop ideotypes that will enable plant breeding to become more effective. In addition to defining crop ideotypes, it is important to quantify the value of specific ideotype traits for specific target production environments so that breeders can decide the cases where these traits should and should not be selected and the priority to give to them. Procedures must be available for selecting and transferring ideotype traits that are efficient (i.e., where large numbers of genotypes can be screened per day, per person, and per dollar spent) and reliable. If a screening method that is both efficient and highly reliable is not available, then it may be possible to use two different screening methods at two different stages in the breeding program. For example, an efficient indirect screening method, such as DNA-marker-assisted selection, could be used in early generations when large numbers of plants must be screened, followed by the use of a direct and more reliable, but less efficient, screening method in more advanced generations.

Ideotype approaches should be regarded as complementing, not replacing, traditional approaches to plant breeding that place strong emphasis on empirical performance-testing procedures (Rasmusson, 1987). Ideotype approaches should be closely integrated within active plant breeding programs (Jackson et al., 1996).

Care must be taken in defining ideotypes because no ideotype is perfect. Often, different ideotypes will be needed for different target production environments, and the effectiveness of ideotypes may change with time as climates and biotic and sociocultural conditions change. An ideotype with inadequate or incorrect components will slow down breeding programs. In earlier years, the International Rice Research Institute (IRRI) took an ideotype approach to developing rice cultivars for irrigated environments that was highly successful (Evans and Fischer, 1999) and resulted in the development of a new cultivar, IR8 in 1966, which was very productive and adopted by many farmers. Further breeding with the same ideotype did not result in increases in yield potential, so, in 1990, the IRRI rice breeding program began pursuing a new ideotype for tropical japonica-type rices to be grown under irrigated conditions. As of 1998, breeding at IRRI based on the new rice ideotype had not been effective, and further modifications to the ideotype were being made (Peng et al., 1999). As understanding of crop physiology increases, so, too, should crop ideotypes evolve and change.

13.1 DEFINING CROP IDEOTYPE TRAITS

For optimal field environments, several factors strongly influence plant biomass production, and they are summarized in Equation 4.2 in Chapter 4. Key factors are the length of the growing season and the extent of ground cover as they influence the seasonal interception of photosynthetically active photons of light. The model described by Equation 4.2 was expanded to consider the grain yield of cereals and grain legumes (Equation 4.3). A key factor in this model is the length of the period when photosynthesis significantly contributes carbohydrate to the developing grain, which often is the period of reproductive activity. Developmental responses of crops to photoperiod and/or temperature are described in Chapter 6. They determine when the crop will flower, the length of the reproductive period, and whether the crop cycle fits into the available season (Bunting, 1975). The impact of earliness on pest and disease problems also must be considered (Chapter 12).

Another key factor that strongly influences grain yield under optimal conditions is the proportion of carbohydrate that is partitioned to grain (*HI*). Increases in grain yield through plant breeding have resulted from increases in *HI* for many crop species (Evans, 1993), but increased partitioning to grain can enhance problems due to herbivorous pests (Chapter 12) and weeds. Clearly, there are optimal *HIs* above which grain yield will no longer increase, and crop management problems will increase. However, it has been hypothesized that the value of *HI* which is optimal may continue to increase as the atmospheric carbon dioxide concentration increases, and that further increases in grain yield may be obtained through further increases in *HI* (Hall and Allen, 1993; Hall and Ziska, 2000; Hall, 2011). Cultivars with higher *HI* tend to be more dwarfed and, as discussed in Chapter 2, may benefit from adjustments in crop management, such as the use of narrower rows and higher plant population densities.

The efficiency by which intercepted photons are converted into plant biomass through photosynthesis and effects of respiration (*Q*) can influence grain yield. An example was presented in Chapter 4 of how *Q* of cowpea may be increased by breeding cultivars with pods that are within rather than above the leaf canopy (Kwapata et al., 1990). It also is possible that *Q* of maize canopies may have been increased by selecting for smaller tassels that intercept fewer photons, thereby having less effect on the photosynthesis of leaves that are below them (Evans and Fischer, 1999). The extent to which *Q* can be increased through selecting for leaf traits that influence photosynthesis is controversial (Evans, 1993). However, some of the progress that has been made in increasing potential productivity of wheat, cotton, and soybean has been positively associated with stomatal conductance and photosynthetic rates (Condon and Hall, 1997; Fischer et al., 1998; Lu et al., 1998; Morrison et al., 1999). This issue can be complex and involve emergent properties, in that mesophyll photosynthetic capacity can be positively associated with maximal stomatal conductance in a process that appears to involve long-term (days) regulation (Schulze and Hall, 1982).

It has been hypothesized that the photosynthetic systems of current C_3 and C_4 crop cultivars may not be well adapted to the current atmospheric carbon dioxide concentration of 400 parts per million or the expected increases to 700 parts per million likely to occur in this century due to inadequate photosynthetic capacity and inadequate carbohydrate sink size (Hall and Allen, 1993). Consequently, selection for increased maximal photosynthesis per unit leaf area but reduced leaf area and increased sink size and harvest index may be warranted.

In developing wheat varieties for optimal environments, it has been suggested that breeders should select for increased numbers of grains/ha. This might be achieved by genetically manipulating plant development to lengthen the phases associated with the determination of grain numbers. However, Sinclair and Jamieson (2006) point out that there is little proof of the assertion that grain number determines yield other than empirical correlations between the number of grains and yield. From a review and analysis of the literature, Sinclair and Jamieson (2006) hypothesize that both yield and grain number are influenced by resource accumulation and use by the crop. Through evolution, annual crops developed mechanisms that regulate the number of grains produced to match the supplies of carbohydrate and protein available to fill the grains. Sinclair and Jamieson (2006) argue that nitrogen might be a key resource limiting grain yield (and grain number) of wheat. Alternatively, it can be argued that carbohydrate coming from photosynthesis could be a key resource that limits grain yield of wheat under optimal conditions. Note that there are complex interactions among photosynthesis and nitrogen uptake and use as emergent properties limiting yield.

The exploitation of emergent properties by plant breeding is constrained by difficulties in selecting for them and the limited information concerning their mechanisms. Some emergent properties may be related to the mathematical phenomenon called *chaos*, which shows that simple processes can have complex consequences, and they may also be related to the possibility that complex systems may have simple rules that guide them (Cohen and Stewart, 1994). For example, the logistic difference equation is a simple equation that is widely used to simulate the changes of organism populations with time, yet this equation in defined circumstances can produce chaotic changes in populations (Gleick, 1987). Chaotic systems also can exhibit some consistent behavior with respect to attractors, which are regions more frequently occupied by the system (Gleick, 1987). The complex morphology of a fern leaf can be simulated by the iteration of a simple set of rules (Gleick, 1987). These simulated leaves, and also real fern leaves and root systems, are fractal in that they have self-similarity, which means a degree of symmetry across different scales, and can be described by a parameter the fractal dimension, that is, an emergent property (Gleick, 1987). I recommend that those readers who are not familiar with chaos and fractals should first read Gleick (1987) and then Cohen and Stewart (1994) to begin to appreciate why the reductionist approach to science can never explain all aspects of the world we live in and to gain more insight into emergent properties.

A model of cowpea grain yield under heat stress (Equation 5.1) was presented in Chapter 5. This is the traditional yield component model, which has not been of much use to plant breeding programs for optimal environments due to the occurrence of negative correlations among the various components. In optimal environments, selecting genotypes with more pods usually results in selection for genotypes with smaller pods and fewer seeds per pod. For cowpea and some other crop species under hot conditions, these negative correlations do not occur because the system is strongly limited by the capacity of the reproductive sink (Ismail and Hall, 1998). In these cases, it is possible to select for more flowers/m^2 and more pods per flower and achieve increases in both traits and in grain yield (Hall, 1992). Controlled-environment studies indicate that heat tolerance during reproductive development, which enhances reproductive sink strength, may be even more valuable in future environments with elevated carbon dioxide concentrations that will enhance the photosynthetic source capabilities of C_3 species because heat tolerance may enhance grain yield over a range of temperatures (Hall and Allen, 1993; Hall and Ziska, 2000; Hall, 2011).

For water-limited environments, three models were presented in Chapter 9: Equations 9.1 through 9.3. The third model (Equation 9.3) may be the most useful for guiding breeding programs. However, progress in enhancing adaptation to drought mainly has come from empirical breeding programs based on selecting for yield under a range of commercial rain-fed environments (Blum, 1988; Cisse et al., 1995, 1997). The third model suggests that grain yield under water-limited conditions may be enhanced by selecting for traits that enhance the amount of soil water available to plants (i.e., deeper roots), water-use efficiency, and the partitioning of carbohydrate to grain. But, as discussed in Chapter 9, there are some negative correlations among these traits that seriously constrain this approach. For all of these traits, there is an optimal value that is adaptive because the traits have costs to adaptation as well as benefits. An example is deeper rooting, which has a carbohydrate *cost* to the plant in its construction and maintenance.

In breeding for adaptation to drought, it is important to recognize that semiarid environments have highly variable rainfall that produces different types of drought. Different ideotypes may be needed to accommodate the different types of drought. Examples are provided of the development of cowpea and rice cultivars for water-limited rain-fed target production zones.

In general, rice is sensitive to drought, and cultivars with greater drought resistance are needed where the crop is grown under water-limited rain-fed conditions. A manual has been developed to provide advice on breeding rice for drought-prone environments (Fischer et al., 2003). Different types of drought-resistant rice cultivars are needed for rain-fed production, depending on whether droughts occur early or late in the growing season (Fukai et al., 1999). Progress has been made in developing rice cultivars with resistance to late-season drought. The major features of these cultivars are an optimum date of flowering, high potential yield under well-watered conditions, and the ability to maintain high leaf water potential (closer to zero) when drought occurs at flowering (Fukai et al., 1999). The factor responsible for the maintenance of high leaf water potential is not known but could involve deeper roots. Fukai et al. (1999) recommended that breeding include parallel screening in two types of field nurseries. Under well–watered conditions screen rice genotypes for grain yield, high harvest index, intermediate height, and small shoot dry matter at anthesis. In addition, under late-season drought screen the same rice genotypes for grain yield, minimal delay in date of flowering, high leaf water potential, low spikelet sterility, and retention of green leaves. Note that visual screening for absence of leaf rolling may be more efficient than screening for high leaf water potential, and a multi-parameter index may be needed for use in weighting the values of the different traits that are being selected. The plant properties of rice that determine its early-season drought resistance are not the same properties that govern its late-season drought resistance. Resistance during the vegetative stage has been associated with ability to recover from the drought, and little progress has been made in breeding rice cultivars with this type of resistance (Fukai et al., 1999).

Droughts can be so severe in the Sahelian zone of Africa (Figures 10.5 and 10.9) that virtually all crop species produce no grain (Hall, 2017). However, cowpea cultivars that can be productive in these harsh conditions have been developed, but at least two types of cultivars are needed. Short-cycle erect cultivars with vegetative-stage drought resistance sown at close spacing produced more grain than medium-cycle spreading cultivars when the rainy season was short but distinct with no mid-season droughts. In contrast, medium-cycle spreading cowpea cultivars with mid-season drought resistance sown at moderate spacing produced higher yields of grain and hay in moderately wet years and had greater ability to recover from mid-season droughts than the short-cycle more erect cultivars (Thiaw et al., 1993). Growing both types of cowpea cultivars increased the stability of the farming system, which is an important objective for subsistence farmers. Improvements in farming practices also should be considered when new cultivars are developed. For example, field studies in Senegal demonstrated that varietal intercrops consisting of alternating rows of medium-cycle spreading and short-cycle erect cultivars produced more grain and hay and were more stable than sole crops of either cultivar in the very dry environments that often occur in the Sahelian zone (Thiaw et al., 1993).

13.2 TESTING THE VALUE OF CROP IDEOTYPE TRAITS

Prior to widespread use of specific phenological, physiological, or morphological traits in breeding programs, their value must be clearly established. A rigorous test involves the development of pairs of lines with and without the trait but with otherwise similar genetic background (i.e., almost isogenic lines). Ideally, several pairs of isogenic lines should be developed with different genetic backgrounds because the agronomic value of a gene(s) can depend on the other genes present in the genome. For example, a cultivar with deeper roots that accesses more soil water may gain more benefit from the deeper roots if it also has greater water-use efficiency. Clearly, the pairs of isogenic lines should have well-adapted genetic backgrounds that complement the trait being evaluated. Obtaining a comprehensive estimate of the value of the ideotype trait also requires that the different pairs of isogenic lines be evaluated in different target production environments with consideration of potential interactions with pest and disease problems (Chapter 12).

The effects of cereal awns on crop performance have been studied extensively using sets of isogenic lines with different lengths of awns. For both wheat and barley, the effects of awns on grain yield depend on both the genetic (cultivar) background and the environment and differ between the two species (Evans, 1993). For wheat, awns enhance grain yield in water-limited environments but tend to trap water and may accentuate lodging and disease problems and enhance premature sprouting in wet environments (Evans, 1993). Consequently, awns are useful for wheat cultivars grown in many drier environments, combining awns with stems that are not too long may be useful in some wetter environments, and awnless varieties may be most effective in very wet environments. For barley, long awns enhance grain yield in many environments, but multiple awns do not (Rasmusson, 1987).

The effects of reproductive-stage heat-tolerance genes have been studied with cowpea using six pairs of lines either having or not having a set of heat-tolerance genes but with different well-adapted genetic backgrounds that are similar for each member of each pair (Ismail and Hall, 1998). These studies demonstrated that in subtropical environments with long days, the set of heat-tolerance genes does not influence grain yield in cool environments but progressively enhances grain yield when temperatures become hotter than a threshold level. These studies also showed that the heat-tolerant cowpea lines were dwarfed, with the strongest effects at hotter temperatures. These heat-tolerant cultivars were most effective when grown on narrower rows (Ismail and Hall, 2000). It is not yet known whether the dwarfing is a pleiotropic consequence of the heat-tolerance genes or due to close genetic linkage. Pleiotropy is where a single gene has multiple effects on a phenotype and must be considered when determining the value of a gene for use in breeding. The heat-tolerance genes enhance early pod production and thereby cause a greater diversion of carbohydrates from vegetative tissues, which may be responsible for the dwarfing. In contrast, if the association is due to close genetic linkage, it may be possible to separate the effects and breed cowpea plants that have reproductive-stage heat tolerance and are not dwarfed. Cowpea lines have been found that have heat tolerance during floral bud development but are not dwarfed (Ismail and Hall, 1999). This indicates that the dwarfing in the heat-tolerant lines which have been developed might be due to genetic linkage between the gene conferring heat tolerance during early floral bud development and a gene causing dwarfing.

The six pairs of cowpea lines also have been evaluated in hot tropical environments. In this case, the heat-tolerant and heat-susceptible lines had similar grain yields. Studies with cowpea lines in a hot glasshouse under long- and short-day conditions (Ehlers and Hall, 1998) provide a possible explanation for the contrasting results from the hot tropical and hot subtropical environments. In the glasshouses, the heat-tolerance genes were effective under both long-day and short-day conditions, whereas the heat-susceptible lines only were damaged under long-day conditions. Consequently, the heat-tolerance genes could be effective under both subtropical and tropical conditions. Under tropical conditions, day lengths can either be long enough to trigger photoperiod effects in SD cultivars or be too short, depending on the latitude and sowing date. An alternative explanation for the contrasting results is that the genetic lines used in the tests conducted in the tropical field conditions were not ideal. The genetic backgrounds of the pairs of lines that were used were adapted to the subtropical test environments but were not well-adapted to the tropical conditions in that the plants suffered considerable damage from various organisms, causing pod rots under both wet and dry conditions (Hall, 2011). I observed that the heat-tolerant lines growing in Africa produced many pods but many of the pods did not develop and produce grain, and they exhibited an unusually high level of shriveling, presumably due to pod rot diseases.

13.3 SELECTING AND TRANSFERRING CROP IDEOTYPE TRAITS

In classical plant breeding, it is often necessary to screen large numbers of plants when searching for parents with desirable ideotype traits for use in crosses. Also, many plants must be screened while breeding to transfer the trait into an adapted and commercially acceptable genetic background. Some morphological traits can be identified visually in an efficient manner so that large numbers of plants can be screened in a short time period, such as for the presence of awns or dwarfing. For other traits, direct screening is not efficient, such as for rooting depth or water-use efficiency or stomatal conductance or photosynthetic capacity. In cases such as these, it is useful to develop an efficient indirect screen that can be used in the initial stages of the breeding program where large numbers of plants must be screened. A screen for speed of rooting has been developed, based on placing an herbicide layer deep in the soil (Robertson et al., 1985). When roots contact the herbicide, the plant reacts, exhibiting leaf chlorosis. Cultivars that showed symptoms sooner were shown, in subsequent experiments, to extract more moisture deep in the soil, indicating they had developed deeper roots than cultivars showing symptoms later. Prolific root hair formation may be simply inherited (Hochmuth et al., 1985) and readily screened for with a non-destructive growth pouch technique (Omwega et al., 1988). Prolific root hair formation can enhance phosphate acquisition in low-phosphate soils (Gahoonia et al., 1997; Gahoonia and Nielsen, 1997).

Intrinsic water-use efficiency can be selected for indirectly, based on the stable carbon isotope composition of leaves, as was discussed in Chapters 8 and 9. Genotypic differences in stomatal conductance can be detected indirectly by measuring their effects on the difference between canopy and air temperature, as was discussed in Chapter 10. When temperature differences are measured with genetic lines growing under limited water that is stored deep in the soil, lines with cooler leaf temperatures would have higher stomatal conductances, but this may be due to the fact that they have deeper roots. In this case, the method may be indirectly screening for either deeper roots or reduced stomatal-closing signals from roots rather than differences in properties of the guard cells.

Genotypic differences in photosynthetic capacity may be detected indirectly by selecting for leaves that have more dry weight per unit leaf area, because they usually have higher levels of photosynthetic enzymes and photosystem components per unit leaf area. Some aspects of chilling tolerance may be detected indirectly by examining electrolyte leakage from seeds or leaf discs under chilling temperatures (Blum, 1988).

Chilling tolerance during emergence in warm-season crops may be selected by screening for both slow electrolyte leakage from seeds and the presence of a specific dehydrin protein in the seed (Ismail et al., 1997). A role for the dehydrin protein in chilling tolerance was indicated by studies with almost isogenic lines (Ismail et al., 1999).

Non-destructive techniques that are effective with single plants or seeds are much more useful to breeding than techniques that require a plant population and stable genetic lines, such as the herbicide root screen and the canopy temperature screen. Non-destructive single-plant screens can be applied in the first segregating generation (i.e., the F_2), where selection pressure can be applied to large numbers of genotypes. Genetic-line screens only can be applied to fairly stable lines at more advanced generations and thus to fewer genotypes. The single-seed test for the presence of a specific dehydrin is particularly powerful because it can be made on a small piece of the cotyledon (Ismail et al., 1999), and the remainder of the seed will germinate and produce a plant that can be either self-pollinated or cross-pollinated to produce seed.

Genetic engineering is useful in ideotype breeding in that, in principle, single genes can be transferred into a cultivar without changing the genetic background, thereby creating an isogenic pair of lines in one step and in a relatively short time. In practice, however, many transformed plants must be created and evaluated because factors such as the placement and manner of gene insertion can influence its expression.

The traditional backcrossing procedure for creating almost-isogenic lines requires many plant generations and several years, but it can be accelerated by using DNA markers to select for the genetic background of the recurrent parent. DNA markers also can be used to develop efficient indirect selection methods. In this case, DNA markers are needed that are closely linked to the trait of interest. Indirect selection using DNA markers is powerful where it can be used for the non-destructive screening of single plants in the first segregating generation. Once a set of stable lines has been bred with selection based on indirect screening procedures, one should either conduct studies to confirm that the desired trait is present, using a more reliable direct screening procedure, or proceed directly to performance trials under field conditions in the target production environment. Marker-assisted backcrossing was particularly useful in transferring a submergence tolerance gene (*SUB*1) into eight rice mega varieties (Mackill et al., 2012).

13.4 PERSPECTIVES FOR THE FUTURE USE OF CROP IDEOTYPES IN PLANT BREEDING

It is useful to place concepts concerning approaches for breeding improved cultivars in an evolutionary context. In several cases, drought resistance of annual plants has been enhanced by selecting genotypes that have more open stomata in water-limited environments. Examples of this are where wheat lines were selected for lower canopy temperatures or where wheat and cowpea lines were selected for greater carbon isotope discrimination (Condon and Hall, 1997) and where wheat and sorghum lines were selected for enhanced osmotic adjustment. The mechanisms of this effect could have been either indirect selection for deeper roots, which, when accessing wetter soil, could produce less stomatal-closing signals or selection for root systems that have a higher threshold for producing stomatal-closing signals and exhibit stomatal closure later when drought becomes more extreme. Clearly, there is a limit to the extent that drought resistance can be enhanced by selecting for plants with either deeper rooting or less *conservative* stomatal responses, because these traits have *costs* as well as *benefits* to adaptation and an optimal level of these traits is needed. It also appears that the natural selection that produced the progenitors of annual crop species favored plants with *conservative* stomatal closing responses in water-limited environments. Presumably, natural selection favored plants that produced at least some viable seeds in most years, which implies that the soil seed bank may be very ineffective in these environments for seed lots that are older than one year (Condon and Hall, 1997).

Reproductive processes, especially production of pollen and pollination, are extremely sensitive to several stresses such as frosts, chilling, high temperatures, and drought. The effects are so severe that they only can be partially explained on the basis of the adaptive advantage from maintaining a balance between photosynthetic source and sink, so that the fewer seeds that are produced are plump and viable. Consequently, reproduction appears to be a weak link in the evolution of flowering plants that was perpetuated by evolution, maybe due to some of the advantages arising from sex. It results in genetic recombination and independent re-assortment of parental genes generating diversity and novel combinations of traits. A potential solution to this problem of the sensitivity of crop plants to stresses during pollination is to develop cultivars that do not require sexual processes while growing in farmers' fields.

Crop plants with an appropriate type of apomixis would be able to produce viable seed from maternal tissue without requiring either meiosis of the embryo mother cell or pollen production and pollination of the embryo or endosperm (Khush, 1994). Advantages of cultivars with this type of apomixis, which were discussed by Jefferson (1993), are described below.

1. Apomixis could confer resistance to stresses that damage pollination and other aspects of sexual reproduction.
2. Apomixis would fix hybridity. F_1 hybrids would have true-breeding seeds, which would make possible use of hybrid vigor in the many crop species where use of hybrid seed is currently difficult or not economically possible. Farmers could reuse seed produced by apomictic hybrid cultivars in that the seed would retain its hybrid vigor. This would provide a significant advantage to poor farmers but they would still need to purchase high-quality seed after a few years of reusing seed from their own fields. The quality of seed declines from year to year due to the presence of seed-borne pathogens and contamination by the presence of off-type plants in the farmers' fields or mechanical mixing during harvesting or processing of seed in warehouses.

3. Crops currently propagated vegetatively, such as most Irish potato and banana, would benefit from apomictic cultivars because they could be propagated by seed.
4. Breeding programs would be accelerated by using apomictic lines because they would confer the ability to immediately fix heterozygous genotypes.

Sex would still be needed to permit the continual breeding of improved cultivars. This could be achieved by using facultative (switchable) apomixis systems, where the default state is apomictic, and the sexual state can be switched on by the breeder by either spraying the plants with a specific chemical or by some other procedure such as growing the plants in a specific environment. For example, there is a male sterility in rice that can be switched off by growing plants in short-day cooler environments (Yuan et al., 1993), and similar types of heat plus long-day-induced male sterility have been observed in some other species such as cowpea (Hall, 1992).

Several different plant species have the genes needed to develop facultative apomictic breeding systems in crop plants (Khush, 1994). For example, much research has been devoted to developing pearl millet cultivars that are apomictic (Dujardin and Hanna, 1994). Through genetic engineering, it may be possible to create and transfer the *cassette* of genes needed for facultative apomixis into crop cultivars (Jefferson, 1993). Crop cultivars with facultative apomixis would result in a revolution in plant biology, plant breeding, and crop production. Presumably, breeding to develop facultative apomictic breeding systems will have to be done by the public sector since the development of systems that can produce F_1 hybrids with true-breeding seeds would not be attractive to commercial plant breeding programs.

When designing plant breeding programs, it is necessary to consider future changes in environment because these programs usually take several years to produce new cultivars, and the new cultivars must be adapted to the conditions that prevail after they are released. Current rapid increases in atmospheric [CO_2] should be considered because they may result in a need for breeding cultivars of C_3 species that have both greater sink strength to balance the increases in photosynthesis that are likely to occur and the components of leaf photosynthesis that can most effectively use the elevated [CO_2] (Hall and Allen, 1993; Hall and Ziska, 2000; Hall, 2011).

Changes in sociocultural conditions also must be considered. In many *developed* societies, crop cultivars and management methods have been developed that reduce the amount of hand labor required in crop production. This trend should continue in those societies where laborers are in short supply, and it should be used to provide some jobs that require a higher skill level and are more desirable to laborers. However, technologies that reduce the amount of necessary hand labor may not be desirable in those societies where suitable alternative employment opportunities are not yet available for those people currently working as farm laborers.

Changes are occurring due to the advent of genetic engineering. Genetic engineering has broad potential for improving agriculture and changing the types of products that it provides to society. The varieties produced by genetic engineering are as safe as those produced by conventional plant breeding as long as the traits that are bred in by either method are safe from a human and an ecological standpoint (Stewart, 2004).

The reactions of different segments of society to the potential risks and benefits of genetic engineering are mixed, however, with some polarization of opinions that reduces opportunities for enhancing mutual understanding and compromise. More complete understanding of the potential risks and benefits of genetic engineering requires comprehensive scientific analysis. The principles described in this book should be useful when conducting part of this analysis in that they can be used to estimate the improvements in productivity and the adaptation of crop plants that can be achieved through genetic engineering and mainstream plant breeding.

ADDITIONAL READING

Blum, A. 1988. *Plant Breeding for Stress Environments.* CRC Press, Boca Raton, FL, p. 223.

Cohen, J. and I. Stewart. 1994. *The Collapse of Chaos: Discovering Simplicity in a Complex World.* Viking Penguin, New York, p. 495.

Donald, C. M. 1968. The breeding of crop ideotypes. *Euphytica* 17: 385–403.

Evans, L. T. 1993. *Crop Evolution, Adaptation and Yield.* Cambridge University Press, Cambridge, UK, p. 500.

Fischer, R. A., D. Rees, K. D. Sayre, Z.-M. Lu, A. G. Condon, and A. Larque Saavedra. 1998. Wheat yield progress associated with higher stomatal conductance and photosynthetic rate, and cooler canopies. *Crop Sci.* 38: 1467–1475.

Fukai, S., G. Pantuwan, B. Jongdee, and M. Cooper. 1999. Screening for drought resistance in rainfed lowland rice. *Field Crops Res.* 64: 61–74.

Gleick, J. 1987. *Chaos: Making a New Science.* Viking Penguin, New York, p. 352.

Hall, A. E. and L. H. Allen, 1993. Designing cultivars for the climatic conditions of the next century, in D. R. Buxton et al. (Eds.), *International Crop Science I.* Crop Science Society of America, Madison, WI, pp. 291–297.

Hall, A. E. and L. H. Ziska. 2000. Crop breeding strategies for the 21st century, in K. R. Reddy and H. F. Hodges (Eds.), *Climate Change and Global Crop Productivity.* CABI Publishing, New York, pp. 407–423.

Hamblin, J. 1993. The ideotype concept: Useful or outdated, in D. R. Buxton et al. (Eds.), *International Crop Science I.* Crop Science Society of America, Madison, WI, pp. 589–597.

Ismail, A. M. and A. E. Hall. 1998. Positive and potential negative effects of heat-tolerance genes in cowpea. *Crop Sci.* 38: 381–390.

Jefferson, R. A. 1993. Beyond model systems—New strategies, methods and mechanisms for agricultural research. *Ann. New York Acad. Sci.* 700: 53–73.

Rasmusson, D. C. 1987. An evaluation of ideotype breeding. *Crop Sc.* 27: 1140–1146.

Reynolds, M. P., E. Acevedo, K. D. Sayre, and R. A. Fischer. 1994a. Yield potential in modern wheat varieties: Its association with a less competitive ideotype. *Field Crops Res.* 37: 149–160.

Sedgley, R. H. 1991. An appraisal of the Donald ideotype after 21 years. *Field Crops Res.* 26: 93–112.

Thiaw, S., A. E. Hall, and D. R. Parker. 1993. Varietal intercropping and the yields and stability of cowpea production in semiarid Senegal. *Field Crops Res.* 33: 217–233.

References

Adolf, V. I., S.-E. Jacobsen, and S. Shabala. 2013. Salt tolerance mechanisms in quinoa (*Chenopodium quinoa* Willd.). *Environ. Exper. Bot.* 92: 43–54.

Agrawal, A. A., C. Laforsch, and R. Tollrian. 1999. Transgenerational induction of defences in animals and plants. *Nature* 401: 60–63.

Ahmed, F. E. and A. E. Hall. 1993. Heat injury during early floral bud development in cowpea. *Crop Sci.* 33: 764–767.

Ahmed, F. E., A. E. Hall, and D. A. DeMason. 1992. Heat injury during floral development in cowpea (*Vigna Unguiculata*, Fabaceae). *Amer. J. Bot.* 79: 784–791.

Ahmed, F. E., A. E. Hall, and M. A. Madore. 1993a. Interactive effects of high temperature and elevated carbon dioxide concentration on cowpea (*Vigna unguiculata* [L.] Walp.). *Plant, Cell Environ.* 16: 835–842.

Ahmed F. E., R. G. Mutters, and A. E. Hall. 1993b. Interactive effects of high temperature and light quality on floral bud development in cowpea. *Austral. J. Plant Physiol.* 20: 661–667.

Al-Khatib, K. and G. M. Paulsen. 1999. High-temperature effects on photosynthetic processes in temperate and tropical cereals. *Crop Sci.* 39: 119–125.

Andersen, P. C., B. V. Brodbeck, and R. F. Mizell III. 1992. Feeding by the leafhopper *Homalodisca coagulata*, in relation to xylem fluid chemistry and tension. *J. Insect Physiol.* 38: 611–622.

Aparicio, N., D. Villegas, J. Casadesus, J. L. Araus, and C. Royo. 2000. Spectral vegetation indices as nondestructive tools for determining durum wheat yield. *Agron. J.* 92: 83–91.

Araus, J. L. 1996. Integrative physiological criteria associated with yield potential, pp. 150–164 in M. P. Reynolds, S. Rajaram, and A. McNab (Eds.), *Increasing Yield Potential in Wheat: Breaking the Barriers.* CIMMYT, Mexico, D.F.

Argyris, J., M. J. Truco, O. Ochoa, S. J. Knapp, D. W. Still, G. M. Lenssen, J. W. Schut, R. W. Michelmore, and K. J. Bradford. 2005. Quantitative trait loci associated with seed and seedling traits in *Lactuca. Theor. Appl. Genet.* 111: 1365–1376.

Arp, W. J. 1991. Effects of source-sink relations on photosynthetic acclimation to elevated CO_2. *Plant, Cell Environ.* 14: 869–875.

Austin, R. B. 1999. Yield of wheat in the United Kingdom: Recent advances and prospects. *Crop Sci.* 39: 1604–1610.

Baker, J. T., L. H. Jr. Allen, K. J. Boote, P. Jones, and J. W. Jones. 1989. Responses of soybean to air temperature and carbon dioxide concentration. *Crop Sci.* 29: 98–105.

Baldocchi, D. and S. Wong. 2008. Accumulated winter chill is decreasing in the fruit growing regions of California. *Climate Change* 87 (Suppl. 1): S153–S166.

Barbour, M. M., R. A. Fischer, K. D. Sayre, and G. D. Farquhar. 2000. Oxygen isotope ratio of leaf and grain material correlates with stomatal conductance and grain yield in irrigated wheat. *Austr. J. Plant Physiol.* 27: 625–637.

Bartkowski, E. J., D. R. Buxton, F. R. H. Katterman, and H. W. Kircher. 1977. Dry seed fatty acid composition and seedling emergence of Pima cotton at low soil temperatures. *Agron. J.* 69: 37–40.

Barton, J. E. and M. Dracup. 2000. Genetically modified crops and the environment. *Agron. J.* 92: 797–803.

Bates, L. M. and A. E. Hall. 1981. Stomatal closure with soil water depletion not associated with changes in bulk leaf water status. *Oecologia* 50: 62–65.

Bedi, S. and A. S. Basra. 1993. Chilling injury in germinating seeds: Basic mechanisms and agricultural implications. *Seed Sci. Res.* 3: 219–229.

Björkman, O. and S. B. Powles. 1981. Leaf movement in the shade species *Oxalis oregana*. I. Response to light level and light quality. *Carnegie Inst. Washington Year Book* 80: 59–62.

Blum, A. 1988. *Plant Breeding for Stress Environments.* CRC Press, Boca Raton, FL, p. 223.

Blum, A., N. Klueva, and H. T. Nguyen. 2001. Wheat thermotolerance is related to yield under heat stress. *Euphytica* 117: 117–123.

Blum, A., K. F. Schertz, R. W. Toler, R. I. Welch, D. T. Rosenow, J. W. Johnson, and L. E. Clark. 1978. Selection for drought avoidance in sorghum using aerial infrared photography. *Agron J.* 70: 472–477.

Board, J. E. and M. L. Peterson. 1980. Management decisions can reduce blanking in rice. *California Agricul.* 34 (11–12): 5–7.

Bolaños, J. and G. O. Edmeades. 1996. The importance of the anthesis-silking interval in breeding for drought tolerance in tropical maize. *Field Crops Res.* 48: 65–80.

Borthwick, H. A. and W. W. Robbins. 1928. Lettuce seed and its germination. *Hilgardia* 3: 275–305.

Bostock, R. M. 1999. Signal conflicts and synergies in induced resistance to multiple attackers. *Physiol. Mol. Plant Pathol.* 55: 99–109.

Boyer, J. S. 1995. *Measuring the Water Status of Plants and Soils.* Academic Press, San Diego, CA, p. 178.

Boyle, M. G., J. S. Boyer, and P. W. Morgan. 1991. Stem infusion of liquid culture medium prevents reproductive failure of maize at low water potential. *Crop Sci.* 31: 1246–1252.

Bradford, K. J. 1994. Water stress and water relations of seed development: A critical review. *Crop Sci.* 34: 1–11.

Bray, E. A. 1993. Molecular responses to water deficit. *Plant Physiol.* 103: 1035–1040.

Breman, H. and C. T. de Wit. 1983. Rangeland productivity and exploitation in the Sahel. *Science* 221: 1341–1347.

Brouwer, R. 1962. Distribution of dry matter in the plant. *Netherlands J. Agric. Sci.* 10: 361–376.

Brown, J. C., C. R. Weber, and B. E. Caldwell. 1967. Efficient and inefficient use of iron by two soybean genotypes and their isolines. *Agron. J.* 59: 459–462.

Bunting, A. H. 1975. Time, phenology and the yields of crops. *Weather* 30: 312–325.

Caffrey, E. A., V. Fonesca, and A. C. Leopold. 1988. Lipid-sugar interactions: Relevance to anhydrous biology. *Plant Physiol.* 86: 754–758.

Campbell, G. S. 1977. *An Introduction to Environmental Biophysics.* Springer-Verlag, New York, p. 159.

Campbell, S. A. and T. J. Close. 1997. Dehydrins: Genes, proteins, and associations with phenotypic traits. *New Phytol.* 137: 61–74.

Canny, M. J. 1995. A new theory for the ascent of sap-cohesion supported by tissue pressure. *Annals Bot.* 75: 343–357.

Chapman, S. C. and G. O. Edmeades. 1999. Selection improves drought tolerance in tropical maize populations: II. Direct and correlated responses among secondary traits. *Crop Sci.* 39: 1315–1324.

Chrispeels, M. J. and D. E. Sadava. 1994. *Plants, Genes and Agriculture.* Jones and Bartlett Publishers, Boston, MA, p. 478.

Cisse, N., M. Ndiaye, S. Thiaw, and A. E. Hall. 1995. Registration of "Mouride" cowpea. *Crop Sci.* 35: 1215–1216.

Cisse, N., M. Ndiaye, S. Thiaw, and A. E. Hall. 1997. Registration of "Melakh" cowpea. *Crop Sci.* 37: 1978.

Clark, R. B. and R. R. Duncan. 1993. Selection of plants to tolerate soil salinity, acidity and mineral deficiencies, pp. 371–379 in D. R. Buxton et al. (Eds.), *International Crop Science I.* Crop Science Society of America, Madison, WI.

Close, T. J. 1996. Dehydrins: Emergence of a biochemical role of a family of plant dehydration proteins. *Physiol. Plant* 97: 795–803.
Close, T. J. 1997. Dehydrins: A commonality in the response of plants to dehydration and low temperature. *Physiol. Plant* 100: 291–296.
Cochemé, J. and P. Franquin. 1967. An agroclimatology survey of a semi-arid area in Africa south of the Sahara. *Technical Note, World Meteorological Organization* No. 86, p. 136.
Cohen, J. and I. Stewart. 1994. *The Collapse of Chaos: Discovering Simplicity in a Complex World.* Viking Penguin, New York, p. 495.
Coleman, J. S., C. G. Jones, and V. A. Krischik. 1992. Phytocentric and exploiter perspectives of phytopathology. *Adv. Plant Pathol.* 8: 149–195.
Condon, A. G. and A. E. Hall. 1997. Adaptation to diverse environments: Variation in water-use efficiency within crop species, pp. 79–116 in L. E. Jackson (Ed.), *Ecology in Agriculture.* Academic Press, San Diego, CA.
Cornish, K., J. W. Radin, E. L. Turcotte, Z. Lu, and E. Zeiger. 1991. Enhanced photosynthesis and stomatal conductance of pima cotton (*Gossypium barbadense* L.) bred for increased yield. *Plant Physiol.* 97: 484–489.
Coviella, C. E. and J. T. Trumble. 1998. Effects of elevated atmospheric carbon dioxide on insect-plant interactions. *Conserv. Biol.* 13: 700–712.
Cowan, I. R. 1977. Stomatal behavior and environment. *Adv. In Bot. Res.* 4: 117–228.
Cowan, I. and G. D. Farquhar. 1977. Stomatal function in relation to leaf metabolism and environment, pp. 471–505 in *Integration of Activity in the Higher Plant. Society for Expermiental Biology Symposia.* XXXI, Cambridge University Press, Cambridge, UK.
DaMatta, F. M. 2004. Ecophysiological constraints on the production of shaded versus unshaded coffee: A review. *Field Crops Res.* 86: 99–114.
Dancette, C. and A. E. Hall. 1979. Agroclimatology applied to water management in the Sudanian and Sahelian zones of Africa, pp. 98–118 in A. E. Hall, G. H. Cannell, and H. W. Lawton (Eds.), *Agriculture in Semi-Arid Environments.* Springer-Verlag, New York.
Dawson, T. E. 1993. Water sources of plants as determined from xylem-water isotopic composition: Perspectives on plant competition, distribution, and water relations, pp. 465–496 in J. R. Ehleringer, A. E. Hall, and G. D. Farquhar (Eds.), *Stable Isotopes and Plant Carbon-Water Relations.* Academic Press, San Diego, CA.
Dicke, M., A. A. Agrawal, and J. Bruin. 2003. Plants talk, but are they deaf? *Trends Plant Sci.* 8: 403–405.
Dobermann, A., D. Dawe, R. P. Roetter, and K. G. Cassman. 2000. Reversal of rice yield decline in a long-term continuous cropping experiment. *Agron. J.* 92: 633–643.
Dolferus, R., X. Ji, and R. A. Richards. 2011. Abiotic stress and control of grain number in cereals. *Plant Sci.* 181: 331–341.
Donald, C. M. 1968. The breeding of crop ideotypes. *Euphytica* 17: 385–403.
Doorenbos, J. and W. O. Pruitt. 1977. Crop water requirements. *FAO Irrigation and Drainage Paper* No. 24 (revised), FAO, United Nations, Rome, Italy, p. 144.
Dow El-Madina, I. M. and A. E. Hall. 1986. Flowering of contrasting cowpea (*Vigna unguiculata* [L.] Walp.) genotypes under different temperatures and photoperiods. *Field Crops Res.* 14: 87–104.
Drake, B. G., P. W. Leadley, W. J. Arp, D. Nassiry, and P. S. Curtis. 1989. An open top chamber for field studies of elevated atmospheric CO_2 concentration on saltmarsh vegetation. *Funct. Ecol.* 3: 363–371.
Dujardin, M. and W. W. Hanna. 1994. Transfer of alien chromosome-carrying gene for apomixis to cultivated *Pennisetum,* pp. 31–34 in G. S. Kush (Ed.), *Apomixis: Exploiting Hybrid Vigor in Rice.* International Rice Research Institute, Manila, Philippines.
Dundas, I., K. B. Saxena, and D. E. Byth. 1981. Microsporogenesis and anther wall development in male-sterile and fertile lines of pigeon pea (*Cajanus cajan* L. Millsp.). *Euphytica* 30: 431–435.

Duvick, D. N. and K. G. Cassman. 1999. Post-green revolution trends in yield potential of temperate maize in the north-central United States. *Crop Sci.* 39: 1622–1630.

Eastin, J. D., R. M. Castleberry, T. J. Gerik, J. H. Hutquist, V. Mahalaksmi, V. B. Ogunela, and J. R. Rice. 1983. Physiological aspects of high temperature and water stress, pp. 91–112 in C. D. Raper and P. J. Kramer (Eds.), *Crop Reactions to Water and Temperature Stresses in Humid, Temperate Climates.* Westview Press, Boulder, CO.

Edmeades, G. O., J. Bolaños, S. C. Chapman, H. R. Lafitte, and M. Bänziger. 1999. Selection improves drought tolerance in tropical maize populations: I. Gains in biomass, grain yield, and harvest index. *Crop Sci.* 39: 1306–1315.

Ehleringer, J. R. 1993. Carbon and water relations in desert plants: An isotopic perspective, pp. 155–172 in J. R. Ehleringer, A. E. Hall, and G. D. Farquhar (Eds.), *Stable Isotopes and Plant Carbon-Water Relations.* Academic Press, San Diego, CA.

Ehleringer, J. and O. Björkman. 1977. Quantum yields for CO_2 uptake in C_3 and C_4 plants. *Plant Physiol.* 59: 86–90.

Ehleringer, J. R. and R. K. Monson. 1993. Evolutionary and ecological aspects of photosynthetic pathway variation. *Annu. Rev. Ecol. Syst.* 24: 411–439.

Ehlers, J. D. and A. E. Hall. 1996. Genotypic classification of cowpea based on responses to heat and photoperiod. *Crop Sci.* 36: 673–679.

Ehlers, J. D. and A. E. Hall. 1998. Heat tolerance of contrasting cowpea lines in short and long days. *Field Crops Res.* 55: 11–21.

Ehlers, J. D., A. E. Hall, P. N. Patel, P. A. Roberts, and W. C. Matthews. 2000. Registration of "California Blackeye 27" cowpea. *Crop Sci.* 40: 854–855.

Ehlers, J. D., B. L. Sanden, C. A. Frate, A. E. Hall, and P. A. Roberts. 2009. Registration of "California Blackeye 50" cowpea. *J. of Plant Registrations* 3: 236–240.

Elawad, H. O. A. and A. E. Hall. 2002. Registration of "Ein El Gazal" cowpea. *Crop Sci.* 42: 1745–1746.

Elfving, D. C., M. R. Kaufmann, and A. E. Hall. 1972. Interpreting leaf water potential measurements with a model of the soil-plant-atmosphere continuum. *Physiol. Plant* 27: 161–168.

Epstein, E. 1972. *Mineral Nutrition of Plants: Principles and Perspectives.* John Wiley & Sons, New York, p. 412.

Epstein, E. and J. D. Norlyn. 1977. Seawater based crop production: A feasibility study. *Science* 197: 249–251.

Evans, J. R. 1998. Photosynthetic characteristics of fast- and slow-growing species, pp. 101–119 in H. Lambers, H. Poorter, and M. M. I. Van Vuuren (Eds.), *Inherent Variation in Plant Growth. Physiological Mechanisms and Ecological Consequences.* Backhuys Publishers, Leiden, The Netherlands.

Evans, L. T. 1993. *Crop Evolution, Adaptation and Yield.* Cambridge University Press, Cambridge, UK, p. 500.

Evans, L. T. 1998. *Feeding the Ten Billion—Plants and Population Growth.* Cambridge University Press, Cambridge, UK, p. 247.

Evans, L. T. and R. A. Fischer. 1999. Yield potential: Its definition, measurement, and significance. *Crop Sci.* 39: 1544–1551.

Farooq, M., H. Bramley, J. A. Palta, and K. H. M. Siddique. 2011. Heat stress to wheat during reproductive and grain-filling phases. *Crit. Rev. Plant Sci.* 30: 1–17.

Farooq, M., T. Aziz, A. Wahid, and K. H. M. Siddique. 2009. Chilling tolerance in maize: Agronomic and physiological approaches. *Crop Pasture Sci.* 60(6): 501–516.

Farquhar, G. D. 1978. Feedforward responses of stomata to humidity. *Austral. J. Plant Physiol.* 5: 787–800.

Farquhar, G. D. and J. Lloyd. 1993. Carbon and oxygen isotope effects in the exchange of carbon dioxide between terrestrial plants and the atmosphere, pp. 47–70 in J. R. Ehleringer, A. E. Hall, and G. D. Farquhar (Eds.), *Stable Isotopes and Plant Carbon-Water Relations.* Academic Press, San Diego, CA.

Farquhar, G. D., M. H. O'Leary, and J. A. Berry. 1982. On the relationship between carbon isotope discrimination and the intercellular carbon dioxide concentration in leaves. *Austr. J. Plant Physiol.* 9: 121–137.

Farquhar, G. D., S. von Caemmerer, and J. A. Berry. 1980. A biochemical model of photosynthetic CO_2 assimilation in leaves of C_3 species. *Planta* 149: 78–90.

Farrar, J. and S. Gunn. 1998. Allocation: Allometry, acclimation—And alchemy?, pp. 183–198 in H. Lambers, H. Poorter, and M. M. I. Van Vuuren (Eds.), *Inherent Variation in Plant Growth. Physiological Mechanisms and Ecological Consequences.* Backhuys Publishers, Leiden, The Netherlands.

Fehr, W. R. 1982. Control of iron-deficiency chlorosis in soybeans by plant breeding. *J. Plant Nutrition* 5: 611–621.

Fischer, K. S., R. Lafite, S. Fukai, G. Atlin, and B. Hardy. 2003. *Breeding Rice for Drought-Prone Environments.* International Rice Research Institute, Los Baños, Philippines, p. 98.

Fischer, R. A. 1980. Influence of water stress on crop yield in semiarid regions, pp. 323–339 in N. C. Turner and P. J. Kramer (Eds.), *Adaptation of Plants to Water and High Temperature Stress.* John Wiley & Sons, New York.

Fischer, R. A. 1985. Number of kernels in wheat crops and the influence of solar radiation and temperature. *J. Agric. Sci. Camb.* 105: 447–461.

Fischer, R. A., D. Byerlee, and G. O. Edmeades. 2014. *Crop Yields and Global Food Security: Will Yield Increase Continue to Feed the World?* ACIAR Monograph No. 158 Australian Centre for International Agricultural Research, Canberra, Australia, p. 634.

Fischer, R. A., D. Rees, K. D. Sayre, Z.-M. Lu, A. G. Condon, and A. Larque Saavedra. 1998. Wheat yield progress associated with higher stomatal conductance and photosynthetic rate, and cooler canopies. *Crop Sci.* 38: 1467–1475.

Flower, D. J. and M. M. Ludlow. 1986. Contribution of osmotic adjustment to the dehydration tolerance of water-stressed pigeonpea (*Cajanus cajan* [L.] millsp.) *Plant, Cell Environ.* 9: 33–40.

Foy, C. D. 1988. Plant adaptation to acid, aluminum-toxic soils. *Commun. Soil Sci. Plant Anal.* 19: 959–987.

Fukai, S., G. Pantuwan, B. Jongdee, and M. Cooper. 1999. Screening for drought resistance in rainfed lowland rice. *Field Crops Res.* 64: 61–74.

Gahoonia, T. S. and N. E. Nielsen. 1997. Variation in root hairs of barley cultivars doubled soil phosphorous uptake. *Euphytica* 98: 177–182.

Gahoonia, T. S., D. Care, and N. E. Nielsen. 1997. Root hairs and phosphorous acquisition of wheat and barley cultivars. *Plant Soil* 191: 181–188.

Gifford, R. M. 1974. A comparison of potential photosynthesis, productivity and yield of plant species with different photosynthetic metabolism. *Austral. J. Plant Physiol.* 1: 107–117.

Gifford, R. M. 1986. Partitioning of photosynthate in the development of crop yield, pp. 535–549 in W. J. Lucas and J. Cronshaw (Eds.), *Phloem Transport.* Alan R. Liss, New York.

Gleick, J. 1987. *Chaos: Making a New Science.* Viking Penguin, New York, p. 352.

Glenn, E. P., J. W. Leary, M. C. Watson, T. L. Thompson, and R. O. Kuehl. 1991. *Salicornia bigelovii* Torr.: An oilseed halophyte for seawater irrigation. *Science* 251: 1065–1067.

Graham, P. and C. P. Vance. 2000. Nitrogen fixation in perspective: An overview of research and extension needs. *Field Crops Res.* 65: 93–106.

Grantz, D. A., J. I. MacPherson, W. J. Massman, and J. Pederson. 1994. Study demonstrates ozone uptake by SJV crops. *California Agriculture* 48(1): 9–12.

Grattan, S. R. and E. V. Maas. 1988. Effect of salinity on phosphate accumulation and injury in soybean. I. Role of substrate Cl and Na. *Plant Soil* 109: 65–71.

Greaves, J. A. 1996. Improving suboptimal temperature tolerance in maize—The search for variation. *J. Exp. Bot.* 47: 307–323.

Gurley, W. B. 2000. HSP101: A key component for the acquisition of thermotolerance in plants. *Plant Cell* 12: 457–460.
Gurney, W. S. C. and R. M. Nisbet. 1998. *Ecological Dynamics.* Oxford University Press, Oxford, UK, p. 335.
Gwathmey, C. O. and A. E. Hall. 1992. Adaptation to midseason drought of cowpea genotypes with contrasting senescence traits. *Crop Sci.* 32: 773–778.
Hall, A. E. 1979. A model of leaf photosynthesis and respiration for predicting carbon dioxide assimilation in different environments. *Oecologia* 143: 299–316.
Hall, A. E. 1982a. Mathematical models of plant water loss and plant water relations, pp. 231–261 in O. L. Lange et al. (Eds.), *Encyclopedia of Plant Physiology, Physiological Plant Ecology,* vol. 12B. Springer-Verlag, New York.
Hall, A. E. 1982b. Humidity and plant productivity, pp. 23–40 in M. Rechigl (Ed.), *CRC Handbook of Agricultural Productivity,* Vol. I: Plant Productivity. CRC Press, Boca Raton, FL.
Hall, A. E. 1992. Breeding for heat tolerance. *Plant Breed. Rev.* 10: 129–168.
Hall, A. E. 1993a. Physiology and breeding for heat tolerance in cowpea, and comparison with other crops, pp. 271–284 in C. G. Kuo (Ed.), *Adaptation of Food Crops to Temperature and Water Stress,* Publ. No. 93-410. Asian Vegetable Research and Development Center, Shanhua, Taiwan.
Hall, A. E. 1993b. Is dehydration tolerance relevant to genotypic differences in leaf senescence and crop adaptation to dry environments?, pp. 1–10 in T. J. Close and E. A. Bray (Eds.), *Plant Responses to Cellular Dehydration during Environmental Stress.* Current Topics in Plant Physiology Vol. 10. American Society of Plant Physiologists, Rockville, MD.
Hall, A. E. 1999. Cowpea, pp. 355–373 in D. L. Smith and C. Hamel (Eds.), *Crop Yield Physiology and Processes.* Springer-Verlag, Berlin, Germany.
Hall, A. E. 2011. Breeding cowpea for future climates, pp. 340–355 in S. S. Yadav, R. J. Redden, J. L. Hatfield, H. Lotze-Campen, and A. E. Hall (Eds.), *Crop Adaptation to Climate Change.* John Wiley & Sons, Chichester, UK.
Hall, A. E. 2012. Heat stress, pp. 118–131 in S. Shabala (Ed.), *Plant Stress Physiology.* CABI, Wallingford, UK.
Hall, A. E. 2017. *Sahelian Droughts: A Partial Agronomic Solution.* Nova Science Publishers, New York, p. 216.
Hall, A. E. and L. H. Allen, Jr. 1993. Designing cultivars for the climatic conditions of the next century, pp. 291–297 in D. R. Buxton et al. (Eds.), *International Crop Science I.* Crop Science Society of America, Madison, WI.
Hall, A. E. and P. N. Patel. 1987. Cowpea improvement for semi-arid regions of Sub-Saharan Africa, pp. 279–290 in J. M. Menyonga, T. Bezuneh, and A. Youdeowei (Eds.), *Food Grain Production in Semi-Arid Africa.* OAU/STRC-SAFGRAD, Ougadougou, Burkina Faso.
Hall, A. E. and E.-D. Schulze. 1980. Stomatal responses to environment and a possible interrelation between stomatal effects on transpiration and CO_2 assimilation. *Plant, Cell Environ.* 3: 467–474.
Hall, A. E. and L. H. Ziska. 2000. Crop breeding strategies for the 21st century, pp. 407–423 in K. R. Reddy and H. F. Hodges (Eds.), *Climate Change and Global Crop Productivity.* CABI Publishing, New York.
Hall, A. E., M. M. A. Khairi, and C. W. Asbell. 1977. Air and soil temperature effects on flowering of citrus. *J. Amer. Soc. Hort. Sci.* 102: 261–263.
Hall, A. E., E.-D. Schulze, and O. L. Lange. 1976. Current perspectives of steady-state stomatal responses to environment, pp. 168–188 in O. L. Lange, L. Kappen, and E.-D. Schulze (Eds.), *Water and Plant Life Problems and Modern Approaches,* Ecological Studies 19. Springer-Verlag, New York.

Hall, A. E., B. B. Singh, and J. D. Ehlers. 1997. Cowpea breeding. *Plant Breeding Rev.* 15: 215–274.

Hall, A. E., R. A. Richards, A. G. Condon, G. C. Wright, and G. D. Farquhar. 1994. Carbon isotope discrimination and plant breeding. *Plant Breeding Rev.* 12: 81–113.

Hall, A. J., F. Vilella, N. Trapani, and C. Chimenti. 1982. The effects of water stress and genotype on the dynamics of pollen-shedding and silking in maize. *Field Crops Res.* 5: 349–363.

Hamblin, J. 1993. The ideotype concept: Useful or outdated, pp. 589–597 in D. R. Buxton et al. (Eds.), *International Crop Science I.* Crop Science Society of America, Madison, WI.

Harmens, H., C. M. Stirling, C. Marshall, and J. F. Farrar. 2000. Does down-regulation of photosynthetic capacity by elevated CO_2 depend on N supply in *Dactylis glomerata*? *Physiol. Plant* 108: 43–50.

Hendrey, G. R. and B. A. Kimball. 1994. The FACE program. *Agric. For. Meteorol.* 70: 3–14.

Herrero, M. P. and R. R. Johnson. 1981. Drought stress and its effects on maize reproductive systems. *Crop Sci.* 21: 105–110.

Hidema, J., T. Kumagai, J. C. Sutherland, and B. M. Sutherland. 1997. Ultraviolet B-sensitive rice cultivar deficient in cyclobutyl pyrimidine dimer repair. *Plant Physiol.* 113: 39–44.

Hillel, D. 1971. *Soil and Water Physical Principles and Processes.* Academic Press, New York, p. 288.

Hochmuth, G. J., W. H. Gabelman, and G. C. Gerloff. 1985. A gene affecting tomato root morphology. *HortScience* 20: 1099–1101.

Hodgson. A. S., J. F. Holland, and P. Rayner. 1989. Effects of field slope and duration of furrow irrigation on growth and yield of six grain-legumes on a waterlogging-prone vertisol. *Field Crops Res.* 22: 165–180.

Hodgson, A. S., J. F. Holland, and E. F. Rogers. 1992. Iron deficiency depresses growth of furrow irrigated soybean and pigeon pea on vertisols of northern N.S.W. *Aust. J. Agric. Res.* 43: 635–644.

Hong, S., U. Lee, and E. Vierling. 2003. Arabidopsis hot mutants define multiple functions required for acclimation to high temperatures. *Plant Physiol.* 132: 757–767.

Hubick, K. T. and G. D. Farquhar. 1989. Carbon isotope discrimination and the ratio of carbon gained to water lost in barley cultivars. *Plant, Cell Environ.* 12: 795–804.

Ishimaru, T., H. Hirabayashi, M. Ida, T. Takai, Y. A. San-Oh, S. Yoshinaga, I. Ando, T. Ogawa, and M. Kondo. 2010. A genetic resource for early-morning flowering trait of wild rice *Oryza officinalis* to mitigate high temperature-induced spikelet sterility at anthesis. *Ann. Bot.* 106: 515–520.

Ismail, A. M. 2013. Flooding and submergence tolerance, pp. 269–289 in C. Kole (Ed.), *Genomics and Breeding for Climate-Resilient Crops*, Vol. 2. Springer-Verlag, Berlin, Germany.

Ismail, A. M. and A. E. Hall. 1998. Positive and negative effects of heat-tolerance genes in cowpea. *Crop Sci.* 38: 381–390.

Ismail, A. M. and A. E. Hall. 1999. Reproductive–stage heat tolerance, leaf membrane thermostability and plant morphology in cowpea. *Crop Sci.* 39: 1762–1768.

Ismail, A. M. and A. E. Hall. 2000. Semidwarf and standard-height cowpea responses to row spacing in different environments. *Crop Sci.* 40: 1618–1623.

Ismail, A. M. and T. Horie. 2017. Genomics, physiology, and molecular breeding approaches for improving salt tolerance. *Annu. Rev. Plant Biol.* 68: 405–434.

Ismail, A. M., A. E. Hall, and E. A. Bray. 1994. Drought and pot size effects on transpiration efficiency and carbon isotope discrimination of cowpea accessions and hybrids. *Aust. J. Plant Physiol.* 21: 23–35.

Ismail, A. M., A. E. Hall, and T. J. Close. 1997. Chilling tolerance during emergence of cowpea associated with a dehydrin and slow electrolyte leakage. *Crop Sci.* 37: 1270–1277.

Ismail, A. M., A. E. Hall, and T. J. Close. 1999. Allelic variation of a dehydrin gene cosegregates with chilling tolerance during seedling emergence. *Proc. Natl. Acad. Sci. USA* 23: 13569–13573.

Ismail, A. M., A. E. Hall, and J. D. Ehlers. 2000. Delayed-leaf-senescence and heat-tolerance traits mainly are independently expressed in cowpea. *Crop Sci.* 40: 1049–1055.

Isom, W. H. and G. F. Worker. 1979. Crop management in semi-arid environments, pp. 199–223 in A. E. Hall, G. H. Cannell, and H. W. Lawton (Eds.), *Agriculture in Semi-Arid Environments.* Springer-Verlag, New York.

Jackson, M. B., R. W. Brailsford, and M. A. Else. 1993. Hormones and plant adaptation to poor aeration: A review, pp. 231–243 in C. G. Kuo (Ed.), *Adaptation of Food Crops to Temperature and Water Stress,* Publ. No. 93-410. Asian Vegetable Research and Development Center, Shanhua, Taiwan.

Jackson, P., M. Robertson, M. Cooper, and G. Hammer. 1996. The role of physiological understanding in plant breeding; from a breeding perspective. *Field Crops Res.* 49: 11–37.

Jarvis, P. G. and K. G. McNaughton. 1986. Stomatal control of transpiration: Scaling up from leaf to region. *Adv. Ecol. Res.* 15: 1–49.

Jefferson, R. A. 1993. Beyond model systems—New strategies, methods and mechanisms for agricultural research. *Ann. New York Acad. Sci.* 700: 53–73.

Jones, H. G. 1992. *Plants and Microclimate*, 2nd ed. Cambridge University Press, Cambridge, UK, p. 428.

Kaeppler, H. F. 2000. Food safety assessment of genetically modified crops. *Agron. J.* 92: 793–797.

Kasperbauer, M. J. 1987. Far-red light reflection from green leaves and effects on phytochrome-mediated assimilate partitioning under field conditions. *Plant Physiol.* 85: 350–354.

Kasperbauer, M. J. 2000. Strawberry yield over red versus black plastic mulch. *Crop Sci.* 40: 171–174.

Kasperbauer, M. J. and P. G. Hunt. 1998. Far-red light affects photosynthate allocation and yield of tomato over red mulch. *Crop Sci.* 38: 970–974.

Keller, E. and K. L. Steffen. 1995. Increased chilling tolerance and altered carbon metabolism in tomato leaves following application of mechanical stress. *Physiol. Plant* 93: 519–525.

Khush, G. S. (Ed.). 1994. *Apomixis: Exploiting Hybrid Vigor in Rice.* International Rice Research Institute, Manila, Philippines, p. 78.

Kim, Y.-U., B.-S. Seo, D.-H. Choi, H. Y. Ban, and B.-W. Lee. 2017. Impact of high temperatures on the marketable tuber yield and related traits of potato. *European J. Agron.* 89: 46–52.

Kimball, M. H., W. L. Sims, and J. E. Welch. 1967. Climatographs for head lettuce in western producing areas. *California Agricul.* 21(4): 3–4.

Kirchhoff, W. R., A. E. Hall and W. H. Isom. 1989a. Phenotypic expression of a chlorophyll mutant in cowpea (*Vigna unguiculata*): Environmental influences and effects on productivity. *Field Crops Res.* 21: 19–28.

Kirchhoff, W. R., A. E. Hall, and M. L. Roose. 1989b. Inheritance of a mutation influencing chlorophyll content and composition in cowpea. *Crop Sci.* 29: 105–108.

Kirchhoff, W. R., A. E. Hall, and W. W. Thomson. 1989c. Gas exchange, carbon isotope discrimination, and chloroplast ultrastructure of a chlorophyll-deficient mutant of cowpea. *Crop Sci.* 29: 109–115.

Kochian, L. V. 1995. Cellular mechanisms of aluminum toxicity and resistance in plants. *Annu. Rev. Plant Physiol. Plant Mol. Biol.* 46: 237–260.

Kononowicz, A. K., K. G. Raghothama, A. M. Casas, M. Reuveni, A. A. Watad, D. Liu, R. A. Bressan, and P. M. Hasegawa. 1993. Osmotin: Regulation of gene expression and function, pp. 144–158 in T. J. Close and E. A. Bray (Eds.), *Plant Responses to Cellular Dehydration during Environmental Stress.* Current Topics in Plant Physiology Vol. 10. American Society of Plant Physiologists, Rockville, MD.

Koster, K. L. and A. C. Leopold. 1988. Sugars and desiccation tolerance in seeds. *Plant Physiol.* 88: 829–832.

Koster, K. L., M. A. Tengbe, V. Furtula, and E. A. Nothnagel. 1994. Effects of low temperature on lateral diffusion in plasma membranes on maize (*Zea mays* L.) root cortex protoplasts: Relevance to chilling sensitivity. *Plant, Cell Environ.* 17: 1285–1294.

Kovach, D. A. and K. J. Bradford. 1992. Temperature dependence of viability and dormancy of *Zizania palustris* var. *interior* seeds stored at high moisture contents. *Ann. Bot.* 69: 297–301.

Kramer, P. J. and J. S. Boyer. 1995. *Water Relations of Plants and Soils.* Academic Press, San Diego, CA, p. 495.

Kwapata, M. B., A. E. Hall, and M. A. Madore. 1990. Response of contrasting vegetable-cowpea cultivars to plant density and harvesting of young pods. II. Dry matter production and photosynthesis. *Field Crops Res.* 24: 11–21.

Lambers, H., F. S. Chapin III, and T. L. Pons. 1998. *Plant Physiological Ecology.* Springer-Verlag, New York, p. 540.

Lampinen, M. J. and T. Noponen. 2003. Thermodynamic analysis of the interaction of the xylem water and phloem sugar solution and its significance for the cohesion theory. *J. Theor. Biol.* 224: 285–298.

Leakey, A. D. B., E. A. Ainsworth, C. J. Bernacchi, A. Rogers, S. P. Long, and D. R. Ort. 2009. Elevated CO_2 effects on plant carbon, nitrogen and water relations: Six important lessons from FACE. *J. Exp. Bot.* 60: 2859–2876.

Lee, G. J. and E. Vierling. 2000. A small heat shock protein cooperates with heat shock protein 70 systems to reactivate a heat-denatured protein. *Plant Physiol.* 122: 189–197.

Letourneau, D. K. 1997. Plant-arthropod interactions in agroecosystems, pp. 239–290 in L. E. Jackson (Ed.), *Ecology in Agriculture.* Academic Press, San Diego, CA.

Lin, W., L. H. Ziska, O. S. Namuco, and K. Bai. 1997. The interaction of high temperature and elevated CO_2 on photosynthetic acclimation of single leaves of rice in situ. *Physiol. Plant* 99: 178–184.

Loomis, R. S. and J. S. Amthor. 1996. Limits to yield revisited, pp. 78–89 in M. P. Reynolds, S. Rajaram, and A. McNab (Eds.), *Increasing Yield Potential in Wheat: Breaking the Barriers.* CIMMYT, Mexico, D.F.

Loomis, R. S. and J. S. Amthor. 1999. Yield potential, plant assimilatory capacity, and metabolic efficiencies. *Crop Sci.* 39: 1584–1596.

Loomis, R. S. and D. J. Connor. 1992. *Crop Ecology.* Cambridge University Press, Cambridge, UK, p. 538.

Loomis, R. S., R. Rabbinge, and E. Ng. 1979. Explanatory models in crop physiology. *Ann. Rev. Plant Physiol.* 30: 339–367.

Loomis, R. S., W. A. Williams, and A. E. Hall. 1971. Agricultural productivity. *Ann. Rev. Plant Physiol.* 22: 431–468.

L6pez Peirera, M., N. Trápani, and V. O. Sadras. 2000. Genetic improvement of sunflower in Argentina between 1930 and 1995. Part III. Dry matter partitioning and grain composition. *Field Crops Res.* 67: 215–221.

L6pez Pereira, M., V. O. Sadras, W. Batista, J. J. Casal, and A. J. Hall. 2017. Light-mediated self-organization of sunflower stands increases oil yield in the field. *Proc. Natl. Acad. Sci. USA* 114: 7975–7980.

Lu, Z., R. G. Pearcy, C. O. Qualset, and E. Zeiger. 1998. Stomatal conductance predicts yields in irrigated Pima cotton and bread wheat grown at high temperatures. *J. Exp. Bot.* 49: 453–460.

Lu, Z., J. W. Radin, E. L. Turcotte, R. Percy, and E. Zeiger. 1994. High yields in advanced lines of Pima cotton are associated with higher stomatal conductance, reduced leaf area and lower leaf temperature. *Physiol. Plant* 92: 266–272.

Ludlow, M. M. 1993. Physiological mechanisms of drought resistance, pp. 11–34 in T. J. Mabry, H. T. Nguyen, R. A. Dixon, and M. S. Bonness (Eds.), *Biotechnology for Aridland Plants.* IC^2 Institute, University of Texas at Austin.

Ludlow, M. M. and R. C. Muchow. 1990. Critical evaluation of traits for improving crop yields in water-limited environments. *Adv. Agron.* 43: 107–153.

Luedeling, E., E. H. Girvetz, M. A. Semenov, and P. H. Brown. 2011. Climate change affects winter chill for temperate fruit and nut trees. *PLos ONE* 6(5): e20155.

Lyons, J. M. 1973. Chilling injury in plants. *Ann. Rev. Plant Physiol.* 24: 445–466.

Maas, E. V. 1986. Salt tolerance of plants. *Appl. Agricult. Res.* 1: 12–26.

Maas, E. V. and S. R. Grattan. 1999. Crop yields as affected by salinity, pp. 55–108 in *Agricultural Drainage,* Agronomy Monograph no. 38. American Society of Agronomy, Madison, WI.

Mackill, D. J., A. M. Ismail, U. S. Singh, R. V. Labios, and T. R. Paris. 2012. Development and rapid adoption of submergence-tolerant (*Sub*1) rice varieties. *Adv. Agronomy* 115: 299–352.

Maestrie-Valero, J. F., L. Testi, M. A. Jiménez-Bello, J. R. Castel, and D. S. Intrigliolo. 2017. Evapotranspiration and carbon exchange in a citrus orchard using eddy covariance. *Irrig. Sci.* 35: 397–408.

Mahall, B. E. and R. M. Callaway. 1991. Root communication among desert shrubs. *Proc. Natl. Acad. Sci. USA* 88: 874–876.

Mahall, B. E. and R. M. Callaway. 1992. Root communication mechanisms and intra community distributions of two Mojave Desert shrubs. *Ecology* 73: 2145–2151.

Marfo, K. O. and A. E. Hall. 1992. Inheritance of heat tolerance during pod set in cowpea. *Crop Sci.* 32: 912–918.

Masle, J. 1992. Genetic variation in the effects of root impedance on growth and transpiration rates of wheat and barley. *Austral. J. Plant Physiol.* 19: 109–125.

Masle, J. and J. B. Passioura. 1987. The effect of soil strength on the growth of young wheat plants. *Austral. J. Plant Physiol.* 14: 643–656.

Massman, W. J. 1998. A review of the molecular diffusivities of H_2O, CO_2, CH_4, CO, O_3, SO_2, NH_3, N_2O, NO, and NO_2 in air O_2 and N_2 near STP. *Atmos. Environ.* 32: 1111–1127.

Matsui, T., O. S. Namuco, L. H. Ziska, and T. Horie. 1997. Effects of high temperature and CO_2 concentration on spikelet fertility in indica rice. *Field Crops Res.* 51: 213–219.

Mayer, A. M. and A. Poljakoff-Mayber. 1989. *The Germination of Seeds*, 4th ed. Pergamon Press, Oxford, England, p. 270.

McCutchan, H. and K. A. Shackel. 1992. Stem-water potential as a sensitive indicator of water stress in prune trees (*Prunus domestica* L. cv. French). *J. Amer. Soc. Hort. Sci.* 117: 607–611.

McNeil, S. D., M. L. Nuccio, and A. D. Hanson. 1999. Betaines and related osmoprotectants. Targets for metabolic engineering of stress resistance. *Plant Physiol.* 120: 945–949.

Meinzer, F. C. and D. A. Grantz. 1990. Stomatal and hydraulic conductance in growing sugarcane—Stomatal adjustment to water transport capacity. *Plant Cell Environ.* 13: 383–388.

Meinzer, F. C., D. A. Grantz, and B. Smit. 1991. Root signals mediate coordination of stomatal and hydraulic conductance in growing sugarcane. *Austral. J. Plant Physiol.* 18: 329–338.

Miflin, B. J. 2000. Crop biotechnology. Where now? *Plant Physiol.* 123: 17–27.

Morrison, M. J., H. D. Voldeng, and E. C. Cober. 1999. Physiological changes from 58 years of genetic improvement of short-season soybean cultivars in Canada. *Agron. J.* 91: 685–689.

Moya, T. B., L. H. Ziska, O. S. Namuco, and D. Olsyk. 1998. Growth dynamics and genotypic variation in tropical, field grown paddy rice (*Oryza sativa* L.) in response to increasing carbon dioxide and temperature. *Global Change Biology* 4: 645–656.

Mulholland, B. J., A. Hussain, C. R. Black, I. B. Taylor, and J. A. Roberts. 1999. Does root-sourced ABA have a role in mediating growth and stomatal responses to soil compaction in tomato (*Lycopersicon esculentum*)? *Physiol. Plant* 107: 267–276.

Munns, R., S. Husain, A. R. Rivelli, R. A. James, A. G. Condon, M. P. Lindsay, E. S. Lagudah, D. P. Schachtman, and R. A. Hare. 2002. Avenues for increasing the salt tolerance of crops, and the role of physiologically based selection traits. *Plant Soil* 247: 93–105.

Murakami, Y., M. Tsuyama, Y. Kobayashi, H. Kodama, and K. Iba. 2000. Trienoic fatty acids and plant tolerance of high temperature. *Science* 287: 476–479.

Murdock, W. 1990. World hunger and population, pp. 3–20 in C. R. Carroll, J. H. Vandermeer, and P. M. Rosset (Eds.), *Agroecology*. McGraw-Hill, New York.

Mutters, R. G. and A. E. Hall. 1992. Reproductive responses of cowpea to high temperature during different night periods. *Crop Sci.* 32: 202–206.

Mutters, R. G., L. G. R. Ferreira, and A. E. Hall. 1989a. Proline content of the anthers and pollen of heat-tolerant and heat-sensitive cowpea subjected to different temperatures. *Crop Sci.* 29: 1497–1500.

Mutters, R. G., A. E. Hall, and P. N. Patel. 1989b. Photoperiod and light quality effects on cowpea floral development at high temperatures. *Crop Sci.* 29: 1501–1505.

Ng, E. L. 2017. Planet plastic. *The Scientist* 31(6): 41–47.

Nielsen, C. L. and A. E. Hall. 1985a. Responses of cowpea (*Vigna unguiculata* [L.] Walp.) in the field to high night temperature during flowering. I. Thermal regimes of production regions and field experimental system. *Field Crops Res.* 10: 167–179.

Nielsen, C. L. and A. E. Hall. 1985b. Responses of cowpea (*Vigna unguiculata* [L.] Walp.) in the field to high night temperatures during flowering. II. Plant responses. *Field Crop Res.* 10: 181–196.

Nishida, I. and N. Murata. 1996. Chilling sensitivity in plants and cyanobacteria: The crucial contribution of membrane lipids. *Annu. Rev. Plant Physiol. Plant Mol. Biol.* 47: 541–568.

Noble, C. L. and M. E. Rogers. 1992. Arguments for the use of physiological criteria for improving the salt tolerance in crops. *Plant Soil* 146: 99–107.

Omwega, C. O., I. J. Thomason, and P. A. Roberts. 1988. A non-destructive technique for screening bean germplasm for resistance to *Meloidogyne incognita*. *Plant Dis.* 72: 970–972.

Onwueme, I. C. and S. A. Adegoroye. 1975. Emergence of seedlings from different depths following high temperature stress. *J. Agr. Sci., Camb.* 84: 525–528.

Passioura, J. B. 1973. Sense and nonsense in crop simulation. *J. Australian Inst. Agric. Sci.* 39: 181–183.

Passioura, J. B. 1982. The role of root system characteristics in the drought resistance of crop plants, pp. 71–82 in *Drought Resistance in Crops with Emphasis on Rice*. International Rice Research Institute, Los Banos, Philippines.

Passioura, J. B. and R. J. Stirzaker. 1993. Feedforward responses of plants to physically inhospitable soil, pp. 715–719 in D. R. Buxton et al. (Eds.), *International Crop Science I.* Crop Sciences Society of America, Madison, WI.

Patel, P. N. and A. E. Hall. 1986. Registration of snap-cowpea germplasms. *Crop Sci.* 26: 207–208.

Peng, S., K. G. Cassman, S. S. Virmani, J. Sheehy, and G. S. Kush. 1999. Yield potential trends of tropical rice since the release of IR8 and the challenge of increasing rice yield potential. *Crop Sci.* 39: 1552–1559.

Peng, S., J. Huang, J. E. Sheehy, R. C. Laza, R. M. Visperas, X. Zhong, G. S. Centeno, G. S. Kush, and K. G. Cassman. 2004. Rice yields decline with higher night temperature from global warming. *Proc. Natl. Acad. Sci. USA* 101: 9971–9975.

Petrie, C. L. and A. E. Hall. 1992. Water relations in cowpea and pearl millet under soil water deficits: I. Contrasting leaf water relations. *Austral. J. Plant Physiol.* 19: 577–589.

Petrie, C. L., Z. J. Kabala, A. E. Hall, and J. Simunek. 1992. Water transport in an unsaturated medium to roots with different local geometries. *Soil Sci. Soc. Amer. J.* 56: 1686–1694.

Poorter, H. and A. van der Werf. 1998. Is inherent variation in RGR determined by LAR at low irradiance and by NAR at high irradiance? A review of herbaceous species, pp. 309–336 in H. Lambers, H. Poorter, and M. M. I. Van Vuuren (Eds.), *Inherent Variation in Plant Growth. Physiological Mechanisms and Ecological Consequences.* Backhuys Publishers, Leiden, The Netherlands.

Popelka, J. C., S. Gollasch, A. Moore, L. Molvig, and T. J. V. Higgins. 2006. Genetic transformation of cowpea (*Vigna unguiculata* L.) and stable transmission of the transgenes to progeny. *Plant Cell Rep.* 25: 304–312

Porch, T. G., and A. E. Hall. 2013. Heat tolerance, pp. 167–202 in C. Kole (Ed.), *Genomics and Breeding for Climate-Resilient Crops*, Vol. 2. Springer-Verlag, Berlin, Germany.

Powles, S. B. and O. Björkman. 1981. Leaf movement in the shade species *Oxalis oregana.* II. Role in protection against injury by intense light. *Carnegie Inst. Washington Year Book* 80: 63–66.

Prasad, P. V. V., P. Q. Craufurd, R. J. Summerfield, and T. R. Wheeler. 2000. Effects of short periods of heat stress on flower production, pod yield and yield components of groundnut (*Arachis hypogaea* L.). *J. Exp. Bot.* 51: 777–784.

Purseglove, J. W. 1968. *Tropical Crops: Dicotyledons 1.* John Wiley & Sons, New York, p. 332.

Queitsch, C., S. Hong, E. Vierling, and S. Lindquist. 2000. Heat shock protein 101 plays a crucial role in thermotolerance in arabidopsis. *Plant Cell* 12: 479–492.

Radin, J. W., L. L. Reaves, J. R. Mauney, and O. F. French. 1992. Yield enhancement in cotton by frequent irrigations during fruiting. *Agron. J.* 84: 551–557.

Rasmusson, D. C. 1987. An evaluation of ideotype breeding. *Crop Sci.* 27: 1140–1146.

Reddy, B. V. S., A. A. Kumar, S. Ramesh, and P. S. Reddy. 2011. Sorghum genetic enhancement for climate change adaptation, pp. 326–339 in S. S. Yadav, R. J. Redden, J. L. Hatfield, H. Lotze-Campen, and A. E. Hall (Eds.), *Crop Adaptation to Climate Change.* John Wiley & Sons, Chichester, UK.

Reddy, K. R., H. F. Hodges, and J. M. McKinion. 1997. A comparison of scenarios for the effect of global climate change on cotton growth and yield. *Austr. J. Plant Physiol.* 24: 707–713.

Reynolds, M. P. and E. E. Ewing. 1989. Effects of high air and soil temperature stress on growth and tuberization in *Solanum tuberosum. Annals Bot.* 64: 241–247.

Reynolds, M. P., E. Acevedo, K. D. Sayre, and R. A. Fischer. 1994a. Yield potential in modern wheat varieties: Its association with a less competitive ideotype. *Field Crops Res.* 37: 149–160.

Reynolds, M. P., M. Balota, M. I. B. Delgado, I. Amani, and R. A. Fischer. 1994b. Physiological and morphological traits associated with spring wheat yield under hot, irrigated conditions. *Austral. J. Plant Physiol.* 21: 717–730.

Rhoades, J. D., A. Kandiah, and A. M. Mashali. 1992. The use of saline waters for crop production. FAO Irrigation and Drainage Paper No. 48. FAO, United Nations, Rome, Italy, p. 133.

Richards, L. A. (Ed.). 1954. Diagnosis and improvement of saline and alkali soils. *USDA Agricultural Handbook* No. 60. U.S. Government Printing Office, Washington, DC, p. 160.

Richards, R. A. 1995. Improving crop production on salt-affected soils: By breeding or management? *Expl. Agric.* 31: 395–408.

Richards, R. A. and J. B. Passioura. 1989. A breeding program to reduce the diameter of the major xylem vessel in the seminal roots of wheat and its effects on grain yield in rainfed environments. *Aust. J. Agric. Res.* 40: 943–950.

Richards, R. A., G. J. Rebetzke, A. G. Condon, and A. F. van Herwaarden. 2002. Breeding opportunities for increasing the efficiency of water use and crop yield in temperate cereals. *Crop Sci.* 42: 111–121.

Roberts, E. H. 1988. Temperature and seed germination, pp. 109–132 in S. P. Long and F. I. Woodward (Eds.), *Plants and Temperature,* Symposium of the Society for Experimental Biology, Number XXXXII. The Company of Biologists, Cambridge, UK.

Robertson, B. M., A. E. Hall, and K. W. Foster. 1985. A field technique for screening for genotypic differences in root growth. *Crop Sci.* 25: 1084–1090.

Rosenberg, N. J. 1974. *Micro-Climate the Biological Environment.* John Wiley & Sons, New York, p. 315.

Sabehat, A., D. Weiss, and S. Lurie. 1998. Heat-shock proteins and cross-tolerance in plants. *Physiol. Plant* 103: 437–441.

Saini, H. S., M. Sedgley, and D. Aspinall. 1984. Developmental anatomy in wheat of male sterility induced by heat stress, water deficit or abscisic acid. *Aust. J. Plant Physiol.* 11: 243–253.

Saxena, M. C. 1987. Agronomy of chickpeas, pp. 207–232 in M. C. Saxena and K. B. Singh (Eds.), *The Chickpea.* CAB Int., Wallingford, UK.

Schenk, H. J., R. M. Calloway, and B. E. Mahall. 1999. Spatial root segregation: Are plants territorial? *Adv. Ecol. Res.* 28: 145–180.

Schläppl, M. R., A. K. Jackson, G. C. Eizenger, A. Wang, C. Chu, Y. Shi, N. Shimoyama, and D. L. Boykin. 2017. Assessment of five chilling tolerance traits and GWAS mapping in rice using the USDA mini-core collection. *Front. Plant Sci.* 8: 957.

Schneider, S. H. 1989. The changing climate. *Scientific American* 261: 70–79.

Schulze, E.-D. and A. E. Hall. 1982. Stomatal responses, water loss and CO_2 assimilation rates of plants in contrasting environments, pp. 180–230 in O. L. Lange et al. (Eds.), *Encyclopedia of Plant Physiology, Physiological Plant Ecology,* vol. 12B. Springer-Verlag, New York.

Sedgley, R. H. 1991. An appraisal of the Donald ideotype after 21 years. *Field Crops Res.* 26: 93–112.

Seneweera, S. and R. M. Norton. 2011. Plant responses to increased carbon dioxide, pp. 198–217 in S. S. Yadav, R. J. Redden, J. L. Hatfield, H. Lotze-Campen, and A. E. Hall (Eds.), *Crop Adaptation to Climate Change.* John Wiley & Sons, Chichester, UK.

Shackel, K. A. and A. E. Hall. 1979. Reversible leaflet movements in relation to drought adaptation of cowpeas, *Vigna unguiculata* (L.) Walp. *Austral. J. Plant Physiol.* 6: 265–276.

Shackel, K. A., H. Ahmadi, W. Biasi, R. Buchner, D. Goldhamer, S. Gurusinghe, J. Hasey et al. 1997. Plant water status as an index of irrigation need in deciduous fruit trees. *HortTechnology* 7: 23–29.

Shannon, M. C. 1997a. Adaptation of plants to salinity. *Adv. Agron.* 60: 75–120.

Shannon, M. C. 1997b. Genetics of salt tolerance in higher plants, pp. 265–289 in P. K. Jaiwal, R. P. Singh, and A. Gulati (Eds.), *Strategies for Improving Salt Tolerance in Higher Plants.* Oxford & IBH Publishing, New Delhi, India.

Shpiler, L. and A. Blum. 1991. Heat tolerance for yield and its components in different wheat cultivars. *Euphytica* 51: 257–263.

Sinclair, T. R. 1994. Limits to crop yield?, pp. 509–532 in K. J. Boote, J. M. Bennett, T. R. Sinclair, and G. M. Paulsen (Eds.), *Physiology and Determination of Crop Yield.* Crop Science Society of America, Madison, WI.

Sinclair, T. R. 1998. Historical changes in harvest index and crop nitrogen accumulation. *Crop Sci.* 38: 638–643.

Sinclair, T. R. and C. T. de Wit. 1975. Photosynthate and nitrogen requirements for seed production by various crops. *Science* 189: 565–567.

Sinclair, T. R. and T. Horie. 1989. Leaf nitrogen, photosynthesis, and crop radiation use efficiency: A review. *Crop Sci.* 29: 90–98.

Sinclair, T. R. and P. D. Jamieson. 2006. Grain number, wheat yield, and bottling beer: An analysis. *Field Crops Res.* 98: 60–67.

Sinclair, T. R., L. C. Purcell, and C. H. Sneller. 2004. Crop transformation and the challenge to increase yield potential. *Trends Plant Sci.* 9(2): 70–75.

Singh, R. P., P. V. Vara Prasad, A. K. Sharma, and K. Raja Reddy. 2011. Impacts of high-temperature stress and potential opportunities for breeding, pp. 166–185 in S. S. Yadav, R. J. Redden, J. L. Hatfield, H. Lotze-Campen, and A. E. Hall (Eds), *Crop Adaptation to Climate Change*. John Wiley & Sons, Chichester, UK.

Slattery, R. A., A. VanLoocke, C. J. Bernacchi, X. Zhu, and D. R. Ort. 2017. Photosynthesis, light use efficiency and yield of reduced-chlorophyll soybean mutants in field conditions. *Front. Plant Sci.* 8: 1–19.

Smith, C. M. 1989. *Plant Resistance to Insects: A Fundamental Approach*. John Wiley & Sons, New York, p. 286.

Smith, H. 1995. Physiological and ecological function within the phytochrome family. *Annu. Rev. Plant Physiol. Plant Mol. Biol.* 46: 289–315.

Smucker, A. J. M. and R. R. Allmaras. 1993. Whole plant responses to soil compaction, pp. 727–731 in D. R. Buxton et al. (Eds.), *International Crop Science I*. Crop Sciences Society of America, Madison, WI.

Soman, P. and J. M. Peacock. 1985. A laboratory technique to screen seedling emergence of sorghum and pearl millet at high soil temperature. *Expl. Agr.* 21: 335–341.

Srinivasan, A., N. P. Saxena, and C. Johansen. 1999. Cold tolerance during early reproductive growth of chickpea (*Cicer arietinum* L.): Genetic variation in gamete development and function. *Field Crops Res.* 60: 209–222.

Stavarek, S. J. and D. W. Rains. 1984. The development of tolerance to mineral stress. *HortScience* 19: 377–382.

Stewart, C. N. Jr. 2004. *Genetically Modified Planet: Environmental Impacts of Genetically Engineered Plants*. Oxford University Press, Oxford, UK, p. 240.

Teramura, A. H., L. H. Ziska, and A. E. Sztein. 1991. Changes in growth and photosynthetic capacity of rice with increased UV-B radiation. *Physiol. Plant* 83: 373–380.

Thiaw, S. and A. E. Hall. 2004. Comparison of selection for either leaf-electrolyte-leakage or pod set in enhancing heat tolerance and grain yield of cowpea. *Field Crop Res.* 86: 239–253.

Thiaw, S., A. E. Hall, and D. R. Parker. 1993. Varietal intercropping and the yields and stability of cowpea production in semiarid Senegal. *Field Crops Res.* 33: 217–233.

Thomashow, M. F. 1998. Role of cold-responsive genes in plant freezing tolerance. *Plant Physiol.* 118: 1–7.

Turk, K. J., A. E. Hall, and C. W. Asbell. 1980. Drought adaptation of cowpea. I. Influence of drought on seed yield. *Agron. J.* 72: 413–420.

Veremis, J. C. and P. A. Roberts. 2000. Diversity of heat-stable genotype specific resistance to *Meloidogyne* in Maranon races of *Lycopersicon peruvianum* complex. *Euphytica* 111: 9–16.

Vertucci, C. W. and E. E. Roos. 1990. Theoretical basis of protocols for seed storage. *Plant Physiol.* 94: 1019–1023.

Vertucci, C. W. and E. E. Roos. 1993. Theoretical basis of protocols for seed storage II. The influence of temperature on optimal moisture levels. *Seed Sci. Res.* 3: 201–213.

Vertucci, C. W., E. E. Roos, and J. Crane. 1994. Theoretical basis of protocols for seed storage III. Optimum moisture contents for pea seeds stored at different temperatures. *Ann. Bot.* 74: 531–540.

Vierling, E. 1991. The roles of heat shock proteins in plants. *Annu. Rev. Plant Physiol. Plant Mol. Biol.* 42: 579–620.

Warrag, M. O. A. and A. E. Hall. 1983. Reproductive responses of cowpea to heat stress: Genotypic differences in tolerance to heat at flowering. *Crop Sci.* 23: 1088–1092.

Warrag, M. O. A. and A. E. Hall. 1984a. Reproductive responses of cowpea (*Vigna unguiculata* [L.] Walp.) to heat stress. I. Responses to soil and day air temperatures. *Field Crops Res.* 8: 3–16.

Warrag, M. O. A. and A. E. Hall. 1984b. Reproductive responses of cowpea (*Vigna unguiculata* [L.] Walp.) to heat stress. II. Responses to night air temperatures. *Field Crops Res.* 8: 17–33.

Wei, C., M. T. Tyree, and E. Steudle. 1999. Direct measurement of xylem pressure in leaves of intact maize plants. A test of the cohesion-tension theory taking hydraulic architecture into consideration. *Plant Physiol.* 121: 1191–1205.

Welbaum, G. E. and F. C. Meinzer. 1990. Compartmentation of solutes and water in developing sugarcane stalk tissue. *Plant Physiol.* 93: 1147–1153.

Welch, R. M. and R. D. Graham. 1999. A new paradigm for world agriculture: Meeting human needs, productive, sustainable, nutritious. *Field Crops Res.* 60: 1–10.

Westgate, M. E. and J. S. Boyer. 1986. Reproduction at low silk and pollen water potentials in maize. *Crop Sci.* 26: 951–956.

Wheeler, D. M. 1995. Relative aluminium tolerance of 10 species of Graminae. *J. Plant Nutr.* 18: 2305–2312.

Wheeler, D. M., D. C. Edmeades, R. A. Christie, and R. Gardner. 1992. Effect of aluminum on the growth of 34 plant species: A summary of results obtained in low ionic strength solution culture. *Plant Soil* 146: 61–66.

Wilson, G. L., P. S. Raju, and J. M. Peacock. 1982. Effect of soil temperature on seedling emergence in sorghum. *Indian J. Agr. Sci.* 52: 848–851.

Wong, S. C., I. R. Cowan, and G. D. Farquhar. 1978. Leaf conductance in relation to assimilation in *Eucalyptus pauciflora* Sieb. Ex Spreng: Influence of irradiance and partial pressure of carbon dioxide. *Plant Physiol.* 62: 670–674.

World Bank. 2010. World development report 2010: Development and climate change. The World Bank, Washington, DC, p. 439.

Xiong, D., X. Ling, J. Huang, and S. Peng. 2017. Meta-analysis and dose-response analysis of high temperature effects on rice yield and quality. *Environ. Exp. Bot.* 141: 1–9.

Yadav, S. S., R. J. Redden, J. L. Hatfield, H. Lotze-Campen, and A. E. Hall (Eds.). 2011. *Crop Adaptation to Climate Change.* John Wiley & Sons, Chichester, UK, p. 595.

Yuan, S.-C., Z.-G. Zhang, H.-H. He, H.-L. Zen, K.-Y. Lu, J.-H. Lian, and B.-X. Wang. 1993. Two photoperiod-reactions in photoperiod-sensitive genic male-sterile rice. *Crop Sci.* 33: 651–660.

Zimmermann, U., F. C. Meinzer, R. Benkert, J. J. Zhu, H. Schneider, G. Goldstein, E. Kuchenbrod, and A. Haase. 1994. Xylem water transport: Is the available evidence consistent with the cohesion theory? *Plant, Cell Environ.* 17: 1169–1181.

Ziska, L. H. and A. E. Hall. 1983a. Seed yields and water use of cowpeas *Vigna unguiculata* (L.) Walp., subjected to planned-water-deficit irrigation. *Irrig. Sci.* 3: 237–245.

Ziska, L. H. and A. E. Hall. 1983b. Soil and plant measurements for determining when to irrigate cowpeas *Vigna unguiculata* (L.) Walp., grown under planned-water-deficits. *Irrig. Sci.* 3: 247–257.

Ziska, L. H., A. E. Hall, and R. M. Hoover. 1985. Irrigation management methods for reducing water use of cowpea (*Vigna unguiculata* [L.] Walp.) and lima beans (*Phaseolus lunatus* L.) While maintaining seed yield at maximum levels. *Irrig. Sci.* 6: 223–239.

Appendix

PLANT SPECIES NAMES AND PAGE REFERENCES

Alfalfa	(*Medicago sativa* L.), 14, 152, 186, 191t, 193t, 198t, 207
Almond	(*Prunus dulcis* Mill. D. A. Webb), 139, 168, 192t, 193t
Amaranth, grain	(*Amaranthus* sps.), 47t
Ambrosia	(*Ambrosia dumosa* Payne), 17, 216
Apple	(*Malus* sps.), 17, 87, 89, 98
Apricot	(*Prunus armeniaca* L.), 88, 157, 192t, 193t, 198t
Arabidopsis	(*Arabidopsis thaliana*), 76
Artichoke	(*Cyndra scolymus* L.), 191t, 198t
Asparagus	(*Asparagus officinalis* L. Myrsiphyllum Willd.), 190, 191t, 198t, 199t
Aspen	(*Populus tremuloides* Michx.), 49
Avocado	(*Persea americana* Mill.), 64, 73, 180, 193, 198t, 200
Bananas	(*Musa* sps.), 87, 157, 182, 225
Barley	(*Hordeum vulgare* L.), 47t, 64t, 76, 79, 89, 153, 191t, 193t, 196, 198t, 199t, 201, 221
Bean, common	(*Phaseolus vulgaris* L.), 20, 47t, 64t, 80, 83–84, 179, 191t, 192–193, 198t, 199t
Bermuda grass	(*Cynodon dactylon* L. Pers.), 190, 191t
Blackberry	(*Rubus* sp.), 192t, 198t
Bluegrass, Kentucky	(*Poa pratensis* L.), 198t
Bottle tree	(*Brachychiton australis*), 143
Boysenberry	(*Rubus ursinus* Cham. and Schlechtend), 192t
Broccoli	(*Brassica oleraceae* L.), 191t, 198t
Cabbage	(*Brassica oleraceae* L.), 74, 191t, 198t, 207–208
Cactus, prickly pear	(*Opuntia basilaris* Engelm. and Bigel), 47t, 180t
Canola	(*Brassica napus* L.), 64t, 190, 191t
Carrot	(*Daucus carota* L.), 71, 99, 192t, 198t
Cassava	(*Manihot esculaenta* Crantz), 47, 201, 208
Cattail	(*Typha latifolia* L.), 128
Cauliflower	(*Brassica oleraceae* L.), 193t, 198t
Celery	(*Apium graveolens* L.), 99, 191t, 198t
Cherry	(*Prunus avium* L.), 198t
Clovers	(*Trifolium* sps.), 67, 191t, 198t
Cocoa	(*Theobroma cacao* L.), 49, 66, 180
Coffee	(Arabica *Coffea arabica* L. and Robusta *C. canephora* Pierre), 49–50, 88, 151, 182

Sudan grass	(*Sorghum bicolor* L. Moench), 190, 191t
Sugar beet	(*Beta vulgaris* L.), 47t, 60, 99, 190, 191t, 193t, 194, 198t
Sugar cane	(*Saccharum* sps.), 47t, 104, 142, 191t
Sunflower	(*Helianthus annuus* L.), 44, 47t, 64t, 106, 191t, 193t, 196–197, 198t
Sweet corn	(*Zea mays* L.), 97, 102, 211
Tea	(*Camellia sinensis* L. O. Kuntze), 157
Tepary bean	(*Phaseolus acutifolius* A. Gray), 64t, 84
Tobacco	(*Nicotiana tabacum* L.), 64t, 77, 198t, 208
Tomato	(*Lycopersicon esculentum* Mill.), 27, 29, 47t, 64t, 73, 76, 79, 191t, 193t, 198t, 211
Tomato, wild	(*Lycopersicon preuvianum*), 211
Triticale	(x *Triticosecale* Wittmack), 64t, 196
Turnip	(*Brassica rapa* L.), 64t, 192t, 198t
Vetch	(*Vicia benghalensis* L.), 64t, 198t
Walnut, black	(*Juglans nigra* L.), 17
Walnut, English	(*Juglans regia* L.), 139, 168, 198t
Wheat	(bread *Triticum aestivum* L. emend. Thell and durum *Triticum turgidum* L.), 2–3, 5, 20–21, 27, 47t, 49, 64t, 73–74, 77, 99, 137, 155, 161, 191t, 196–197, 198t, 199t, 215, 218, 221
Wheatgrass, tall	(*Agropyron elongatum* (Host) P. Beauv. subsp. *ruthenicum* Beldie), 190, 191t
Wild rice	(*Zizania aquatica* L. native species and *Z. palustris* L. cultivated species), 66, 81
Zucchini squash	(*Cucurbita pepo* L.), 191

Index

Note: Tables are indicated by bold page references; figures, photos and/or illustrations are indicated by an *italicized number.*

T

Printed in the United States
by Baker & Taylor Publisher Services